Materials, Techniques, and Ecological Impact of Sustainable Prototyping

Romdhane Ben Khalifa
https://orcid.org/0000-0002-9171-9584
National Engineering School of Gabes, Tunisia

IGI Global
Scientific Publishing
Publishing Tomorrow's Research Today

Vice President of Editorial Melissa Wagner
Director of Acquisitions Mikaela Felty
Director of Book Development Jocelynn Hessler
Production Manager Mike Brehm
Cover Design Jose Rosado

Published in the United States of America by
IGI Global Scientific Publishing
701 East Chocolate Avenue
Hershey, PA, 17033, USA
Tel: 717-533-8845 | Fax: 717-533-7115
Website: https://www.igi-global.com E-mail: cust@igi-global.com

Library of Congress Cataloging-in-Publication Data

LCCN: 2025034551 (CIP Data Pending)
ISBN13: 979-8-3373-5473-6
Isbn13Softcover: 979-8-3373-5474-3
EISBN13: 979-8-3373-5475-0

British Cataloguing in Publication Data
A Cataloguing in Publication record for this book is available from the British Library.

Table of Contents

Detailed Table of Contents

This chapter explores sustainable prototyping through additive manufacturing (AM) of superalloys, particularly in aerospace and energy sectors. It begins with an overview of superalloy material science and its environmental implications, then examines AM technologies that offer efficiency and reduced ecological impact. By customizing AM parameters and applying lifecycle assessments, both performance and sustainability can be optimized. Topology optimization enhances material efficiency, while strategies like waste minimization and powder recycling support circular economy goals. The chapter emphasizes scalability and evolving standards as key factors for broader industry adoption, supported by case studies showing real-world environmental and economic benefits. Future innovation paths, including AI-integrated manufacturing and eco-friendly alloy development, are discussed as part of a holistic approach that unites technical advancement with ecological responsibility.

This chapter explores the revolutionary concept of 4D printing, a transformative extension of traditional additive manufacturing that incorporates the dimension of time. Unlike 3D printing, which constructs objects layer by layer, 4D printing leverages smart materials that can alter their shape and properties in response to various external stimuli such as temperature, moisture, and light. This adaptive capability, often referred to as programmable matter (PM), opens new avenues for designing materials that can autonomously adjust to their environment, thereby enhancing functionality and performance. The chapter discusses the principles of 4D printing, its diverse applications across industries and the implications for sustainable development. Furthermore, it addresses the challenges faced in material innovation, design processes, and manufacturing integration. Through a comprehensive review of ongoing research and future prospects, this chapter highlights the potential of 4D printing to reshape manufacturing paradigms and contribute to the creation of dynamic responsive products

The transition towards renewable and sustainable energy systems requires not only the deployment of clean energy technologies but also the maximization of energy efficiency and resource optimization throughout generation, conversion, distribution, and end-use. This chapter explores strategies, technologies, and methodologies to enhance energy efficiency and optimize resource utilization in renewable energy systems, including solar, wind, biomass, geothermal, and hydropower. Key focus areas include design optimization, energy storage integration, digitalization through smart grids, advanced control systems, and the circular economy in material usage. The chapter also discusses case studies, future opportunities, and the role of artificial intelligence (AI) in improving performance and sustainability outcomes

In the framework of ecologically sustainable development, this research paper explores the crucial relationship between financial literacy and the use of green energy. The study looks into how institutional and individual financial knowledge affects choices about energy-efficient technology, sustainable financial products, and investments in renewable energy. This paper highlights important gaps in green energy financial literacy that impede the shift to sustainable energy systems by thoroughly analyzing the body of existing literature, actual data, and theoretical frameworks. The study uses a mixed-methods approach, combining qualitative interviews with policymakers and industry professionals with quantitative surveys of 2,500 respondents from a range of demographic categories. Important findings show that although 78% of respondents are generally aware of the advantages of green energy, only 32% of them are sufficiently knowledgeable about green investments, tax breaks, financing options, and long-term cost-benefit evaluations.

Preface

INTRODUCTION

The book "Materials, Techniques, and Ecological Impact of Sustainable Prototyping" outlines a new climate action strategy for the design and manufacturing sectors, which are known for their significant environmental impact. Construction activities generate major carbon emissions, and prototype development creates substantial waste. The book proposes combining material development, advanced manufacturing, and environmental assessment to produce sustainable products across industries, including product development and urban planning. The use of low-energy materials, recycled plastics, byproduct-enhanced concrete, certified timber, and bio-waste-derived composites, matches traditional performance while lowering lifecycle emissions. These materials follow circular economy principles: using sustainable resources, modular designs for disassembly, and biodegradables that maximize material life and reduce waste. The book offers techniques such as digital twin simulations, sustainable cutting tools, material recycling, and clean energy to minimize prototyping costs. By employing bio-based and recycled materials and encouraging construction and product reuse, the approach keeps waste out of landfills and reduces emissions. The text highlights ecological and socio-economic benefits, such as cost reductions, regulatory compliance, improved image, and innovations in renewable energy and AI-driven urban planning. It delivers practical methods to benefit engineers, researchers, business owners, and policymakers. The book comprehensively covers 3D printing, materials science, ecology, and sustainable investment strategies, drawing on interdisciplinary perspectives for actionable solutions. It supports transformation by urging early lifecycle assessments, biomaterials education, and balancing speed with sustainability in prototyping. The book teaches readers to view prototyping as central to environmental sustainability and to develop sustainable design for the future.

OBJECTIVES

The book Materials, Techniques, and Ecological Impact of Sustainable Prototyping provides a transformative framework for product development and construction, centered on sustainable prototyping as a complete system. The main goal is to replace traditional, linear development with circular-economy systems that extend material lifespans and reduce waste. By integrating environmental, social, and economic factors into each development stage, the book equips designers, engineers, and organizations to optimize resource use and minimize harm. It analyzes existing prototyping methods, revealing inefficiencies and environmental impacts, and demonstrates the urgent need for system change. The proposed theoretical framework integrates sustainable materials, green design principles, modern manufacturing, clean energy, and waste management solutions, enabling the selection of products that meet both performance needs and sustainability goals. The book prioritizes life-cycle assessment and environmental impact evaluation from a project's start, using digital tools to maximize resource efficiency and enable end-of-life planning before manufacturing. Practical operational methods are presented for businesses to decrease waste, recycle materials, and support adaptive reuse in construction. Ultimately, sustainable prototyping systems not only offer environmental benefits but also drive social and economic advantages lowering operational costs, improving energy performance, boosting corporate image, and ensuring regulatory compliance. These integrated benefits help organizations remain competitive and aligned with the UN Sustainable Development Goals. By uniting diverse expertise and promoting systems thinking, the book lays out a clear path for creating sustainable products and buildings that serve today's needs without compromising those of future generations, supporting global sustainability objectives across industries.

TARGET AUDIENCE

The book Materials, Techniques, and Ecological Impact of Sustainable Prototyping serves multiple academic disciplines by bringing together researchers who aim to protect the environment while developing new design approaches and construction solutions. The book delivers comprehensive information through its structured format, which helps professionals across different industrial sectors.

The main audience is academic staff in engineering, architecture, materials science, and environmental studies. The book provides complex theoretical information, lifecycle assessment techniques, and strategies for sustainable prototyping and the circular economy, including how 3D printing and bio-fabrication reduce environmental damage. Environmental scientists and sustainability researchers find supporting data for waste-reduction, carbon-emissions, and resource-conservation policy research.

The book offers designers and engineers practical guidance on using sustainable materials—renewable, recycled, and recyclable and on designing products for disassembly, repair, and end-of-life recycling. It presents energy-efficient manufacturing, waste-reduction strategies, and closed-loop systems for professionals in product design, manufacturing, and construction, including lifecycle assessment methods and sustainable selection criteria.

The built environment sector, including contractors, architects, managers, and construction firms, benefits from frameworks for reducing environmental impact from design to demolition. BIM and digital fabrication professionals use tools to integrate digital tools with performance and lifecycle assessment, supporting project managers in waste management and sustainability programs.

The main audience includes business owners, entrepreneurs, and decision-makers seeking market leadership through sustainable innovation. The book shows how sustainable prototyping lowers costs by improving resource management, reducing expenses, and enhancing brand value. It supports executives in manufacturing, construction, real estate, and consumer goods to link sustainability with regulations, stakeholder needs, and financial goals.

The book provides guidance for manufacturing specialists, production planners, and supply chain managers on sustainable production, energy, and material efficiency. It presents advanced approaches, such as 3D printing, bio fabrication, and additive manufacturing, that generate less waste than subtractive methods. Quality assurance and process engineers get tools to use lifecycle assessment frameworks to maximize manufacturing efficiency.

The book guides environmental consultants, officers, and ESG professionals in environmental evaluation, waste reduction, and sustainability integration. It includes details on renewable energy, carbon footprint analysis, and circular economy frameworks for building sustainability plans. Policy advisors and regulators get frameworks for standards and green building regulations.

The book equips educators in design, engineering, architecture, and environmental studies to teach sustainable prototyping. Students gain essential knowledge for successful careers. The book serves as both a reference and educational resource, connecting theory to practice and preparing professionals to prioritize sustainability.

The book provides urban planners, regional developers, and infrastructure specialists with insights into sustainable development, urban transformation, and smart-city digital twin technology. It demonstrates how to link construction methods with solar systems for urban or rural planning. Frameworks support infrastructure resilience, climate adaptation, and community development at scale.

Green investment professionals, sustainability investors, and financial analysts studying ESG investments find information on new sustainable technologies, market trends, and financial aspects. The book supports financial institutions with sections on green investment decisions, financial literacy, and sustainability cost-benefit analysis. It helps asset managers understand sustainable business models that align with sustainability-focused strategies.

The book helps government agencies, policymakers, and regulators create standards, incentives, and sustainable development regulations for construction and manufacturing. Municipal leaders gain operational methods for achieving climate goals through case studies and best practices. International development organizations and SDG-focused bodies seek resource-efficiency solutions for global implementation.

The book offers supply chain and procurement specialists methods for selecting sustainable materials, sources, and logistics to improve the environment. Corporate purchasing teams learn to build sustainable value chains by choosing materials, requiring supplier disclosure, and planning product disposal.

Design software, BIM tool, and lifecycle platform developers, as well as digital fabrication creators, find sustainability-focused requirements and examples. AI and machine learning experts learn about optimization systems, digital twins, and predictive modeling for sustainable prototyping.

The book addresses interdisciplinary readers since sustainable prototyping requires knowledge spanning materials science, engineering, design, environmental science, business, policy, and finance. It unites technical, managerial, policy, and investment perspectives through a complete analysis of materials, techniques, ecological impacts, artificial intelligence, digital twins, renewable energy, urban development, and financial literacy. Readers gain science-based solutions to create sustainable prototypes supporting global sustainability goals.

THE CHALLENGES

The field of sustainable prototyping encounters various obstacles that require solutions from both industry professionals and this research document. Stakeholders need to identify all obstacles that affect product development and construction operations to achieve sustainability.

The main obstacle to sustainable material adoption is their high prices, limited availability, and insufficient performance. Increasing prices prevent manufacturers and construction companies from using sustainable materials. These cost more than conventional materials. Designers must balance sustainable materials with operational needs. Sustainable materials should perform at least as well as conventional ones in all environments. The book needs to explain how different material streams impact recycled material quality. It should describe the supply chain risks bio-based materials face. The book also needs to outline how scientists develop new sustainable materials that fulfill environmental standards and market needs. The environmental impacts of new materials throughout their life cycles remain unclear. Researchers lack complete information on embodied energy consumption, water usage, and disposal requirements. Organizations face major operational and technical challenges when adopting sustainable practices. These methods often require distinct approaches compared to existing manufacturing systems. The speed, material selection, and production capacity of additive manufacturing and 3D printing still fall short of traditional manufacturing. Adopting bio-fabrication, sustainable molding, and other advanced techniques demands new equipment, employee training, and process optimization. This results in significant expenses and operational disruptions. The book needs to demonstrate that sustainable technologies often remain unaffordable for most businesses. Implementation speed depends on sector, product intricacy, and company readiness. Digital tools like BIM, lifecycle assessment software, and digital twins require organizations to build new infrastructure and acquire specialized expertise. Organizational change management systems are hard for small firms to implement. Evaluating environmental effects and conducting lifecycle assessments face significant challenges. There are no established measurement protocols, and data remains insufficient for assessment. The book supports early tool adoption, but practitioners encounter varying standards between regions. They also lack sufficient methods for developing new sustainable solutions. Designers struggle to predict environmental impacts and create circular product designs. Complete information on disposal scenarios for new materials and products is lacking. Integrating sustainability into product development is difficult. Different design platforms, assessment tools, and information systems have poor interoperability. Current educational programs do not provide enough training. Designers and engineers lack essential knowledge of sustainable materials, lifecycle thinking, and the circular economy. High costs of sustainable prototyping equipment, training, materials, and process development slow technology adoption. The book shows that sustainable practices can reduce operational expenses. However, small and medium-sized enterprises struggle to afford initial costs for equipment, training, and materials. Organizations face challenges in measuring financial returns from sustainability initiatives. Benefits often appear in the long term, as market value and brand reputation, rather than in direct savings. Requirements to cut expenses hinder adoption. In markets focused on price, or in industries without sustainable product value, affordable options often triumph over environmental sustainability. Sustainable prototyping adoption meets barriers because countries have varying and shifting environmental rules. Organizations operating globally must handle different material standards, lifecycle assessment methods, and green building requirements. The lack of third-party verification and weak enforcement of standards allows greenwashing, eroding customer trust. Regulation fails to keep up with new solutions, as laws change only after technology advances. Implementing sustainable prototyping requires organizations to change cultural values, supply chain operations, and standards. This often meets opposition from established practices and powerful interests. Construction and manufacturing have developed over decades of linear production. Complete redesign of design, procurement, production, and waste management is required. Achieving sustainability across supply chains demands trust-based partnerships. Full transparency and mutual understanding between stakeholders are necessary. The market has not made sustainability a key factor in customer decisions, which restricts sustainable product growth. The book provides essential frameworks and strategies. Yet, implementing sustainable prototyping at an industrial scale faces many obstacles. Industrial-scale adoption requires substantial financial resources for new facilities, training, and supply network changes. Running dual systems during transition causes inefficiencies and higher expenses. Small businesses and developing countries often lack resources to build full sustainability systems. This widens the performance gap compared to competitors. Developing sustainable claims is challenging because they are complex. Lifecycle assessment results are inconsistent because different methods and boundary conditions yield varying outcomes. This makes comparisons difficult. The book needs to explain that full environmental accounting from cradle to grave is costly and may produce unexpected results. For example, transporting recycled materials globally can lead to more emissions than sourcing conventional materials locally. Organizations struggle to monitor and present their real achievements. Verification systems lack standardization and development. The rapid pace of technological advancement complicates the identification of lasting, sustainable methods. Market adoption of new materials and processes is uncertain. Better alternatives may emerge before widespread acceptance, making current investments useless. The book needs to understand that predicting sustainability innovation is difficult, as new technologies keep appearing.

OVERVIEW OF THE BOOK

Materials, Techniques, and Ecological Impact of Sustainable Prototyping is a twelve-chapter look at how design and manufacturing can shift toward sustainability. It examines materials, techniques, and ecological impacts together. Each chapter covers a different, linked part of prototyping.

Chapter 1 establishes prototyping as a full life-cycle system that embeds environmental, social, and economic principles into development. It criticizes the linear, subtractive methods that consume many resources and generate construction and demolition waste and emissions. This chapter creates a combined framework that links materials, eco-design, manufacturing, energy-saving, and waste management. Key plans include choosing recycled materials, using additive processes to reduce waste, designing modular, reusable parts, conducting life-cycle checks, and using digital tools from the start. Results show lower effects, costs, improved reputation, and support for development goals.

Chapter 2 explores emerging technologies—Internet of Things (IoT), artificial intelligence (AI), and digital twin technology—within frameworks like the circular economy, eco-design, and lifecycle management. It shows how new technologies and environmental awareness enable the development of efficient, environmentally friendly products. The chapter emphasizes the importance of material selection, energy reduction, and waste minimization in production and use. Case studies and analytical frameworks demonstrate application in sustainable design. Trends and ethical considerations guide policy innovation for future development.

Chapter 3 sits at the intersection of Industry 4.0 and sustainability goals. It explores sustainable manufacturing as a blend of technological advancement with ecological and economic aims. The chapter critiques the environmental degradation caused by manufacturing. It shows how digital technologies, such as artificial intelligence, IoT, cloud computing, and data analytics, optimize operations, decentralize supply chains, and enable circular models through reuse, remanufacturing, and recycling. Sustainable manufacturing uses closed-loop systems and eco-design principles. It applies additive manufacturing (3D printing) to reduce waste, robotics for production, and digital twin technology for process optimization. The chapter examines blockchain for supply chain tracking and transparency. Blockchain verifies and shares sustainability claims.

Chapter 4 uses bi-stacked Long Short-Term Memory (LSTM) neural networks to capture patterns in energy use and emissions. The methodology predicts carbon outputs. It notes that rapid urbanization and climate change add to the pressure on cities. Literature review and case studies from Singapore, Helsinki, and Shanghai support its points. Digital twin technology enables oversight and uses information for real-time system optimization. It helps achieve energy efficiency, reduce carbon footprints, improve resilience, and support governance. The analysis looks at obstacles, including ethical ones, to DTT implementation. It also describes opportunities like intelligence deployment, sector collaboration, and standardization. The chapter ends with steps for planners, policymakers, and practitioners to build smarter, more resilient, and more equitable urban futures.

Chapter 5 explores sustainable prototyping through manufacturing superalloys, key materials for aerospace and energy. It links material science with sustainability goals. The chapter shows how customized manufacturing parameters and lifecycle are explored in Chapter 5, which explores sustainable prototyping in manufacturing superalloys, essential in aerospace and energy. By linking material science to sustainability goals, the chapter describes how customized manufacturing parameters and lifecycle assessments improve performance and lower impact. It covers topology optimization, waste minimization, and powder recycling to support the circular economy, referencing case studies that show economic benefits and future innovations such as AI in manufacturing and eco-friendly alloy development. mart materials adapt to their environment and improve how objects function. 4D printing offers new ways to design adaptable things. The chapter explains how 4D printing works and shows its industrial applications. It reviews how 4D printing affects the environment and points out issues with materials and design integration. The chapter reviews current research and future ideas. It says 4D printing can change how we make things and help create responsive, eco-friendly products. The chapter gives a clear picture of 4D printing.

Chapter 6 explores the revolutionary concept of 4D printing, a transformative extension of traditional additive manufacturing that incorporates the dimension of time. Unlike 3D printing, which constructs objects layer by layer, 4D printing leverages smart materials that can alter their shape and properties in response to various external stimuli such as temperature, moisture, and light. This adaptive capability, often referred to as programmable matter (PM), opens new avenues for designing materials that can autonomously adjust to their environment, thereby enhancing functionality and performance. The chapter discusses the principles of 4D printing, its diverse applications across industries and the implications for sustainable development. Furthermore, it addresses the challenges faced in material innovation, design processes, and manufacturing integration. Through a comprehensive review of ongoing research and future prospects, this chapter highlights the potential of 4D printing to reshape manufacturing paradigms and contribute to the creation of dynamic responsive products

Chapter 7 looks at the shift toward sustainable energy systems. It shows how to make the most of energy efficiency. This chapter shows how to use resources to generate, shift, transmit, and wield power. It explores plans and tools for sources such as solar wind, biomass, geothermal, and hydropower. It keeps the focus on design, adding energy storage via grids, smart control systems, and circular-economy ideas for material use. I see case studies that show use. The discussion of opportunities and intelligence's role in improving performance and sustainable outcomes gives forward-looking ideas.

Chapter 8 examines the economic and policy dimensions required to accelerate the transition to renewable and sustainable energy. It explains how consumer awareness and knowledge of energy investments can be enhanced and analyzes how consumers make investment decisions, including prevailing investment behavior patterns. The chapter reviews policy measures that encourage the adoption of renewable energy and emphasizes that a successful energy transition requires more than technological and engineering solutions. It also depends on investment literacy, active consumer engagement, and supportive policy frameworks for renewable and sustainable energy.

Ensure it covers the economic and policy aspects needed to accelerate the transition to renewable and sustainable energy. It explains how to raise consumer knowledge about energy investments and examines how consumers decide to invest and the patterns in investment behavior. It reviews the policy actions that promote energy adoption and notes that a successful energy transition requires more than technology and engineering. The chapter says the transition also needs investment literacy, active consumer participation, and policy plans that support renewable and sustainable energy.

Materials, Techniques, and Ecological Impact of Sustainable Prototyping serves as a reference for researchers, scholars, practicing professionals, and graduate students seeking an overview of sustainable prototyping methods and applications. Materials, Techniques, and Ecological Impact of Sustainable Prototyping examines the links among materials science, advanced manufacturing techniques, and ecological assessment. Materials, Techniques, and Ecological Impact of Sustainable Prototyping is a guide. Materials, Techniques, and Ecological Impact of Sustainable Prototyping offers frameworks that help design and production practices advance across many industries and sectors.

The book focuses on quality assurance, process optimization, and the integration of materials and new technologies in additive manufacturing and other fabrication methods. The book mixes theory with practice. The book provides readers with knowledge and tools for using prototyping in real-world situations. The book supports research, industrial manufacturing, construction, urban planning, and policy development.

We thank the specialists and experts who shared their knowledge, experience, and ideas as we created the book. The specialists and the experts added a lot to the content. The specialist's expertise aligned the theory with practice and generated new ideas. We especially thank the researchers who work on materials, manufacturing, digital manufacturing, and ecological impact assessment. The researcher's knowledge shaped the advice. The case studies that appear in the book.

We also thank the peer reviewers and editorial consultants for feedback during the editorial process. The peer reviewers and editorial consultants carefully examined the manuscript. The peer reviewers and editorial consultants analyzed the manuscript. Gave recommendations for improvement. The recommendations strengthened the manuscript, improved its clarity of presentation, refined its accuracy, and ensured it meets academic and professional standards. The iterative refinement process, guided by expert review, preserved the quality of all material. The iterative refinement process also raised the quality while keeping the book’s vision and practical applicability.

I see the publication as a promise of researchers, practitioners, reviewers, and editors to move prototyping forward. The publication uses a clear study to examine challenges, the record of practice, and the open sharing of both wins and limits in modern sustainable design and manufacturing. I think the resource will help the readers. The resource will help research grow, guide work, train a wave of sustainability-focused practitioners, and contribute to the big change toward more responsible, circular, and careful environmental design and manufacturing worldwide.

Romdhane Ben Khalifa
National Engineering School of Gabes, Tunisia

Acknowledgment

This book, *Materials, Techniques, and Ecological Impact of Sustainable Prototyping*, is the result of a shared commitment to rethinking how we design, test, and manufacture ideas in a world of finite resources, and we are grateful for the ongoing collaboration and support from all involved that made this work possible.

First and foremost, we thank all contributing authors for their thoughtful research, creativity, and patience throughout the editorial process. Their chapters reflect a rare balance of technical depth and ecological awareness, showing how sustainable prototyping can be both scientifically rigorous and socially meaningful. Each contribution added a distinct perspective—spanning bio-based materials, low-impact fabrication, circular design strategies, digital workflows, and life-cycle assessment—yet all converged on a common aim: to prototype not only for performance, but for responsibility, highlighting the power of diverse collaboration.

We extend our sincere appreciation to the reviewers whose careful evaluations strengthened the book considerably. Their constructive feedback improved clarity, expanded critical discussion, and ensured that the methodological and environmental claims in each chapter were well supported. Peer review is often invisible labor; we recognize the intellectual generosity behind it and are deeply thankful for their dedication.

We are also grateful to the research communities and laboratories that served as the foundation for many of the ideas presented here. Sustainable prototyping is inherently interdisciplinary, and the dialogue between materials science, mechanical and industrial engineering, architecture, human-computer interaction, and environmental studies continuously enriched this project. We acknowledge the conferences, workshops, and academic networks where early versions of these studies were debated, challenged, and refined, emphasizing the collaborative effort that drives progress in sustainability.

Many thanks go to our home institutions and departments for the resources and encouragement provided during the development of this volume. Support for experimental work, access to fabrication facilities, and the freedom to pursue cross-disciplinary inquiry were essential. We also acknowledge the students and research assistants who contributed to data collection, prototyping trials, and model validation. Their curiosity and persistence energized the project and helped ensure its practical relevance.

We express special gratitude to industry partners, maker spaces, and community organizations that opened their doors and shared their experiences. Sustainable prototyping must translate into real-world practice, and the lessons drawn from applied contexts limitations, opportunities, and unexpected solutions added an important layer of realism to the book. The openness of practitioners to discuss both successes and failures was invaluable.

Our appreciation goes to the editorial and production team for guiding this book from concept to completion with professionalism and care. We thank the publisher for supporting a topic that requires both technical precision and ethical urgency. The steady coordination, detailed feedback, and commitment to quality allowed this volume to take its final shape.

Finally, we acknowledge the broader community of researchers, educators, designers, engineers, and activists working toward more sustainable futures. This book stands on the progress they have already made and seeks to contribute to it in return. The knowledge collected here will encourage further exploration of low-impact materials and methods, more transparent ecological assessment, and prototyping cultures that prioritize regeneration over waste. Your ongoing commitment inspires us all to keep pushing for meaningful change.

To everyone who supported, challenged, and inspired this effort thank you. Your contributions have helped build a resource intended not only to document current advances but also to motivate the next generation of sustainable prototyping research and practice.

Chapter 1
Introduction to Sustainable Prototyping

Romdhane Ben Khalifa
https://orcid.org/0000-0002-9171-9584
National Engineering School of Gabes, Tunisia

ABSTRACT

Sustainable prototyping embeds environmental, social, and economic principles across the prototyping lifecycle, targeting construction's high material, energy, and waste burdens. Conventional linear, subtractive practices overconsume resources, generate vast construction/demolition waste, and emit heavily, requiring circular strategies for sustainable development. This paper develops a framework integrating sustainable materials, eco-design, additive manufacturing, energy-efficient design, and end-of-life/waste management. Strategies include renewable/recycled low-embodied-energy materials; additive processes to cut waste; modular, durable, repairable designs for disassembly/reuse/recycling; and early LCA/digital tools for optimization. Energy efficiency and waste prevention/recycling/adaptive reuse shift construction to circularity. Adoption reduces impacts/costs, boosts sustainability/reputation, guiding greener products/environments.

INTRODUCTION

The new design approach of sustainable prototyping combines environmental thinking with iterative prototype development to create products. The essential process of traditional prototyping for design refinement and functionality testing results in substantial resource use and waste, which creates detrimental environmental effects (Lauff et al., 2018). Sustainable prototyping implements methods that decrease resource usage and minimize waste output while selecting environmentally beneficial

DOI: 10.4018/979-8-3373-5473-6.ch001

materials for its operations (Bamgbade et al., 2017; Gurova et al., 2020). Sustainable development goals become achievable through this method because it supports the current needs of people without harming future generations' ability to fulfill theirs (Bamgbade et al., 2016). Sustainable construction introduces a concept that creates an environment that supports current human requirements while preserving the ability of future populations to meet their needs (Bamgbade et al., 2016). The increasing need to meet social, economic, and organizational targets has led to the construction of buildings and structures (Yunus et al., 2024). Sustainable prototyping plays an essential role in reducing the environmental impacts of construction activities (Yap et al., 2024). The construction industry implements sustainable practices by using recyclable materials and non-toxic elements, as well as energy-efficient building designs and water-saving methods, thereby supporting environmental, social, and economic sustainability (Yunus et al., 2024). The practice aims to develop environmentally friendly and resource-efficient structures that minimize societal and environmental impacts during construction (Neyestani, 2017; Yap et al., 2024).

The increasing environmental impact of standard prototyping methods drives the development of sustainable prototyping practices. The conventional manufacturing process relies on subtractive techniques, which require operators to remove material from solid blocks to achieve the desired product form (Harun et al., 2025). The manufacturing process generates significant waste when working with intricate designs and materials that are difficult to recycle. The iterative testing process in prototyping yields multiple prototype versions, collectively increasing resource usage and waste generation. The design process should start by minimizing waste generation and promoting material reuse before recycling becomes an option. The European Union faces a yearly construction waste problem of 820 million tonnes, so designers need to incorporate solutions during the design phase (Harun et al., 2025). Sustainable prototyping solves these problems by promoting environmentally friendly methods for creating prototypes. The practice of sustainable prototyping comprises three main strategies that integrate material selection with additive manufacturing and design-for-disassembly principles (Manu, 2024). The implementation of sustainable practices in prototyping brings multiple advantages, including reduced environmental impact, lower costs, and improved product functionality. Sustainable prototyping methods stimulate creative thinking by enabling designers to explore new materials and production techniques that integrate environmental responsibility with technological sophistication. The design of products and services should include end-of-life considerations to make their components and materials available for reuse or recycling.

The circular economy model of sustainable prototyping keeps materials active for extended periods to achieve maximum resource efficiency and waste reduction. The design process for end-of-life functionality of prototypes should include features that allow disassembly and enable component and material reuse or recycling (Ho et al., 2019). Sustainable prototyping helps businesses show their environmental responsibility while building better brand value and market leadership. Sustainable prototyping methods stimulate creative development by pushing designers to find innovative materials and production methods that balance environmental responsibility with technological sophistication. The proper disposal of waste materials and recycling operations play a vital role in renovation activities. The proper management of waste during building operations and waste disposal requires strict adherence to established waste management protocols. The implementation of waste prevention alongside material reuse and recycling strategies helps minimize environmental impacts. The building area requires proper waste disposal, as it generates high resource use and CO_2 emissions (HARUN et al., 2025). The construction phase implements green practices, energy conservation methods, ecological thinking, and sustainable development principles. The implementation of stability in construction projects leads to reduced resource usage, enhanced energy performance, decreased operational expenses, and improved public perception. The transition toward sustainable construction practices is driven by three main elements: rising consumer environmental consciousness, enhanced sustainability materials and technologies, and strengthened environmental regulations. Sustainable product design requires analyzing product life cycles, implementing the 6 R's, and developing housing and service infrastructure to achieve sustainability (Gupta et al., 2015). The built environment requires interdisciplinary collaboration and systemic thinking throughout its entire development process. Figure 1 contrasts the traditional linear prototyping path design, build, test, and discard with a sustainable circular lifecycle that reintegrates prototypes through reuse, refurbishment, or recycling. This comparison highlights how closing the loop reduces material consumption, waste generation, and embodied emissions across construction prototyping.

The need to fulfill social, economic, and organizational targets has led to the construction of buildings and structures. Sustainable construction introduced a new approach to building that meets current human needs while securing a future for generations to come. The construction of buildings leads to substantial power usage, water consumption, and natural resource depletion while producing waste that ends up in landfills. According to Kumar et al. (2021), the selection of sustainable materials for green building design should begin with materials that match or outperform conventional materials. The application of green building techniques extends to three main areas: new construction and renovation work, and ongoing facility maintenance. The design of green buildings includes integrated landscaping systems with water-collection infrastructure and solar-power generation capabilities. The creation of enduring values depends on the proper management of energy resources, materials, and water consumption. Indeed, this paper begins by examining the limits of prototyping methods in construction. This paper points out the need for alternatives, underscoring the urgency for innovation. This paper then presents a framework that brings together materials, design principles, production techniques, and waste management strategies. It concludes by discussing how life cycle assessments and digital tools can be effectively used to enhance resource efficiency and build trust in sustainable development, motivating the audience to adopt these approaches.

Figure 1. Traditional vs. sustainable prototyping lifecycle

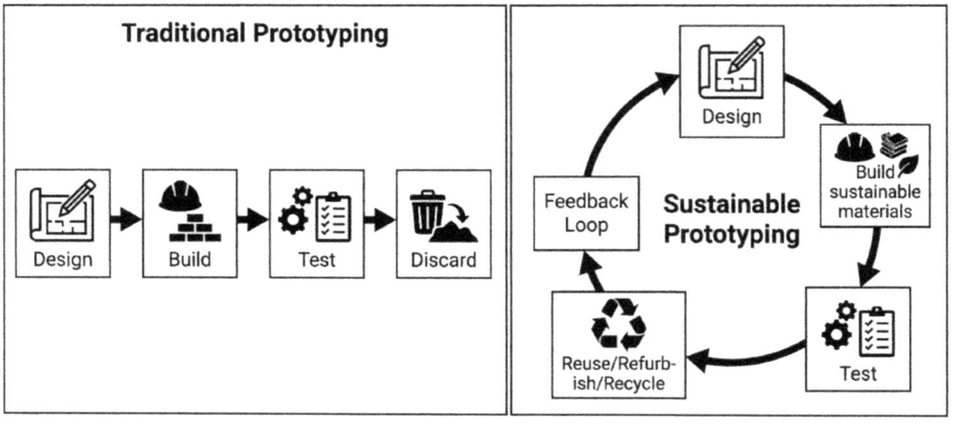

LITERATURE REVIEW

The complete sustainable prototyping system addresses all stages of product development, from material selection and design methods to manufacturing systems and product disposal strategies. The approach recognizes how environmental elements link with social and economic factors to achieve maximum performance in all three dimensions (C, 2014). The construction methods of today cost more than green building solutions, yet they help reduce energy expenses according to Yunus et al. (2024). The delivery of high-performance buildings requires process integration between the planning and design, and construction and operation stages of the facility lifecycle, according to Orsi et al. (2020). Sustainable buildings achieve environmental and human-friendly operation through efficient resource management of energy, water, and materials compared to conventional buildings (Neyestani, 2017). Sustainable architecture achieves environmental benefits through resource-efficient practices that minimize material use, energy and water consumption, and ecosystem preservation (Attiya et al., 2020).

The practice of sustainable prototyping requires designers to understand all material characteristics, including their environmental impacts, functional properties, and market availability. Designers should select renewable, recycled, and recyclable materials while avoiding hazardous substances in their work. The sustainable prototyping approach achieves resource efficiency by reducing material waste and energy use throughout the prototyping process. The transition to environmentally friendly construction requires researchers to conduct thorough evaluations of recycled materials from start to finish (Almusaed et al., 2024). Life cycle assessments, together with comparative studies, demonstrate that sustainable alternatives offer environmental advantages (Abera, 2024).

Design approaches in sustainable prototyping focus on creating products that last longer and can be fixed and updated. The design process should create prototypes that can handle demanding tests and operational conditions to achieve longer service life and decrease replacement requirements. Sustainable prototyping promotes modular designs that enable simple disassembly and component replacement for future maintenance and upgrades. The entire local community participates in traditional house construction through site selection, material choice, and house-building stages, while respecting local traditions and environmental factors (Mariana et al., 2022).

Sustainable prototyping involves developing production methods that reduce environmental harm while minimizing resource use (figure 2). The 3D printing technology known as additive manufacturing shows excellent potential for sustainable prototyping because it produces complex shapes with minimal material loss (Laverne et al., 2019). Designers need to study different manufacturing approaches, including bio-fabrication and sustainable molding, which use sustainable resources while reducing energy usage (Giudice et al., 2020). Construction projects that incorporate sustainability principles lead to reduced resource use, improved energy performance, lower operational expenses, and better public perception.

The implementation of proper waste management systems, together with recycling programs, helps decrease environmental damage. Proper waste management helps cities become cleaner, healthier, and more environmentally friendly. The construction industry now places greater importance on sustainable practices because environmental problems such as pollution and resource depletion continue to grow (Harun et al., 2025). In sustainable prototyping, end-of-life management of prototypes requires strategies to recycle materials effectively, find alternative uses for them, or dispose of them through environmentally friendly methods (Alam et al., 2022; Liu et al., 2020). Designers who choose materials that can be recycled or composted will achieve lower environmental effects during prototype disposal. The practice of adaptive reuse extends the durability of demolition projects while preserving materials and reducing waste. Sustainability needs to become a fundamental element during the planning stage of any project. The implementation of environmental rules by projects will enhance operational efficiency and reduce emissions. The essential steps for waste reduction include recycling programs, fossil fuel reduction, and sustainable procurement practices (Harun et al., 2025). The concept of sustainable development has gained widespread acceptance as a new operational model for both public and private organizations (Chow et al., 2021). Our society needs sustainable development to meet present requirements and future needs worldwide (Despeisse et al., 2012). Project success depends on continuous monitoring and effective communication to maintain both schedule adherence and budget control. Sustainable development works to enhance indoor quality through resource preservation, material recycling, and emission reduction, as well as the protection of both local and global ecosystems.

Figure 2. Sustainability benefits overviewer

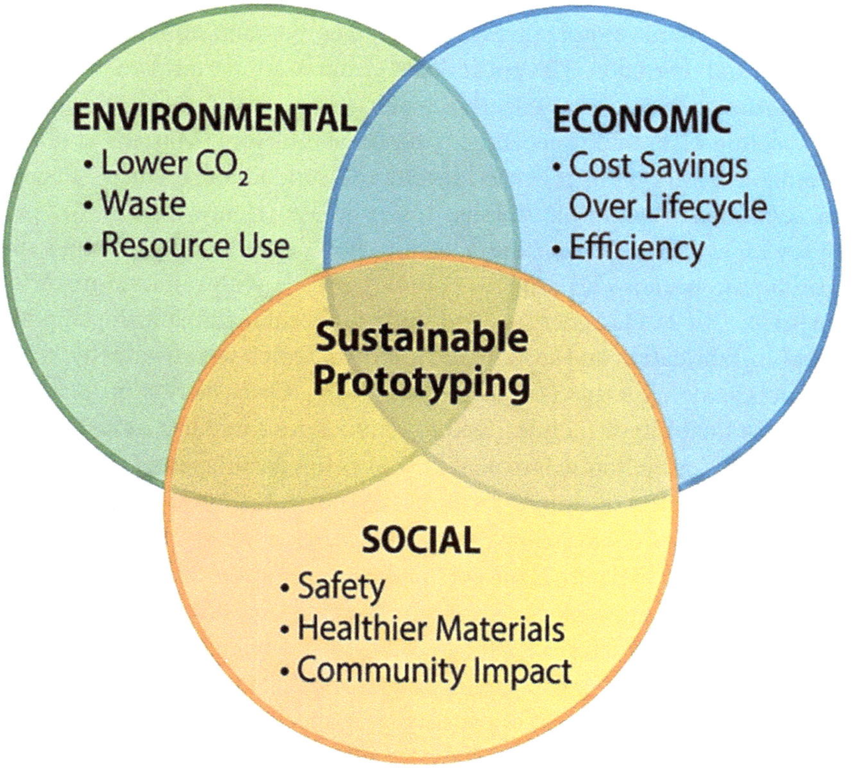

THE PROTOTYPING LIFECYCLE AND SUSTAINABLE STRATEGIES

The integration of sustainable strategies throughout the prototyping lifecycle is paramount for minimizing environmental impact and maximizing resource efficiency (Baldassarre et al., 2020). Sustainable design emphasizes the life-cycle implications of products, systems, and activities, aiming to optimize environmental and social well-being (Faludi et al., 2020). The design phase presents a critical opportunity to incorporate sustainability considerations into the prototyping process (Zajemska et al., 2025). This encompasses selecting environmentally benign materials, optimizing designs for resource efficiency, and considering the end-of-life management of prototypes. Waste should be appropriately managed and disposed of at each stage of the project. Implementing measures such as waste reduction, material reuse, and recycling can mitigate environmental consequences (Harun et al., 2025).

During the prototyping phase, designers should prioritize energy-efficient equipment and processes, minimize waste generation, and implement closed-loop systems for material recovery and reuse. The shift toward sustainable materials requires a comprehensive understanding of material characteristics, including environmental performance, functional properties, and availability. The construction sector can contribute to sustainable development by embracing eco-friendly building techniques and materials. By implementing energy-efficient lighting, HVAC systems, and water-saving fixtures, you can significantly reduce energy consumption and promote water conservation. The evaluation phase should encompass environmental performance assessment, life cycle analysis, and social impact assessment to identify areas for improvement and optimization. The entire construction industry needs to adopt sustainable practices to enhance environmental performance. The efficient utilization of materials, energy, and water is crucial for establishing long-term, durable values. To lessen environmental effects, recycling or the employment of renewable materials can also be beneficial. Environmental challenges such as pollution and resource depletion underscore the importance of a sustainable approach to manufacturing (Harun et al., 2025).

The prototyping process is not just about creating a physical model but also about understanding and addressing the environmental consequences of the process (Alting & Jøgensen, 1993). By embracing sustainable prototyping practices, organizations can reduce their environmental footprint, enhance their brand reputation, and contribute to a more sustainable future (Hussin et al., 2013). By following these guidelines, it is possible to create goods that are both useful and environmentally beneficial. Ecodesign reduces environmental concerns by redesigning manufacturing functions and improving systems (Burke et al., 2021). The application of sustainability in design is significant in creating more ecologically friendly products. Figure 3 synthesizes the core pillars that define sustainable prototyping, bringing together sustainable materials, eco-design, low-waste manufacturing, energy efficiency, waste management, digital tools, and LCA. It shows how these interdependent pillars collectively reduce environmental impact while maintaining prototype performance and feasibility.

Integrating downstream life-cycle data into eco-design tools is essential for achieving truly sustainable product development (Ramani et al., 2010). Adopting eco-design methodologies in the early stages is crucial; however, challenges such as the lack of information models, semantic interoperability, and measurement science can impede the realization of sustainable products and services (Ramani et al., 2010). The building industry must strike a balance among the three core aspects of environmental, social, and economic sustainability to create a sustainable built environment (Hauke et al., 2016). By incorporating stability into construction projects, resource consumption is reduced, energy efficiency is increased, operating costs are lowered, and the public image is improved (Harun et al., 2025; Singh, 2018).

Figure 3. Core pillars of sustainable prototyping

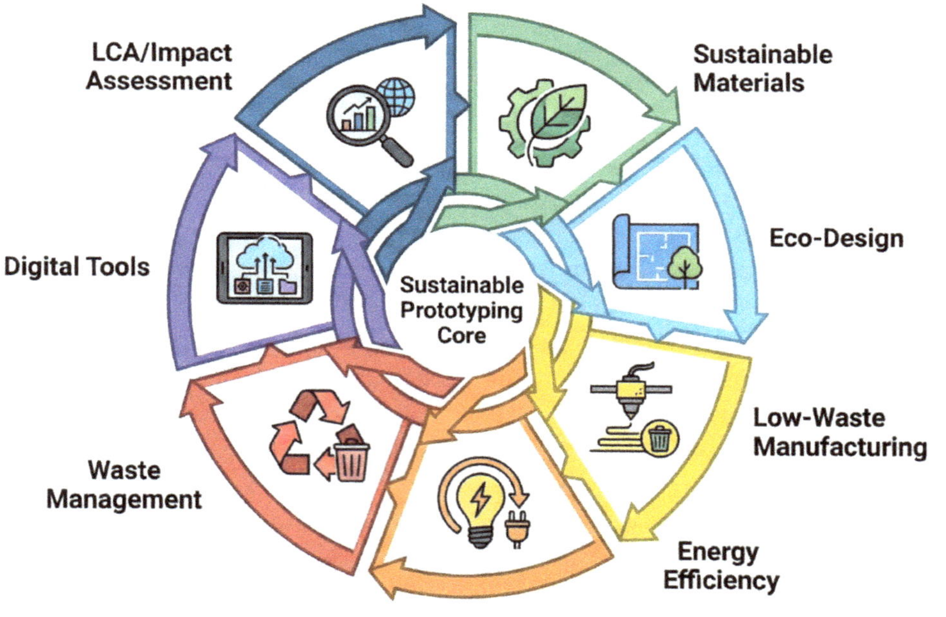

SUSTAINABLE MATERIALS IN PROTOTYPING RESULTS

The selection of materials plays a pivotal role in sustainable prototyping, with a growing emphasis on utilizing materials that minimize environmental impact and promote resource conservation. Sustainable materials are revolutionizing building design and construction, with eco-friendly concrete, FSC-certified wood, and natural insulating materials like cellulose fibers becoming standard in residential and commercial developments (Popescu et al., 2024). Materials like building blocks, flooring, paints, doors, windows, and conventional slabs can be replaced with sustainable alternatives identified through extensive literature reviews (Prajwal et al., 2021). These options help protect the environment, preserve resources, and build a more sustainable future for everybody. Selecting environmentally friendly building products is a great way to improve a building's environmental performance (Singh, 2018).

The adoption of sustainable building materials offers the construction industry a viable avenue to contribute to environmental protection by minimizing environmental impact and advancing sustainable development (Song & Zhang, 2018). Materials with low embodied energy, recycled content, and recyclability are gaining prominence in prototyping applications. The use of green building materials can reduce resource and energy consumption, as well as other environmental consequences associated with building (Firoozi et al., 2024). The use of SCMs in ESG construction is often touted for its environmental and financial benefits, but there are also some opposing arguments to consider.

The construction industry's substantial consumption of resources, including water, timber, and raw materials, and its significant contribution to waste generation and carbon emissions underscore the urgent need for sustainable practices (Tokbolat et al., 2019). Innovative materials that serve as sustainable substitutes for conventional building materials are essential to creating a more sustainable and environmentally responsible construction industry. Projects should adhere to environmental regulations by increasing workplace efficiency and lowering. Figure 4 presents the full prototyping lifecycle and shows the specific sustainable strategies applied at each stage, from eco-design and material selection to low-waste fabrication and end-of-life planning. By combining these actions in one continuous flow, the figure emphasizes that sustainability is achieved through coordinated decisions across the entire process rather than isolated interventions.

Using renewable and recyclable materials is essential for reducing the environmental impact of construction and promoting a circular economy. Adopting strategies such as waste prevention, material reuse, and recycling can help reduce environmental impacts. Sustainable construction techniques increasingly focus on the use of innovative materials to reduce environmental impact, enhance energy efficiency, and improve the overall sustainability of buildings and infrastructure (Iluyomade & Okwandu, 2024).

This involves using materials from sustainable sources, reducing waste, and implementing closed-loop systems to promote resource efficiency. By embracing sustainable practices and innovative materials, the construction industry can mitigate its environmental impact and contribute to a more sustainable future. Adopting ESG principles in the production and use of sustainable construction materials can significantly reduce energy waste. The use of sustainable construction materials can help achieve the UNSDGs by directly and indirectly impacting several goals and indicators, including climate action, affordable and clean energy, responsible consumption and production, and sustainable cities and communities (Yap et al., 2024).

Figure 4. Prototyping lifecycle with embedded sustainable strategies

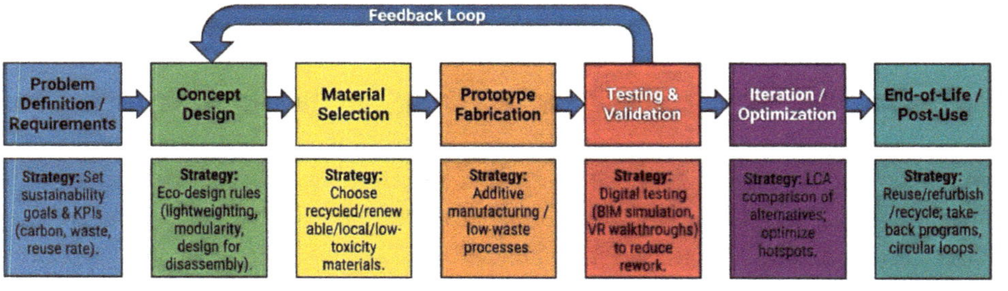

ENERGY USE AND EFFICIENCY

Energy use during prototyping operations determines both environmental pollution levels and resource depletion rates. The construction sector can achieve sustainable development by adopting ESG principles, which help reduce environmental harm caused by energy waste. The construction industry faces severe environmental problems because it requires large amounts of energy and natural resources while generating substantial waste output (Roque et al., 2021). The main objective of energy-efficient building design is to reduce fossil fuel dependence by lowering energy use, improving efficiency, and promoting the adoption of renewable energy.

The building sector focuses its sustainability initiatives on environmental performance assessment of constructed facilities and infrastructure systems. Design strategies should focus on sustainability and environmental responsibility by reducing energy use and promoting renewable materials and energy-conservation practices.

The substantial energy requirements for building material production make sustainable materials that need less production energy an effective way to reduce industrial environmental impact (Silva et al., 2023). Figure 5 maps the relative energy demand across prototyping lifecycle stages, highlighting where consumption is highest (typically material production, fabrication, and repeated iterations). This hotspot view supports targeted efficiency actions, helping teams prioritize interventions that deliver the largest energy and emissions reductions.

The entire building lifecycle requires energy optimization for sustainable construction practices, including the design and construction phases and the operational and demolition stages. The combination of energy-efficient design, construction practices, and material selection enables buildings to use less energy while reducing their environmental impact (Ahn et al., 2010). The construction industry generates 40% of global energy-related carbon dioxide emissions, making it an essential sector for reducing environmental impact (Priya et al., 2025). Adoption of sustainable construction practices and materials enables the built environment to become more sustainable and resilient while reducing energy use, lowering greenhouse gas emissions, and reducing waste production (Yap et al., 2024).

Figure 5. Energy hotspot map across lifecycle stages

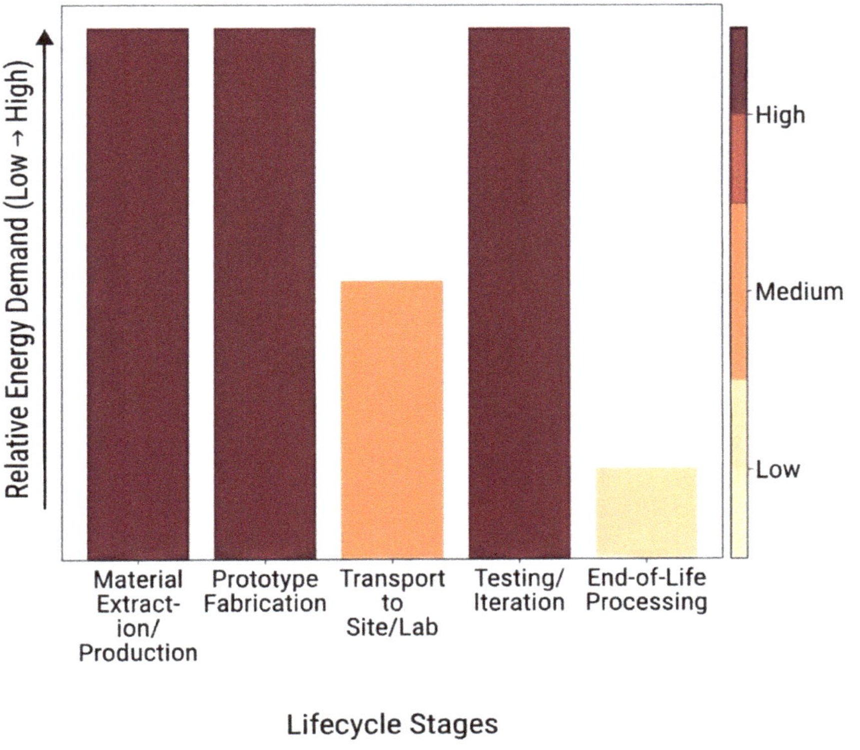

The construction industry must adopt resource-efficient practices and eliminate wasteful operations to achieve sustainable development in building processes. The construction industry can achieve significant reductions in carbon emissions by adopting energy-efficient technologies, optimizing building designs, and integrating renewable energy systems, thereby supporting global climate change mitigation (Yap et al., 2024). Green buildings achieve better environmental performance through the implementation of renewable energy systems, water collection systems, and landscaping elements.

The combination of passive solar heating and cooling systems with high-performance insulation and efficient lighting creates buildings that use less energy and reduce operating expenses. Sustainable construction methods exist to protect both the environment and preserve resources for future generations. Strategies that focus on waste reduction, material reuse, and recycling help reduce environmental impact while maximizing resource efficiency. The practice of waste management and recycling plays a vital role in making construction projects more sustainable. Figure 6 presents the energy pathway in prototyping, tracking how energy enters the process, is consumed in fabrication, transport, and testing, and where losses occur through inefficiencies or idle use. By visualizing these hotspots, the figure clarifies where efficiency strategies and cleaner energy sources can most effectively reduce the prototype's lifecycle impact.

Figure 6. Energy flow in the prototyping process (input- use- loss- output)

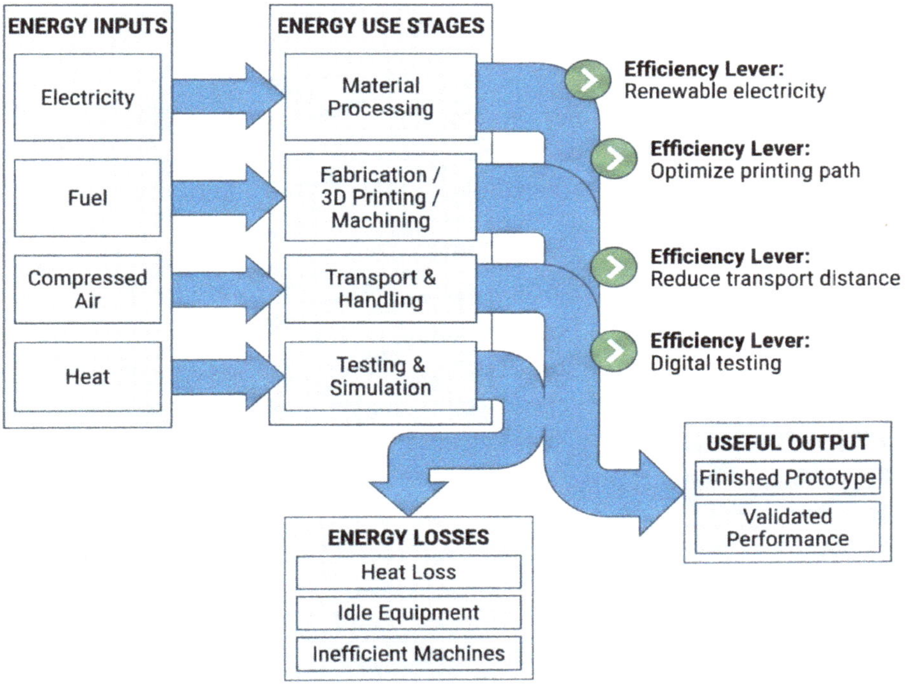

DESIGN FOR DISASSEMBLY AND LIFECYCLE MANAGEMENT

Design for Disassembly and lifecycle management are critical considerations in sustainable prototyping, focusing on the end-of-life phase of products and buildings to minimize waste and maximize resource recovery. By 2050, this transition could result in a 38% decrease in global CO2 emissions from materials, which is equivalent to two billion tons, mainly as a result of the lower need for steel, aluminum, concrete, and plastic (Bovo et al., 2025). Effective waste management entails minimizing waste generation, maximizing material usage, and reducing environmental effects.

The adoption of a circular economy approach in the construction industry necessitates a shift towards designing buildings and infrastructure with disassembly and deconstruction in mind, allowing for the recovery and reuse of materials at the end of their service life. Sustainable construction practices also emphasize durability and adaptability to ensure buildings can withstand evolving needs and environmental conditions over time. Life cycle assessment is a valuable tool for evaluating the environmental impacts of construction materials, products, and systems across their entire life cycle, from raw material extraction to end-of-life disposal. When designing construction projects, it is important to aim for stability by using environmentally friendly methods, conserving energy, thinking ecologically, and planning for the long term (Akadiri et al., 2012).

The integration of sustainable practices into construction projects not only reduces environmental impacts but also offers economic benefits, including lower operating costs, greater energy efficiency, and improved public image. Furthermore, by adopting a holistic approach to waste management, construction projects can minimize environmental effects and conserve resources for future generations. Construction and demolition waste must be handled using novel strategies in accordance with the Waste Framework Directive. Deconstruction, which entails carefully dismantling building parts for reuse, dramatically improves the sustainability of the building sector by replacing the conventional linear material consumption model with a cyclical one in which building components are made to be demountable, reusable, and recyclable (Akbarnezhad et al., 2013).

Effective waste management not only minimizes environmental impact but also contributes to resource conservation and cost savings, making it an essential component of sustainable construction practices. Construction projects that embrace sustainability can also improve indoor air quality, promote occupant health and well-being, and create more comfortable and productive environments. The design and construction phases of a building's life cycle produce construction and demolition waste, while the occupation and demolition phases also contribute to it. Adaptive reuse projects offer a long-term alternative to demolition, helping to conserve resources and reduce waste (Harun et al., 2025).

To ensure the project stays on schedule and within budget, frequent monitoring and communication are essential throughout each stage. Maximizing the amount of waste recycled or reused during refurbishment projects helps reduce the environmental impact of the works (Sung et al., 2011). Prioritizing design stability reduces use, saves energy, lowers costs, and improves public perception. When choosing materials, it is important to consider their environmental qualities and choose options that are better for the environment than what is typically used. By implementing sustainable prototyping practices, construction projects can minimize environmental impacts, conserve resources, and create more sustainable and resilient built environments for future generations. Figure 7 summarizes how Design for Disassembly is embedded across the prototyping lifecycle, linking modular design and reversible connections to circular end-of-life pathways (deconstruction, reuse, refurbishment, and recycling). It highlights the role of lifecycle management in reducing construction waste and in supporting the significant CO_2-reduction potential of circular material use, as projected by (Bovo et al., 2025).

Figure 7. Design for disassembly & lifecycle management in sustainable prototyping (Bovo et al., 2025)

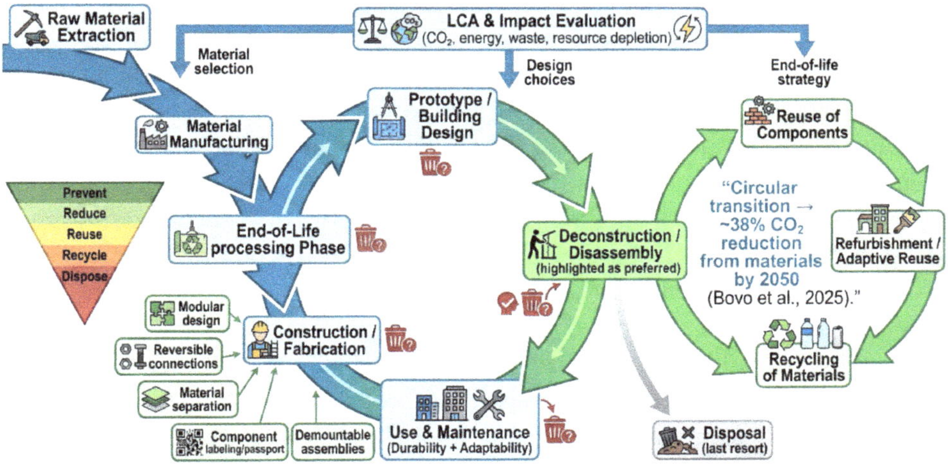

DIGITAL PROTOTYPING AND TOOLS FOR SUSTAINABILITY

Digital prototyping and tools for sustainability represent a transformative approach to design and construction, leveraging advanced technologies to optimize resource utilization, minimize waste, and enhance environmental performance (Kozlovská & Spišáková, 2013).

The selection of materials should prioritize those with minimal environmental impact, opting for renewable, recyclable, and locally sourced materials to reduce transportation emissions and support regional economies.

Digital tools play a crucial role in promoting sustainability by enabling designers and builders to make informed decisions, optimize building performance, and minimize environmental impacts throughout the building lifecycle. The ability to model, simulate, and analyze various design scenarios allows for the identification of potential environmental impacts and the implementation of strategies to mitigate them. Furthermore, integrating digital technologies into construction processes can improve efficiency, reduce waste, and enhance project outcomes. The integration of green practices, energy conservation, ecological considerations, and long-term development goals in overall design leads to reduced resource consumption, increased energy efficiency, lower operating costs, and a better public image (Shafiq & Afzal, 2020).

Digital prototyping tools enable designers and engineers to create virtual models of buildings and infrastructure, allowing for the simulation and analysis of various design scenarios to optimize energy efficiency, minimize material waste, and improve overall sustainability (Gunawardena et al., 2014). Building Information Modeling is a core component of sustainable design, offering capabilities for building performance analysis and simulation, enabling architects and engineers to analyze designs digitally before construction (Zhang & Xiao, 2013). Through digital modeling and simulation, designers can assess the environmental impacts of design options and make informed decisions to minimize adverse effects. BIM tools encourage integrated lifecycle green building management from design, construction, and prefabrication to operation and maintenance (Yang, 2014).

Virtual reality and augmented reality technologies offer immersive experiences that enable stakeholders to visualize and interact with building designs in realistic, intuitive ways, facilitating better communication, collaboration, and decision-making. These cutting-edge technologies can also be used to improve workplace safety, lower emissions, encourage recycling, and reduce the use of fossil fuels (Shafiq & Afzal, 2020). Figure 8 illustrates the integrated digital prototyping workflow, where sustainability goals and material inputs feed a BIM-based digital twin that centralizes the project model. Performance simulations and LCA analyses, combined with VR/AR collaborative reviews, create a continuous feedback loop that optimizes design decisions and reduces physical prototyping waste.

Green BIM considers sustainability in infrastructure development through digital models for design and construction, including assessments of energy flow and solar ecosystems (Bonenberg & Wei, 2015). The integration of BIM with sustainable design principles enables designers to make informed decisions, optimize building performance, and minimize environmental impacts throughout the building lifecycle (Goyal & Singh, 2020). BIM technology helps designers and architects identify and assess sustainable design parameters early, enhancing building sustainability (Taha, 2020).

Figure 8. Integrated flow of digital inputs, BIM-based digital twinning, performance analysis and immersive collaborations tools

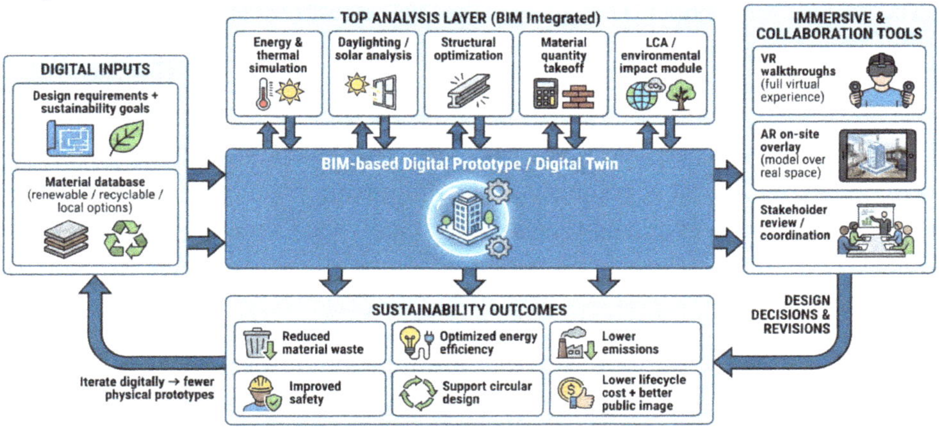

ASSESSMENT METHODS FOR SUSTAINABLE PROTOTYPING

Assessment methods for sustainable prototyping are essential for evaluating the environmental, social, and economic impacts of proposed designs and construction processes. Life Cycle Assessment is a comprehensive approach that evaluates the environmental impacts of a product or system throughout its entire life cycle, from raw material extraction to end-of-life disposal.

LCA helps quantify energy consumption, greenhouse gas emissions, and waste generation, providing valuable insights for identifying opportunities to reduce environmental impacts and improve sustainability performance. LCA has been used to evaluate environmental performance in building projects, incorporating all stages from resource extraction to end-of-life processes (Bamgbade et al., 2017). The complexity of building projects and the availability of resources necessitate realistic, attainable timelines.

Carbon footprint analysis quantifies the greenhouse gas emissions associated with a product, process, or activity, providing a precise, concise measure of its contribution to climate change. By quantifying the carbon footprint, stakeholders can identify opportunities to reduce emissions and mitigate climate change impacts through sustainable design and construction practices. Incorporating stability principles into construction projects enhances resource efficiency, promotes energy conservation, lowers operating expenses, and strengthens public perception.

Material flow analysis examines the flow of materials through a system, from raw material extraction to manufacturing, use, and disposal, to identify opportunities to reduce waste, improve resource efficiency, and promote circular economy principles. By tracking material flows, stakeholders can identify opportunities to reduce material consumption, increase recycling rates, and minimize environmental impacts associated with material production and disposal.

Environmental impact assessment is a systematic process for evaluating the potential environmental consequences of a proposed project or development, considering factors such as air and water quality, biodiversity, and ecosystem services. EIA helps decision-makers understand the potential environmental impacts of a project and identify mitigation measures to minimize negative consequences and promote environmental sustainability.

Evaluating fair labor practices and ensuring safe working conditions are crucial aspects of assessing the sustainability of construction materials (Yap et al., 2024). Furthermore, conducting environmental assessments within the built environment necessitates contextualization specific to construction firms, benefiting both the firms and the environment (Bamgbade et al., 2017). These assessments enable the quantification of emissions linked to different lifecycle stages, building materials, and construction processes (Chaudhary & Akhtar, 2022). Life cycle assessments can be employed to evaluate environmental performance in building projects, encompassing all stages from resource extraction to end-of-life processes (Park, 2025). Moreover, companies can integrate ESG considerations into their selection of construction materials to achieve cost savings and improve operational efficiencies (Yap et al., 2024).

Sustainable construction practices are motivated by concerns over climate change, resource depletion, and social equity (Misnan et al., 2024). Construction projects are adopting sustainability practices to protect the environment, boost profits, and improve customer satisfaction (Liu et al., 2019). Additionally, life-cycle cost analysis can reveal significant savings from reduced maintenance, lower energy consumption, and potential incentives or tax credits (Yap et al., 2024).

IMPLEMENTATION IN PRACTICE: CASE STUDIES

The integration of sustainable design principles, energy-efficient technologies, and eco-friendly materials in green building projects yields successful outcomes that reduce environmental impacts while creating high-performance buildings (Bamgbade et al., 2016). Sustainable design and construction methods in these projects demonstrate their ability to reduce energy consumption, improve water conservation, enhance indoor air quality, and improve occupant health (Ochieng et al., 2014). BIM integration with sustainable design principles enables designers to select optimal building solutions that deliver better performance and environmental protection from start to finish of the building cycle. A study of construction projects showed that stakeholder involvement is a critical factor in successful project execution and enduring sustainability (Siddiqi et al., 2008).

Adaptive reuse projects serve as examples that show how existing buildings can be transformed into new spaces while minimizing waste and resource consumption and protecting cultural heritage. The transformation of existing buildings into contemporary, energy-efficient facilities is made possible by adaptive reuse projects that preserve their original architectural features while reducing the environmental impacts of new construction. The construction process is more successful when projects involve community members from the start, as this leads to better social equity and stronger bonds with residents (Siddiqi et al., 2008). The construction industry needs to prioritize sustainable materials in ESG construction to reduce the environmental impact of buildings while supporting the achievement of the UN Sustainable Development Goals (Yap et al., 2024).

The implementation of modular construction through off-site fabrication brings multiple advantages to projects, including decreased waste generation, accelerated project completion, and enhanced operational efficiency. The construction projects show that modular building methods deliver better performance, lower costs, and reduced environmental impact than conventional construction techniques.

The construction industry in Malaysia faces both benefits and challenges in implementing ESG practices because sustainability targets remain essential for the growth of sustainable building practices, which have increased public understanding of how buildings affect the environment and society (Sulaiman et al., 2024).

The construction industry has seen a significant (Kamali & Hewage, 2016; Misopoulos et al., 2019; Sobin et al., 2010). The worldwide drive for sustainability has become more evident because the COVID-19 pandemic revealed how economic activities affect social structures and environmental systems (Jamaludin et al., 2025).

TRENDS, CHALLENGES, AND FUTURE DIRECTIONS

The construction industry is undergoing an ongoing transformation due to new industry trends, ongoing difficulties, and shifting business priorities (Jamaludin et al., 2025). The construction industry is undergoing a digital transformation through BIM, drones, and IoT sensors, which enhance project management, improve building performance, and enhance team collaboration (Wang & Xue, 2024). The construction industry needs to partner with stakeholders on community-based projects to create positive social impacts from its construction activities (Wang & Xue, 2024). The construction of buildings will depend heavily on renewable materials, renewable energy systems, intelligent building systems, and sustainable supply chain management (Wang & Xue, 2024).

Sustainable practices in the construction industry continue to gain popularity, but multiple barriers prevent their widespread adoption. The construction industry faces multiple barriers to sustainable practice, including employee resistance to change. The construction industry can achieve greater efficiency and waste reduction, and better project results through adoption, due to high initial expenses, restricted funding options, insufficient public knowledge and training, regulatory hurdles, and technological progress.

Multiple essential areas require additional funding and focus to help the construction industry transition more quickly toward sustainable practices. The development of common standards, the sharing of best practices, and the promotion of sustainable innovation depend on stakeholder collaboration among governments, industry associations, and research institutions (Goubran, 2019). Sustainable building practices receive government support through financial incentives, including tax benefits and density rewards, as well as simplified building authorization procedures (Yunus et al., 2024). The built environment reduces carbon emissions by using sustainable materials, such as recycled plastics, timber, and bamboo. Sustainable construction depends on data integration, which requires stakeholders to move away from their conventional information-sharing practices (Khattra et al., 2020). The construction sector needs to establish training programs that teach professionals about sustainable building design, construction, and operations. The adoption of Industry 4.0 technology shows potential for sustainable construction, but its implementation in developing nations faces multiple obstacles during its initial development phase (Lozano-Ramírez & Garcia-Lopez, 2025). The increased focus from regulators and investors on ESG matters will force construction companies to disclose their sustainability initiatives and environmental performance data with greater transparency (Wang & Xue, 2024). The successful implementation of ESG depends on the active involvement of construction industry professionals and their relevant stakeholders, who must reach consensus (Wang & Xue, 2024). The construction industry needs to collaborate with other sectors to achieve better environmental protection by sharing resources and exchanging expertise (Yan, 2015).

The growing market demand for green buildings, together with technological advances and material developments, will continue to accelerate the adoption of sustainable practices. The construction industry requires its own methodology for implementing Green Supply Chain Management, as this concept has proven successful in other sectors. However, it needs to be adapted to the construction industry, a vital sector for reaching sustainability targets while reducing the environmental and social consequences of built environments. (Benachio et al., 2019).

CONCLUSION AND RECOMMENDATIONS

Traditional step-by-step construction methods use large amounts of materials, energy, and water, and produce waste and pollution. Sustainable prototyping is a needed change from those ways. The paper shows that sustainability gains are most significant when sustainable prototyping is built into the prototyping lifecycle rather than added to it. Choosing materials such as recycled, renewable, low-toxicity, and locally sourced options reduces the carbon embodied in the material and supports reuse. Sustainable materials also support use. In my work, I see that eco-design principles, such as modularity, lightweight design, durability, and design for disassembly, reduce waste at the source and keep prototype parts functional after testing. Eco-design principles let you reuse components by throwing them away. I also see that additive manufacturing adds to this shift. Additive manufacturing places material where it is needed, cuts offcuts, and creates shapes that use less material. Additive manufacturing creates geometries that are hard to achieve with traditional manufacturing. Alongside the choices, energy efficiency strategies and structured waste management plans lower the impacts during prototype production and testing. Energy efficiency strategies cut energy use. Structured waste management plans cut waste. I think sustainability works as a system of linked decisions. The integration of life cycle assessment and environmental impact evaluation gives proof of improvement. The integration of life cycle assessment and environmental impact evaluation lets firms compare alternatives, set benchmarks, and refine methods over time. IT sees that digital prototyping tools such as BIM, simulation, and Virtual Reality/Augmented Reality help the design process by reducing repetition and enabling early performance checks, working together, and testing for sustainability. Because of that, sustainable prototyping is not a response; it is also an economic and social chance. Digital prototyping tools lower the cost over the life of the product by making work more efficient and by cutting waste. Digital prototyping tools help the project meet environmental, social, and governance (ESG) impact rules. Digital prototyping tools also support development goals. Construction stakeholders need to speed up adoption. Construction stakeholders should start prototyping in the concept phase. Construction stakeholders should use material selection databases and clear eco-design rules. Construction stakeholders should grow manufacturing with pilot projects. Construction stakeholders should let pilot projects grow step by step. Construction stakeholders should add circular economy targets to project KPIs so that reuse, refurbishment, and recycling become a part of work, not an event. Through an in-depth examination of the existing literature, it becomes apparent that stakeholders in the construction industry can achieve tangible outcomes by adhering to a specific sequence of actions. This structured approach has been shown to yield positive results, underscoring the importance of a methodical and informed strategy in this field. By embracing these key steps, construction stakeholders can set themselves up for success and ultimately drive meaningful progress in their projects. It should establish on-site waste hierarchy practices. Firms need to separate waste, set up take-back agreements, and plan to take prototypes. Firms should treat Life Cycle Assessment (LCA) as a decision tool, not an optional add-on. Firms must invest in the tools. Firms need the training. Firms need to address cost concerns and resistance to change. Firms want the teams to benefit from the computer model decisions that help the environment. The policymakers and industry leaders should encourage prototyping through incentives, green procurement standards, and cross-sector collaboration. The cross-sector collaboration should involve academia, manufacturers, and construction firms. The organizations should track ESG outcomes from prototyping. Report the outcomes transparently. The actions make the proposed framework practical and expandable. The actions enable innovation without sacrificing performance or project goals.

REFERENCES

Abera, Y. A. (2024). Sustainable building materials: A comprehensive study on eco-friendly alternatives for construction. *Composites and Advanced Materials*, *33*, 26349833241255957. Advance online publication. DOI: 10.1177/26349833241255957

Ahn, C. R., Lee, S. H., Peña-Mora, F., & AbouRizk, S. (2010). Toward Environmentally Sustainable Construction Processes: The U.S. and Canada's Perspective on Energy Consumption and GHG/CAP Emissions. *Sustainability (Basel)*, *2*(1), 354–370. DOI: 10.3390/su2010354

Akadiri, P. O., Olomolaiye, P., & Chinyio, E. (2012). Multi-criteria evaluation model for selecting sustainable materials for building projects. *Automation in Construction*, *30*, 113–125. DOI: 10.1016/j.autcon.2012.10.004

Akbarnezhad, A., Ong, K. C. G., & Chandra, L. R. (2013). Economic and environmental assessment of deconstruction strategies using building information modeling. *Automation in Construction*, *37*, 131–144. DOI: 10.1016/j.autcon.2013.10.017

Alam, M., Arifuzzaman, M., Islam, M. K., Alfuhaid, A. F., & Al-Mamun, A. (2022). Sustainable Solution for Deteriorated and Aged RCC Structures: A Review of Buildings, Bridges and Pavements [Review of Sustainable Solution for Deteriorated and Aged RCC Structures: A Review of Buildings, Bridges and Pavements]. IOP Conference Series Earth and Environmental Science, 1026(1), 12009. IOP Publishing. DOI: 10.1088/1755-1315/1026/1/012009

Almusaed, A., Yitmen, İ., Myhren, J. A., & Almssad, A. (2024). Assessing the Impact of Recycled Building Materials on Environmental Sustainability and Energy Efficiency: A Comprehensive Framework for Reducing Greenhouse Gas Emissions. *Buildings (Basel, Switzerland)*, *14*(6), 1566. DOI: 10.3390/buildings14061566

Alting, D. L., & Jøgensen, D. J. (1993). The Life Cycle Concept as a Basis for Sustainable Industrial Production. *CIRP Annals*, *42*(1), 163–167. DOI: 10.1016/S0007-8506(07)62417-2

Attiya, E. M. E., Shebl, M. A., & Nasser, M. M.Eman M. E. AttiyaMohamed A. SheblMaii M. Nasser. (2020). A Comparative Analysis of LEED and GPRS for the Applicability in Egyptian Office Buildings. *International Journal of Engineering Research & Technology (Ahmedabad)*, *3*(3). Advance online publication. DOI: 10.17577/IJERTV9IS030299

Baldassarre, B., Konietzko, J., Brown, P., Calabretta, G., Bocken, N., Karpen, I. O., & Hultink, E. J. (2020). Addressing the design-implementation gap of sustainable business models by prototyping: A tool for planning and executing small-scale pilots. *Journal of Cleaner Production*, *255*, 120295. DOI: 10.1016/j.jclepro.2020.120295

Bamgbade, J. A., Kamaruddeen, A. M., & Nawi, M. N. M. (2017). Towards environmental sustainability adoption in construction firms: An empirical analysis of market orientation and organizational innovativeness impacts. *Sustainable Cities and Society*, *32*, 486–495. DOI: 10.1016/j.scs.2017.04.015

Bamgbade, J. A., Mohammed, K. A., & Nawi, M. N. M. (2016). Assessing the sustainable construction of large construction companies in Malaysia. *AIP Conference Proceedings*, *1761*, 20027. DOI: 10.1063/1.4960867

Benachio, G. L. F., Freitas, M. do C. D., & Tavares, S. F. (2019). Green Supply Chain Management in the Construction Industry: A literature review [Review of Green Supply Chain Management in the Construction Industry: A literature review]. IOP Conference Series Earth and Environmental Science, 225, 12011. IOP Publishing. DOI: 10.1088/1755-1315/225/1/012011

Bonenberg, W., & Wei, X. (2015). Green BIM in Sustainable Infrastructure. *Procedia Manufacturing*, *3*, 1654–1659. DOI: 10.1016/j.promfg.2015.07.483

Bovo, M., Santolini, E., & Barbaresi, A. (2025). An Alternative Modular Wooden System for Fast Assembly/Disassembly of Buildings. *Buildings (Basel, Switzerland)*, *15*(7), 1196. DOI: 10.3390/buildings15071196

Burke, H., Zhang, A., & Wang, J. X. (2021). Integrating product design and supply chain management for a circular economy. *Production Planning and Control*, *34*(11), 1097–1113. DOI: 10.1080/09537287.2021.1983063

Chaudhary, A., & Akhtar, A. (2022). Template for Evaluating Cradle-to-Site Environmental Life Cycle Impacts of Buildings in India. *ACS Environmental Au*, *3*(2), 94–104. DOI: 10.1021/acsenvironau.2c00035

Chow, T. C., Zailani, S., Rahman, M. K., Zhang, Q., Bhuiyan, M. A., & Patwary, A. K. (2021). Impact of sustainable project management on project plan and project success of the manufacturing firm: Structural model assessment. *PLoS One*, *16*(11), e0259819. Advance online publication. DOI: 10.1371/journal.pone.0259819

Despeisse, M., Ball, P., Evans, S., & Levers, A. (2012). Industrial ecology at factory level – a conceptual model. *Journal of Cleaner Production*, *31*, 30–39. DOI: 10.1016/j.jclepro.2012.02.027

Faludi, J., Hoffenson, S., Kwok, S. Y., Saidani, M., Hallstedt, S. I., Telenko, C., & Martinez, V. (2020). A Research Roadmap for Sustainable Design Methods and Tools. *Sustainability (Basel)*, *12*(19), 8174. DOI: 10.3390/su12198174

Firoozi, A. A., Firoozi, A. A., Oyejobi, D. O., Avudaiappan, S., & Flores, E. S. (2024). Emerging Trends in Sustainable Building Materials: Technological Innovations, Enhanced Performance, and Future Directions. *Results in Engineering*, *24*, 103521. DOI: 10.1016/j.rineng.2024.103521

Giudice, F., Barbagallo, R., & Fargione, G. (2020). A Design for Additive Manufacturing approach based on process energy efficiency: Electron beam melted components. *Journal of Cleaner Production*, *290*, 125185. DOI: 10.1016/j.jclepro.2020.125185

Goubran, S. (2019). On the Role of Construction in Achieving the SDGs. *Journal of Sustainability Research*, *1*(2). Advance online publication. DOI: 10.20900/jsr20190020

Goyal, L. K., & Singh, H. (2020). BIM Approach for Sustainable Design of Flat Slab Buildings: A Review [Review of BIM Approach for Sustainable Design of Flat Slab Buildings: A Review]. IOP Conference Series Materials Science and Engineering, 955(1), 12012. IOP Publishing. DOI: 10.1088/1757-899X/955/1/012012

Gunawardena, D., Mendis, P., Ngo, D., Aye, L., & Alfano, J. (2014). Sustainable Prefabricated Modular Buildings. DOI: 10.13140/2.1.4847.3920

Gupta, S., Dangayach, G. S., & Singh, A. K. (2015). Key Determinants of Sustainable Product Design and Manufacturing. *Procedia CIRP*, *26*, 99–102. DOI: 10.1016/j.procir.2014.07.166

Gurova, O., Merritt, T., Papachristos, E., & Vaajakari, J. (2020). Sustainable Solutions for Wearable Technologies: Mapping the Product Development Life Cycle. *Sustainability (Basel)*, *12*(20), 8444. DOI: 10.3390/su12208444

Harun, S. N., Kamaruddin, N., & Choo, J. (2025). INTERIOR DESIGN AND RENOVATION WORKS FOR 2 LEVEL FACTORY OFFICE AT ROVSKI, USJ.

Hauke, B., Kuhnhenne, M., Lawson, M., Veljković, M., & Siebers, R. (2016). What does 'sustainable construction' mean? An overview (p. 1). DOI: 10.1002/9781118740828.ch1

Ho, F. H., Hanim, A.-R. S., Ariffin, R. G. R., Rahman, M. N. A., Hassan, M. F., Arifin, A. M. T., Haq, R. H. A., Ahmad, S., & Abdullah, H. (2019). Analytical analysis for material efficiency practice: An exploratory study of Malaysia manufacturing cases. *Journal of Physics: Conference Series*, *1150*, 12006. DOI: 10.1088/1742-6596/1150/1/012006

Hussin, J. M., Rahman, I. A., & Memon, A. H. (2013). The Way Forward in Sustainable Construction: Issues and Challenges. *International Journal of Advances in Applied Sciences*, *2*(1). Advance online publication. DOI: 10.11591/ijaas.v2i1.1321

Iluyomade, T. D., & Okwandu, A. C.Tosin Daniel IluyomadeAzubuike Chukwudi Okwandu. (2024). Innovative materials in sustainable construction: A review [Review of Innovative materials in sustainable construction: A review]. *International Journal of Science and Research Archive*, *12*(1), 2435–2447. DOI: 10.30574/ijsra.2024.12.1.1048

Jamaludin, A. F., Husain, M. F. H., & Razali, M. N. (2025). Assessing environmental, social and governance of property- listed companies in Malaysia.pdf.

Kamali, M. H., & Hewage, K. (2016). Development of performance criteria for sustainability evaluation of modular versus conventional construction methods. *Journal of Cleaner Production*, *142*, 3592–3606. DOI: 10.1016/j.jclepro.2016.10.108

Khattra, S. K., Singh, H., & Singh, J. (2020). Leveraging the Potential of BIM towards Sustainable Construction. *IOP Conference Series. Materials Science and Engineering*, *955*(1), 12011. DOI: 10.1088/1757-899X/955/1/012011

Kozlovská, M., & Spišáková, M. (2013). Contruction waste generation across construction project life-cycle. *Organization Technology and Management in Construction An International Journal*, *5*(1), 687–695. DOI: 10.5592/otmcj.2013.1.5

Kumar, R., Aggarwal, V., & Surinder, M. G. (2021). Sustainable materials and techniques in affordable high-rise buildings - A case study. E3S Web of Conferences, 309, 1080. DOI: 10.1051/e3sconf/202130901080

Lauff, C., Kotys-Schwartz, D., & Rentschler, M. E. (2018). What is a Prototype? What are the Roles of Prototypes in Companies? *Journal of Mechanical Design*, *140*(6), 061102. Advance online publication. DOI: 10.1115/1.4039340

Laverne, F., Marquardt, R., Segonds, F., Koutiri, I., & Perry, N. (2019). Improving resources consumption of additive manufacturing use during early design stages: A case study. *International Journal of Sustainable Engineering*, *12*(6), 365–375. DOI: 10.1080/19397038.2019.1620897

Liu, G., Zhang, C., Zhao, M., Guo, W., & Luo, Q. (2020). Comparison of Nanomaterials with Other Unconventional Materials Used as Additives for Soil Improvement in the Context of Sustainable Development. Nanomaterials, 11(1), 15. Multidisciplinary Digital Publishing Institute. DOI: 10.3390/nano11010015

Liu, S. H., Rahmawati, Y., & Zawawi, N. A. W. A. (2019). Critical success factors of collaborative approach in delivering sustainable construction. MATEC Web of Conferences, 270, 5003. DOI: 10.1051/matecconf/201927005003

Lozano-Ramírez, N. E., & Garcia-Lopez, N. P. (2025). Mapping Construction 4.0 and Sustainability for Innovation in the Construction Industry of Developing Countries. 1(1). DOI: 10.7771/3067-4883.1512

Manu, B. A. (2024). Innovative Construction Materials: Advancing Sustainability, Durability, Efficiency, and Cost-effectiveness in Modern Infrastructure. *International Journal of Research Publication and Reviews*, *5*(12), 4987–4999. DOI: 10.55248/gengpi.5.1224.0215

Mariana, Y., Suryawinata, B. A., & Peranginangin, E. (2022). Sustainability in West Java traditional house: A case study from Julang Ngapak House. *IOP Conference Series. Earth and Environmental Science*, *998*(1), 12040. DOI: 10.1088/1755-1315/998/1/012040

Misnan, M. S., Ismail, M. Z., & Yan, T. J. (2024). Construction Project Management Issues and Development in Current for Future Construction Project: Challenges and Prospects in Sustainable Project Management. *International Journal of Research and Innovation in Social Science*, *1997*(II), 1997–2011. Advance online publication. DOI: 10.47772/IJRISS.2024.802141

Misopoulos, F., Manthou, V., & Michaelides, Z. (2019). Environmental and Social Sustainability in UK Construction Industry: A Systematic Literature Review. *European Journal of Economics and Business Studies*, *5*(1), 100. DOI: 10.26417/ejes.v5i1.p100-115

Neyestani, B. (2017). A Review on Sustainable Building (Green Building) [Review of A Review on Sustainable Building (Green Building)]. SSRN Electronic Journal. RELX Group (Netherlands). DOI: 10.2139/ssrn.2968885

Ochieng, E. G., Wynn, T. S., Zuofa, T., Ruan, X., & Price, A. D. F.. (2014). Integration of Sustainability Principles into Construction Project Delivery. *J Archit Eng Tech*, *3*(1), 116. DOI: 10.4172/2168-9717.1000116

Orsi, A., Guillén-Guillamón, I., & Pellicer, E. (2020). Optimization of Green Building Design Processes: Case Studies within the European Union. *Sustainability (Basel)*, *12*(6), 2276. DOI: 10.3390/su12062276

Park, J. (2025). A Holistic Approach to Managing Built Assets'. *Life Cycle Carbon Footprint.*, *1*(1). Advance online publication. DOI: 10.7771/3067-4883.1376

Popescu, C., Dissanayake, H., Mansi, E., & Stancu, A. (2024). Eco Breakthroughs: Sustainable Materials Transforming the Future of Our Planet. *Sustainability (Basel)*, *16*(23), 10790. DOI: 10.3390/su162310790

Prajwal, S., Kumar, V., & Patil, N. N. (2021). Implementation of sustainable materials in an existing residential building and comparison. *IOP Conference Series. Materials Science and Engineering*, *1166*(1), 12038. DOI: 10.1088/1757-899X/1166/1/012038

Priya, E., Vasanthi, P., Prabhu, B., & Murugesan, P. (2025). Sawdust as a Sustainable Additive: Comparative Insights into its Role in Concrete and Brick Applications. *Cleaner Waste Systems*, *100286*, 100286. Advance online publication. DOI: 10.1016/j.clwas.2025.100286

Ramani, K., Ramanujan, D., Bernstein, W. Z., Zhao, F., Sutherland, J. W., Handwerker, C. A., Choi, J., Kim, H., & Thurston, D. (2010). Integrated Sustainable Life Cycle Design: A Review [Review of Integrated Sustainable Life Cycle Design: A Review]. Journal of Mechanical Design, 132(9). American Society of Mechanical Engineers. DOI: 10.1115/1.4002308

Roque, A. J., Paleologos, E. K., O'Kelly, B. C., Tang, A. M., Reddy, K. R., Vitone, C., Mohamed, A. O., Koda, E., Goli, V. S. N. S., Vieira, C. S., Fei, X., Sollecito, F., Vaverková, M. D., Plötze, M., Petti, R., Podlasek, A., Puzrin, A. M., Cotecchia, F., Osiński, P., & Singh, D. N. (2021). Sustainable environmental geotechnics practices for a green economy. *Environmental Geotechnics*, *9*(2), 68–84. DOI: 10.1680/jenge.21.00091

Shafiq, M. T., & Afzal, M. (2020). Potential of Virtual Design Construction Technologies to Improve Job-Site Safety in Gulf Corporation Council. *Sustainability (Basel)*, *12*(9), 3826. DOI: 10.3390/su12093826

Siddiqi, K., Chatman, D., & Cook, G. (2008). Role of education and industry towards more sustainable construction. *International Journal of Environmental Technology and Management*, *8*(2/3), 310. DOI: 10.1504/IJETM.2008.017336

Silva, A., Gaspar, F., & Бакатович, A. A. (2023). Composite Materials of Rice Husk and Reed Fibers for Thermal Insulation Plates Using Sodium Silicate as a Binder. *Sustainability (Basel)*, *15*(14), 11273. DOI: 10.3390/su151411273

Singh, C. S. (2018). Green Construction: Analysis on Green and Sustainable Building Techniques. *Civil Engineering Research Journal*, *4*(3). Advance online publication. DOI: 10.19080/CERJ.2018.04.555638

Sobin, N., Molenaar, K. R., & Gransberg, D. D. (2010). Sustainability by..." A Synthesis of Procurement Approaches for High Performance Buildings. Construction Research Congress 2022, 1366. DOI: 10.1061/41109(373)137

Song, Y., & Zhang, H. (2018). Research on sustainability of building materials. *IOP Conference Series. Materials Science and Engineering*, *452*, 22169. DOI: 10.1088/1757-899X/452/2/022169

Sulaiman, N. N., Omar, A. J., Pakir, F., & Mohamad, M. (2024). Evaluating Environmental, Social and Governance (ESG) Practice Among Malaysian Public Listed Construction Companies Using FTSE Russell Rating Model. *International Journal of Sustainable Construction Engineering Technology*, *15*(1). Advance online publication. DOI: 10.30880/ijscet.2024.15.01.003

Sung, H., Kim, K., & Kim, C. K. (2011). Case Study on Selective Demolition Method for Refurbishing Deteriorated Residential Apartments. *Journal of Construction Engineering and Management*, *138*(2), 294–303. DOI: 10.1061/(ASCE) CO.1943-7862.0000424

Taha, F. F. (2020). Adoption of BIM technology to develop sustainable buildings in the Iraqi construction sector. *Indian Journal of Science and Technology*, *13*(15), 1596–1606. DOI: 10.17485/IJST/v13i15.223

Tokbolat, S., Karaca, F., Durdyev, S., & Calay, R. K. (2019). Construction professionals' perspectives on drivers and barriers of sustainable construction. *Environment, Development and Sustainability*, *22*(5), 4361–4378. DOI: 10.1007/s10668-019-00388-3

Wang, J., & Xue, F. (2024). Emerging Trends of ESG in the Construction Sector: A Promising Pathway to Sustainable and Responsible Development. In Lecture notes in operations research (p. 2033). Springer International Publishing. DOI: 10.1007/978-981-97-1949-5_142

Yan, M.-R. (2015). Project-Based Market Competition and Policy Implications for Sustainable Developments in Building and Construction Sectors. *Sustainability (Basel)*, *7*(11), 15423–15448. DOI: 10.3390/su71115423

Yang, Y. (2014). Integrating BIM Technology to Achieve Green Buildings' Sustainable Performance. *Advanced Materials Research*, *1271*. Advance online publication. DOI: 10.4028/www.scientific.net/amr.1073-1076.1271

Yap, C. K., Leow, C. S., & Goh, B. (2024). Sustainable construction materials under ESG: a literature review and synthesis [Review of Sustainable construction materials under ESG: a literature review and synthesis]. MOJ Biology and Medicine, 9(1), 1. MedCrave Group. DOI: 10.15406/mojbm.2024.09.00208

Yunus, J. N., Fahrurazi, L. S., Hamid, M. S. A., Noor, R. N. H. R. M., & Majid, M. R. A. (2024). The Implementation of Sustainability Practices in Construction Site and its Effectiveness in Klang Valley. *International Journal of Academic Research in Business & Social Sciences*, *14*(9). Advance online publication. DOI: 10.6007/IJARBSS/v14-i9/23020

Zajemska, M., Wojtyto, D., Michalik, J., & Berski, S. (2025). Review of Current Trends in Sustainable Construction. *Energies*, *18*(10), 2559. DOI: 10.3390/en18102559

Zhang, S., & Xiao, T. (2013). Building Information Modeling and Sustainable Architecture Design Analysis. DOI: 10.2991/icaicte.2013.15

Chapter 2
Integrating Smart and Sustainable Technologies in Product Design:
Bridging Innovation and Environmental Responsibility

Kaushal Kishore Mishra
https://orcid.org/0000-0001-6466-0575
Chandigarh University, India

Pawan Pant
https://orcid.org/0000-0003-1774-5650
Chandigarh University, India

Azmee Zaheer
https://orcid.org/0000-0002-6281-2608
Chandigarh University, India

Pushpak Sharma
Chandigarh University, India

ABSTRACT

Product design has revolutionised the way businesses integrate innovative and sustainable technologies into product development, bridging the gap between innovation and environmental responsibility. This chapter explores the convergence of advanced technologies, including IoT, AI, and digital twins, with principles of sustainability, such as the circular economy, eco-design, and lifecycle management. It demonstrates how synergies are creating innovative, efficient, and environmen-

DOI: 10.4018/979-8-3373-5473-6.ch002

tally responsive products, highlighting the need to address material choice, energy use, and waste reduction in material production and usage. Cases and frameworks illustrate best practices that stakeholders can use to apply such systems. Finally, it analyses emergent trends and ethical concerns that highlight implications for policy innovation in the pursuit of a sustainable future.

1. INTRODUCTION

Sustainable design with intelligence is transforming how products are designed, driving innovation in environmental stewardship. Considering the significant pressure on companies to address climate and resource scarcity issues, the emerging advanced technologies of IoT, AI, and digital twins have provided unique opportunities to achieve minimal environmental impacts when product performance optimisation is considered a transformative change (Geissdoerfer et al., 2017). These technologies not only make the products functional but also enable real-time monitoring, predictive analytics, and customisation, extend product life cycles, and prevent waste (Kamble et al., 2020). Concomitantly, sustainability principles, such as circular economy frameworks, eco-design, and LCA, will increasingly form the backbone of products designed to be energy-efficient and recyclable, as well as to serve environmental ends (Ellen MacArthur Foundation, 2013).

Synergies between innovation and sustainability can be used in order to revitalise industries across the spectrum - from consumer electronics to automotive and health care, dealing with material choice issues, producing more efficiently with minimal waste; for example, smart manufacturing and IoT/AI control can manage energy use and decrease emissions (Chen et al., 2020). Despite these developments, significant barriers remain, including high implementation costs, lack of standardisation, and ethical concerns regarding data privacy and accessibility (Beier et al., 2020). Overcoming these challenges will require a multifaceted effort that incorporates technological innovation, is supported by policy initiatives, and involves stakeholder collaboration.

This chapter's primary intention is to explore the intersection of innovative and sustainable technologies, examining challenges, current situations, product design applications, and future trends. The aim is to provide design manufacturers and policymakers with guidelines or case-study illustrations of practical efforts to achieve a future full of innovation while maintaining a sustainable system.

2. CONCEPTUAL BACKGROUND

The two related paradigms of technological innovation and environmental sustainability form the foundation for integrating innovative and sustainable technologies into product design. The IoT, AI, and digital twins represent some of the innovative technologies that have enabled real-time data capture, predictive analytics, and adaptive functions not previously seen in product development at this stage (Porter & Heppelmann, 2015). These enable designers and producers to create better, more efficient, and user-centric items that respond to ever-changing conditions. For instance, Internet of Things-enabled sensors enable continuous monitoring throughout the product's lifetime, enabling predictive maintenance and optimising resource utilisation (Chen et al., 2020).

In line with this, sustainability principles such as the circular economy and eco-design have also emerged in response to mounting environmental concerns. The circular economy aims to achieve resource efficiency, waste reduction, and product lifecycle extension through the reuse, recycling, and remanufacturing of products. This is according to the Ellen MacArthur Foundation (2013). Eco-design integrates environmental considerations into product development, resulting in products that are energy-efficient, durable, and environmentally friendly (Bocken et al., 2014). The ecological footprint of the products will be reduced by their design for disassembly and recyclability.

The convergence of these paradigms has the potential for synergy. Innovative technologies can drive sustainable practices more effectively by providing the necessary data for monitoring and improving environmental performance (Geissdoerfer et al., 2017). For example, digital twins can simulate how a product impacts the environment for its lifecycle. This enables designers to make more informed decisions when selecting materials and processes. Still, these approaches entail challenges such as technological complexity and high initial costs and require the collaboration of multidisciplinary practitioners (Beier et al., 2020). The conceptual underpinnings of such theories are, hence, essential to enable their complete applications in product design and development.

3. TECHNOLOGICAL ENABLERS FOR SMART AND SUSTAINABLE PRODUCT DESIGN

Advanced technologies, which have come to represent a significant enabler of innovative functionality and sustainability through integration into product design, will enable the creation of creative, low-impact products that maximise efficiency and end-user satisfaction. Key technological enablers are:

a) Internet of Things (IoT): IoT is widely being used as a key constituent in innovative, sustainable product design. It tracks the real-time performance of the products and their environmental aspects. IoT sensors within products monitor usage behaviours, energy consumption, and rate of wear over time, enabling predictive maintenance and extending the useful life of goods (Porter & Heppelmann, 2015). Furthermore, an IoT-based system can optimise resource utilisation on the shop floor, thereby achieving further waste reductions and energy efficiencies (Chen et al., 2020).
b) Artificial Intelligence (AI) and Machine Learning (ML): Algorithms from AI may optimise the choice of materials, increase energy efficiency, and make forecasts of probable environmental impacts of a product at all stages of its life cycle. For example, artificial intelligence enables the analysis of millions of rows, the identification of patterns, and the proposal of design changes that help create products with a reduced carbon footprint (Kamble et al., 2020). In manufacturing, process efficiency and waste reduction in AI-based systems may be part of an eco-centric approach.
c) Digital Twins: Digital twins generate virtual representations of actual products and are capable of simulating the behaviour of a product design and analysing how this might respond in different conditions. Such digital twin technology would be capable of conducting an LCA because these models enable the direct computation of environmental impacts on design decisions. It is therefore possible to compute carbon emissions through material selection and manufacturing techniques, even when prototyping does not occur in physical space. This could lead to the conservation of energy resources (Tao et al., 2019).
d) Additive Manufacturing (AM): AM is commonly referred to as 3D Printing. This approach enables the designer to conceive a sustainable product by eliminating waste and facilitating localised manufacturing. AM will employ recyclable or biodegradable materials and accommodate on-demand manufacturing with minimal inventory or transportation (Gibson et al., 2015).
e) Advanced Materials and Smart Materials: Advanced materials encompass lightweight composites, biodegradable polymers, energy-efficient coatings, and innovative materials that adapt to their environment, enabling superior product performance with minimal energy consumption. Self-healing materials can extend a product's lifespan by automatically healing minor damage without human intervention (Binnemans et al., 2015).
f) Blockchain Technology: Blockchain enhances the traceability and transparency of products throughout their lifecycle, thereby facilitating sustainable material sourcing and compliance with environmental regulations. This data sharing will help stakeholders securely access information, thereby promoting accountability in sustainable practices (Saberi et al., 2019).

g) Automation and Robotics: Automation and robotics are integral to sustainable manufacturing since they increase precision and reduce waste. AI robots optimise production processes, handle complex assembly tasks, and ensure quality control, thereby reducing defects and conserving resources (Beier et al., 2020).

4. SUSTAINABLE PRODUCT LIFECYCLE MANAGEMENT

Figure 1. Sustainable PLM

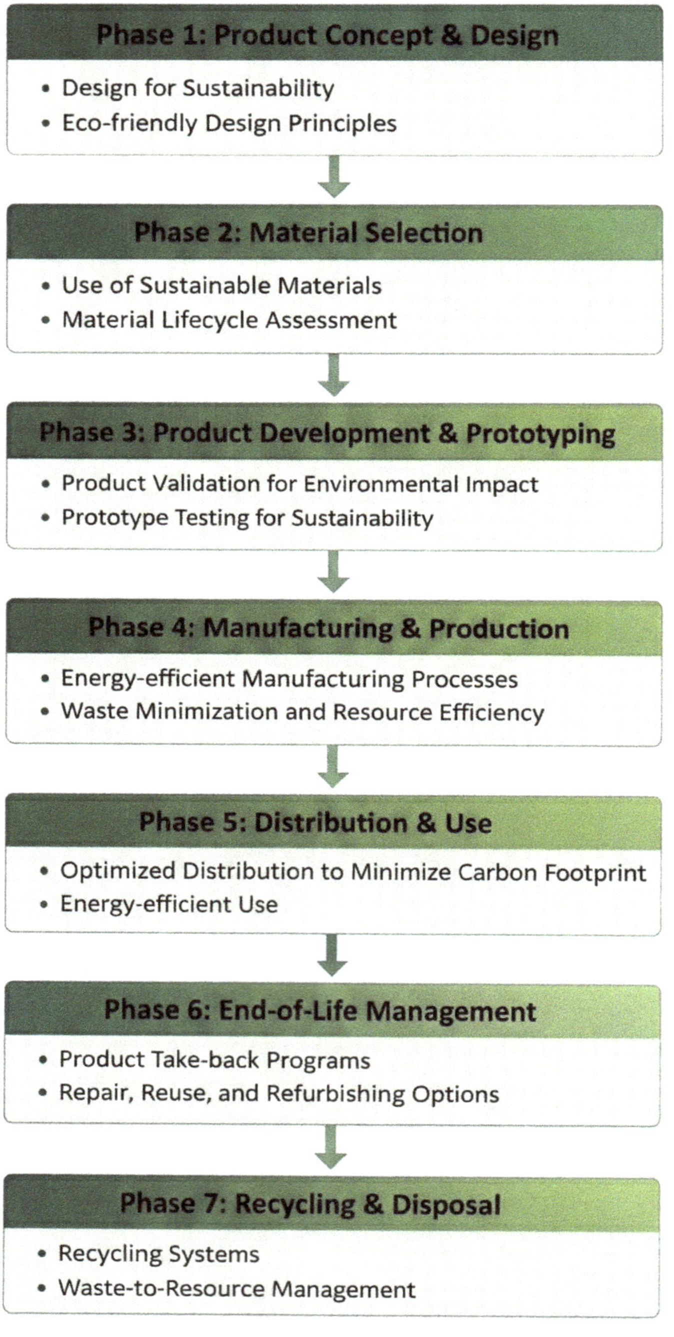

Sustainable Product Lifecycle Management (PLM) (Figure 1) integrates sustainability principles across all stages of the product lifecycle, from design and development to end-of-life. It involves implementing strategies that reduce environmental impacts, optimise resource use, and enhance circularity, helping businesses achieve ecological goals while increasing competitiveness and innovation. The image illustrates a sustainable product lifecycle framework outlining the primary steps from concept and design to recycling and disposal. Below is a further explanation of each phase, accompanied by relevant citations and references.

4.1 Sustainable Product Lifecycle

a) Phase 1: Product Concept & Design

The first step in product development defines the product's purpose, functionality, and sustainability. An essential feature of this step is incorporating eco-friendly design principles that reduce environmental impacts from the start. In other words, considering sustainability in the early stages of the design process can reduce resource usage, increase energy efficiency, and ensure long-term environmental benefits (Pigosso et al., 2015).

A third significant way is design for sustainability, which applies life-cycle thinking to minimise a product's carbon footprint, waste generated, and energy consumption. This method ensures that environmental issues are addressed throughout the product life cycle, from raw material extraction to final disposal (Bocken et al., 2016). The integration of sustainability into the very design philosophy that shapes a product enables these businesses to develop innovative, environmentally conscious products.

b) Phase 2: Material Selection

Material selection is a critical phase in developing a sustainable product because it determines the extent to which the product impacts the environment. In sustainable material development, renewable, biodegradable, or recycled materials are used to avoid resource depletion and generate less waste. The process often finds successful applications, particularly in cases where resources are limited, thanks to products made from sustainable materials such as bamboo, bioplastics, and recycled metals (Ashby, 2012). Another critical approach is Material Lifecycle Assessment (LCA), which evaluates the environmental implications of materials throughout their entire life cycle, from extraction and raw material production to use and end-of-life disposal. LCA helps designers and manufacturers make informed decisions regarding material sourcing for sustainability at all levels, as Guinee (2016) notes.

This integrated approach enables industries to reduce carbon emissions, minimise waste, and improve overall environmental performance.

c) Phase 3: Product Development & Prototyping

Product development and prototyping form the phase in which a product is refined for design while upholding its sustainability credentials. This phase of product development has an essential focus on validating the environmental impact of a product to ensure that its sustainability performance is upheld through established eco-certifications and compliance with ecological legislation. Certifications such as Energy Star, Cradle to Cradle, and LEED serve as verification of a product's conformity to environmental standards, thereby reducing resource consumption, emissions, and waste (Hauschild et al., 2018). Compliance with such regulatory frameworks enhances consumers' trust in the product and positions it in the market as an eco-friendly option.

Prototype testing for sustainability is essential. Products are tested for energy efficiency, durability, and recyclability. The product must then undergo scrutiny at this stage to determine how the raw material and design choice are considered environmentally friendly, ensuring they meet sustainability thresholds before being produced in quantity. Testing thus pinpoints aspects for redesign to reduce energy-guzzling parts, prolong its shelf life, and recycle at the end of its life (Sherwin, 2004). It is possible to integrate sustainability into products by considering it during the prototyping phase, ensuring businesses offer products with features that deliver long-term environmental benefits.

d) Phase 4: Manufacturing & Production

The manufacturing and production phases are crucial to ensuring that sustainability principles are effectively implemented in the final product. There is an excellent focus on energy-efficient manufacturing, where lean and green manufacturing techniques are implemented to optimise energy use and reduce carbon emissions. Strategies include automation, process optimisation, and integration of renewable energy sources to minimise resource consumption and improve production efficiency. More intelligent manufacturing systems, which rely on AI and IoT, also continue to monitor energy usage in real time to prevent waste (Kusiak, 2018). Such techniques enable industries to make optimal reductions in their environmental footprint and operational costs.

Finally, waste minimisation and resource efficiency, which reduce material waste and improve resource recovery strategies, are essential. This would involve adopting the principles of a circular economy: remanufacturing, reusing, and recycling materials to extend their lifecycle. For example, closed-loop manufacturing systems recycle scrap materials and byproducts into production rather than sending them out as waste (Allwood et al., 2011). Sustainable manufacturing frameworks conserve raw materials and promote economic efficiency, ultimately contributing to an environmentally responsible supply chain.

e) Phase 5: Distribution & Use

The distribution and use phase is crucial to ensuring that sustainability efforts extend beyond production into the product's delivery and lifecycle use. A primary focus is on optimised distribution to minimise carbon footprint, which involves implementing green logistics and supply chain optimisation strategies. These encompass fuel-efficient transportation modes, routing optimisation, and alternative energy-fuelled fleets. Additionally, companies are strategically utilising local sourcing, green packaging, and carbon offset programs to make their distribution more sustainable (McKinnon et al., 2015). The integration of Blockchain and AI-driven logistics will further enhance supply chain transparency while reducing inefficiencies.

Another crucial aspect is the energy-efficient use of products, which encourages users to consume them sustainably. It includes low-energy appliances, eco-modes, and innovative technology integration that enable consumers to consume less energy. Consumer education programs and awareness campaigns on sustainable usage also extend the environmental benefits of the product across its lifecycle (Jackson, 2005). Companies are now exploring PaaS models in which the product is leased rather than sold, thereby encouraging longevity and reuse while reducing waste.

f) Phase 6: End-of-Life Management

The end-of-life management stage is a significant one, as products are designed to generate minimal environmental waste and are oriented towards a circular economy. In this respect, a prominent strategy for product take-back involves reverse logistics, with an emphasis on extended producer responsibility. This calls for firms to institute a collection system to recover their used products, such as through material recycling, remanufacturing, or environmentally friendly final disposal. Governments and business enterprises are also establishing regulations to require manufacturers to manage post-consumer waste, thereby reducing landfill inputs and increasing resource recovery (Ghisellini et al., 2016). Companies leading the world also implement Blockchain and AI tracking of returned products, maximising material recovery.

Another is the availability of repair, reuse, and remanufacture options. In this component, it tries to extend the product life cycle by maintaining the original product, remanufacturing, and accessing secondary markets. Modular design encourages consumers to replace only the faulty parts rather than the entire product. Practices that include buy-back programs in second-hand markets and the sale of refurbished items generate economic and environmental value by reducing demand for raw materials (Stahel, 2016). Apple, Dell, and Patagonia have successfully implemented this refurbishment model, reduced overall waste while making items more accessible to budget-conscious consumers.

g) Phase 7: Recycling & Disposal

Recycling and disposal are the final stages in a product's lifecycle, through which waste is disposed of and resources are recovered effectively. Closed-loop recycling, a key feature of a recycling system, helps reduce landfilling of waste products by recycling materials at the end-of-life phase of a product, reprocessing them, and reintegrating them into the manufacturing process. This would ultimately save raw materials, reduce carbon emissions, and curtail the use of virgin resources (Ellen MacArthur Foundation, 2013). Technology investments in companies like chemical recycling and AI-based sorting of waste are also on the rise for improved material recovery and better recycling rates. Governments and international bodies are also establishing recycling targets and offering incentives to companies to promote accountability and consumer participation.

Another key strategy involves waste-to-resource management, which converts waste materials into reusable materials or a source of energy. Examples of this approach include converting biodegradable waste into biofuel, collecting rare metals from electronic devices, and utilising industrial waste in construction. Innovation in the lines of plasma gasification, anaerobic digestion, and capture of carbon added to the potential of waste energy solutions by conserving the natural environment and enhancing economic returns on investment (Korhonen et al., 2018). The successful implementation of these strategies contributes to the circular economy model, where waste is viewed as a resource rather than a liability.

4.2 Technologies Enabling Sustainable PLM

Technologies for Sustainable Product Lifecycle Management (PLM) are advancing toward more efficient, environmentally friendly practices across a product's lifecycle. IoT and Big Data are key enablers, enabling the collection of real-time data on product performance and resource consumption via IoT devices. This helps optimise operations across various lifecycle stages, thereby facilitating the successful management of production at end of life in pursuit of sustainability goals (Chen et al., 2020).

Another critical technology is Digital Twins. These create virtual models of the physical products that simulate different lifecycle scenarios. In this way, manufacturers can make proactive, informed decisions on product design, product performance monitoring, and end-of-life strategies to improve sustainability (Tao et al., 2019). Blockchain technology also provides a means to ensure transparency and traceability throughout the product's lifecycle. It establishes that, through blockchain, the digital record of all sourcing, production, and recycling activities is securely maintained. This supports accountability and compliance with sustainability standards for the circular economy approach (Saberi et al., 2019). These technologies support more sustainable PLM practices through increased efficiency, reduced waste, and enhanced transparency.

4.3 Benefits of Sustainable PLM

Most sustainable practices adopted by Product Lifecycle Management (PLM) offer opportunities for environmental, economic, and social benefits, creating advantages for both the firm and society as a whole. Sound ecological practices influence reductions in greenhouse gas emissions, minimising resource depletion and waste, as products are reused, refurbished, and recycled rather than discarded once their lifecycle has been exhausted (Bocken et al., 2016). Economically, the more efficient use of available resources leads to cost savings, reduced waste generation, and a longer product lifespan. This translates into lower production costs and higher profitability (Geissdoerfer et al., 2017). There is the regulatory side: sustainable PLM will comply with all relevant environmental regulations and standards, such as ISO 14001, which focuses on environmental management systems, and the EU Circular Economy Action Plan, a strategy promoting sustainable production and consumption (European Commission, 2020). Lastly, sustainability efforts benefit customer satisfaction, as companies with an ethical and environmental commitment typically enjoy an enhanced reputation and higher customer loyalty (Niinimäki & Hassi, 2011). All these together make sustainable PLM a strategic approach for businesses looking to succeed in an increasingly dynamic global market.

4.4 Challenges and Solutions

Although there are various advantages to transitioning towards sustainable PLM, the process is accompanied by numerous challenges that firms must overcome to reap the maximum benefits. One of the most significant barriers is the cost of implementing sustainable technologies. It has been considered a capital-intensive area, with investments in IoT, Digital Twins, and Blockchain technologies, thereby discouraging smaller enterprises and, to some extent, medium-sized companies from adopting it (Bocken et al., 2016). Changing the current supply chain setup for circular applications also presents a significant challenge. This mandates a significant scale of redesigns, coordination involving various stakeholders, and changes to logistics and sourcing strategies, which are often complex and time-consuming (Geissdoerfer et al., 2017). Another challenge is that there is limited awareness among people of sustainable design and PLM. Most companies lack the skills and resources to design products that fully align with sustainable principles, thereby limiting their widespread deployment (Niinimäki & Hassi, 2011).

However, various solutions may help overcome such challenges. Governments and regulatory authorities can also play crucial roles. They may provide inducements, grants, or subsidies for green activities to help reduce investment costs (European Commission, 2020). There can be collaboration among various stakeholders, such as suppliers, manufacturers, and regulators, who share their knowledge and resources, which is highly useful in practice for designing and scaling sustainable practices (Bocken et al., 2016). Last but not least, investment in training and capacity building is essential to ensure that employees possess the proper skills and knowledge necessary for sustainable design and PLM, thereby contributing to the long-term success of such initiatives (Geissdoerfer et al., 2017).

5. FRAMEWORK FOR INTEGRATING SMART AND SUSTAINABLE TECHNOLOGIES

Figure 2. Integration of smart and sustainable technologies in product design

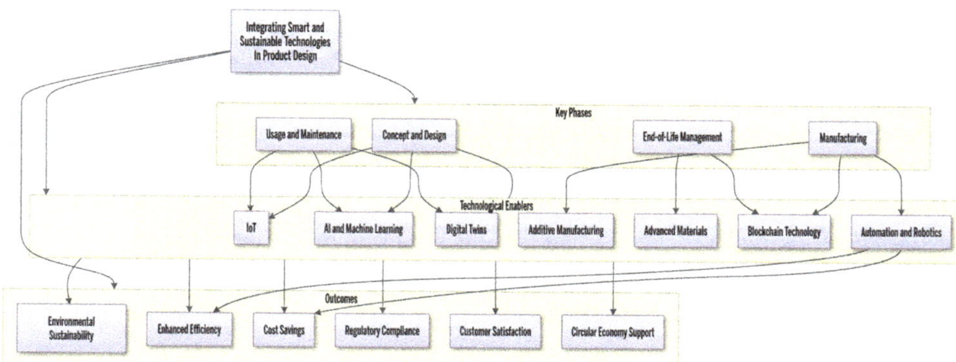

Product design has integrated smart, sustainable technologies, transforming the approach to innovative, efficient, and environmentally responsible products. Using digital tools, automation, and intelligent systems in modern industries has optimised product design, manufacturing, usage, and end-of-life management. Technological advancements improve sustainability by minimising waste, conserving resources, and reducing environmental footprints.

5.1 Innovative and Sustainable Product Design Key Stages

a) Concept and Design

This brings in the initial critical phase where sustainability will be introduced inside the PLM model (Figure 2) in the conception and design phases that present the basis of creating responsible products by being conscious not only on an environmental and ecological level but also on economics with conscious choices initiated from the early stages while maintaining ecological material utilisation and energy-related designs for producing the result. Companies can see that it is only through this manner that products last for long periods. By using many sustainable raw materials, the use of materials is reduced while emphasizing energy-saving features. This stage, which incorporates advanced technologies to optimise the product development process with minimal environmental damage, is considered necessary.

Tools Used:

- Artificial Intelligence & Machine Learning: The higher usage of AI and ML is being employed to make product design more sustainable. AI can predict the efficiency of various materials used in a design and help in selecting cost-effective, eco-friendly ones. Optimising the design process by leveraging AI and ML with historical and real-time data can help identify patterns, leading to more sustainable results. This sets the basis for determining the efficiency of material utilisation and waste prevention and, more importantly, enhancing the efficiency of product design (Bocken et al., 2016). For example, machine learning-based algorithms can help predict the outcomes of various design alternatives based on natural parameters, such as resource consumption and energy usage, as well as recyclability, to inform decisions as early as the design stage.
- Digital Twins – AI and ML: Artificial Intelligence, Machine Learning, and Digital Twins continue to be an increasingly important development in the realisation of sustainable products. AI now predicts the efficient use of resources across various materials, enabling designers to select cost-effective materials for their products in a way that is both cost-effective and beneficial to the ecosystem. AI can also optimise using historical and live data analysis to identify patterns that drive more sustainable processes. These technologies

help determine the best use of materials, minimise waste, and enhance overall design functionality (Bocken et al., 2016). For instance, a machine learning algorithm can predict how different design variants will perform across environmental parameters, such as resource consumption, energy use, and recyclability. This enables designers to choose appropriately from the outset of the design process, one that suits the sustainability goal.

b) Manufacturing

Manufacturing in sustainable PLM is an essential stage for reducing environmental impact, improving operational efficiency, and promoting ethical practices. The manufacturing technique, therefore, incorporates clever production methods. Through innovative production techniques, waste can be minimised and emissions reduced. Additionally, the methods utilise advanced technologies, resulting in more efficient processes, sustainable material use, and energy savings throughout the entire production phase.

Key Technologies:

- Additive Manufacturing (3D Printing) – This is one of the key sustainable manufacturing technologies, also referred to as Additive Manufacturing, also known as 3D Printing. Unlike the subtractive approach in traditional manufacturing, where excess material is literally cut from a much larger block, additive manufacturing adds products layer by layer, using only the material required for the final design, thereby providing significant material savings with virtually no waste. 3D Printing also opens the potential for complex geometries that would otherwise be deemed impossible or very wasteful through traditional methods, making it the most efficient approach to resource use (Ford & Despeisse, 2016). Due to the reduction in material use, 3D Printing serves as a strategic tool to minimise the footprint of production operations.
- Advanced Materials – The use of advanced materials is also a significant contributor to sustainability in manufacturing. Improvements in biodegradable materials and the creation of more effective recycled materials can dramatically reduce the product's environmental footprint. For example, using biodegradable plastics and other sustainable materials can replace conventional materials, thereby reducing waste and making end-of-life management much simpler (Allwood et al., 2011). On the other hand, recycled materials promote circular manufacturing by enabling the production of products from previously used materials, thereby reducing the need for virgin resources.

- Blockchain Technology - Blockchain technology plays a crucial role in sustainability, particularly in the manufacturing process, mainly due to its increased transparency within the supply chain. It enables tracing and verification from source to final product, ensuring that materials are sourced ethically and that all sustainability standards are met. This level of transparency is indispensable for manufacturers interested in ethical behaviour and environmental footprints. Blockchain technology enables compliance with regulatory requirements and fosters accountability across sourcing and production processes (Saberi et al., 2019).
- Automation and Robotics – Automation and Robotics are revolutionising manufacturing by adding precision to the processes while consuming less energy. Automated systems and robotic machines can perform work with greater accuracy and efficiency than human labour. It leads to less waste, lower energy consumption, and fewer emissions. Additionally, robots can operate in hazardous conditions, thereby enhancing worker safety while optimising resource utilisation (Frank et al., 2019). The companies will therefore increase their sustainability initiatives while ensuring that production meets high standards by automating several production processes.

c) Usage and Maintenance

In PLM, the usage and maintenance phase is the most critical because it extends the product's lifespan while ensuring it performs efficiently. This stage, therefore, aims to optimise product efficiency by adopting innovative technologies to monitor and predict issues, thereby enabling maintenance activities that prolong product life and optimally use available resources. Therefore, the manufacturers integrate these technologies during the lifecycle of their products. With those technologies included, such products will provide optimal functionality at all times while having minimal environmental impact.

Improving product efficiency -One of the most important technologies used in this stage is IoT. IoT products integrated into the product enable real-time tracking and monitoring of its performance, providing valuable data on usage patterns, energy consumption, and operational efficiency. It helps manufacturers and users track how efficiently the product is used, allowing it to operate at optimal energy levels. With the early detection of issues like wear and tear or degradation in performance by IoT sensors, timely interventions can be made, and their maintenance can be taken care of, resulting in less downtime and fewer unnecessary repairs, which maximises the operational life of a product. Through regular maintenance tailored to the product's actual conditions, IoT-based systems help minimise waste and energy consumption, making the product sustainable throughout its lifecycle (Lee et al., 2019).

Technologies Used:

IoT (Internet of Things)—IoT can provide real-time insights into how a product is being used, resulting in more efficient energy consumption and reduced unnecessary resource use. For instance, smart thermostats or devices can automatically adjust their operation according to energy usage to optimise energy use and minimise environmental impact. Such data can also be used to inform users on how to increase product efficiency, thereby minimising consumption and waste (Lee et al., 2019).

d) End-of-Life Management

The End-of-Life Management stage is critical, as it ensures the proper disposal or recycling of products in accordance with sustainable development principles. This aspect is based on circular economy thinking, where products can be designed so that they include recyclable elements and can be reused for longer, thus minimising waste to the environment. Successful end-of-life strategies will enable companies to minimise their environmental impacts while recovering valued materials, enhancing the sustainability of a resource-efficient system.

Key Strategies:

- Lifecycle Assessment: Lifecycle Assessment, or LCA, is one of the most widely used approaches within the realm of End-of-Life Management. An LCA provides an integrated process for evaluating all environmental effects resulting from a product's entire life cycle, from the extraction of raw materials through to final disposal. Through this process, the LCA is designed to provide opportunity points for improving environmental performance at every stage of the product life cycle—for instance, reducing energy use in manufacturing or improving recyclability at the end of the product's life. Companies can utilise data-driven decision-making to mitigate adverse environmental impacts and promote sustainability (Hauschild et al., 2018). This approach is imperative for aligning product designs with circular economy principles, utilising available resources effectively, and ensuring their ultimate end-of-life is managed.
- Reverse Logistics – One of the essential strategies of End-of-Life Management is Reverse Logistics. Reverse logistics is the process by which used products are collected from consumers, returned to a manufacturer or recycling facility, and refurbished or recycled for reuse. This is one of the critical strategies that enable the implementation of circular economy activities, in which materials such as metals, plastics, and electronics are not discarded but reused and incorporated into the manufacturing of new products. Reverse logistics

also enables the take-back of products at the end of their useful life, thereby allowing the businesses to recover valuable materials and minimise the environmental impacts of disposal (Govindan et al., 2015). Integrating reverse logistics into the product lifecycle is a strategy companies use to close the supply chain loop, eliminate waste, and advance sustainability and circularity.

5.2 Technological Enablers

a) The reason that smart and sustainable product design will be rooted in Technological Enablers is that, based on the current trend, innovative technologies that increase efficiency, reduce environmental impact, and improve sustainability throughout a product's entire lifespan can now be leveraged to enhance a product's design. Such technologies will maximise the product design and also enhance each stage of production. Usage. end-of-life.

b) IoT & AI: The primary enablers of sustainable product design are IoT and AI, as they enable real-time tracking of product performance and predictive maintenance needs. The IoT devices embedded in the product record usage patterns, energy consumption, and performance, while AI algorithms analyse that data to optimise energy use and predict maintenance needs. This results in minimal downtime, lower energy consumption, and a longer product life, ultimately leading to a more sustainable product life cycle.

c) Digital Twins: The Digital twin approach is a fantastic way of optimising the product's design. It simulates virtual versions of the products and their overall performance over their lifetimes. This simulation of the authentic product using these virtual models would allow designers to test and observe it under different conditions for its intended application. Potential weaknesses and inefficiencies are therefore identified before production begins, helping eliminate material waste.

d) Additive Manufacturing—This is the application of additive manufacturing, better known as 3D Printing. It introduces technology for new product designs, achieved through layerwise buildup. Layer-by-layer helps reduce waste from materials used in production, demonstrating greater efficiency than other manufacturing methods. This results in creating product designs that enable light use and resource-efficient consumption. Ultimately, this could mean that material use and related waste would be minimised during the productive process.

e) Blockchain Technology: Blockchain Technology is an upgrade for supply chain transparency, whereby material sourcing and tracing can be safely verified at each point in a product's life cycle. The blockchain assurance ensures that materials are sourced in an environmentally responsible manner, with reduced risk of unethical practices, thereby ensuring complete adherence to all environmental regulations. A transparent supply chain, facilitated by blockchain, supports responsible sourcing, which in turn leads businesses to sustainability standards.

f) Automation & Robotics—Automation and robotics are widely used in the manufacturing sector. They offer improved precision and reduced energy consumption. Resources will also be optimised. They help save significant energy, produce fewer emissions, and use fewer materials. Thus, the entire manufacturing process becomes sustainable.

5.3 Key Outcomes of Smart and Sustainable Product Design

The key outcomes of innovative and sustainable product design focus on achieving both environmental and economic goals while fulfilling regulatory requirements and meeting customer expectations. These are essential in a transition towards a more sustainable and circular economy that seeks resource efficiency and waste reduction at every step of the product lifecycle.

a) Environmental Sustainability: Smart and sustainable product designs minimise the environmental impact through reduced carbon footprint, lower emissions, and reduced waste. This suggests that energy-efficient materials and processes are adopted and integrated with circular economy principles. To that extent, the whole lifecycle of product design diminishes its ecological footprint. For instance, the Ellen MacArthur Foundation (2013) calculated that such designs could contribute to achieving global sustainability goals, particularly in resource conservation and pollution prevention.

b) Increased Efficiency: Sustainable design maximises resource use. IoT, AI, and automation technologies enhance energy efficiency, optimise production workflows, and ensure resource efficiency. According to McKinsey (2020), innovative design and manufacturing processes not only increase operational efficiency but also reduce resource consumption, thereby lowering costs and speeding up production cycles.

c) Cost reduction: As innovative technologies are added to product designs, material wastage and energy consumption will decrease, just as in additive manufacturing, IoT monitoring, and AI optimisation. Such reductions will lower the manufacturer's operating costs. According to Frank et al. (2019), infrequent repairs and downtime result in minimal resource waste and, therefore, lower overall costs.

d) Regulatory Compliance: The increasing and stringent environmental legislation demands that companies follow the level of sustainable production and its effect on the environment. Sustainable product design enables companies to adhere to environmental policies, including ISO 14001 standards and the EU Circular

Economy Action Plan. According to KPMG (2021), adopting sustainable practices supports compliance and corporate responsibility, thereby boosting reputation and marketability.

e) Customer Satisfaction: Consumers today are becoming increasingly concerned about the environmental impact of the products they purchase. Durable, sustainable, and high-quality products create customer loyalty and satisfaction. Sustainable products designed for durability will enhance consumer trust and ensure a strong brand reputation in the market. According to Stahel (2016), the outcome will be increased demand and customer retention.

f) Circularity: The other fundamental goal of sustainable product design is to fulfil the principle of the circular economy concept, where products have a life to be used over and over by reusing, repairing, or recycling them. By making products easy to disassemble or recyclable, companies can utilise materials for a more extended period while also avoiding the further consumption of newly developed raw material stocks. Lacy and Rutqvist (2015) posit that clever design allows for the practice of a circular economy. This approach yields closed-loop systems, ensuring long-term sustainability through waste reduction and the conservation of natural resources.

6. CASE STUDIES

6.1 Tesla - Electric Vehicles and Sustainable Manufacturing

6.1.1 Background

The firm Tesla, Inc., entered the automotive industry as an electric-car producer with a cleaner manufacturing process. Founded in 2003, Tesla believed it could help the world accelerate its transition to sustainable energy by producing high-performance electric cars that reduce fossil fuel consumption and pave the way for renewable energy. Tesla's commitment to sustainable design extends beyond electric vehicles and into its entire supply chain, manufacturing processes, and product lifecycle management. This case study describes how Tesla applies innovative and sustainable technologies in the design and manufacturing of its products, particularly its electric vehicles, and utilises renewable energy sources.

6.1.2 Tesla Electric Vehicle Design and Development

Tesla's electric cars are designed with sustainability in mind. The company offers a range of high-performance electric vehicles, including the Model S, Model 3, Model X, and Model Y, while pursuing reductions in carbon emissions and improvements in energy efficiency.

a) Battery Technology. Tesla's automobiles use lithium-ion batteries with high energy density and long cycle life. Significant improvements in battery chemistry have enabled electric cars to become more efficient and economical. Tesla Gigafactories are designed to produce batteries at industrial scale while maintaining competitive prices and improving efficiency. The Gigafactories also manufacture the lowest-impact battery cells and utilise renewable energy sources.
b) Energy Efficiency and Range: Tesla EVs' designs are superior in terms of energy efficiency and range. They use aerodynamic design, lightweight materials, and regenerative braking technology to promise an extended range. One of the significant issues that people face when using electric vehicles is range anxiety. The competitor to this one is the Model S Long Range, which offers a driving range of more than 370 miles on a single charge.
c) Autopilot and Smart Features: Tesla vehicles have innovative features, such as high-level ADAS. Autopilot uses machine learning and AI to guide a car through navigation, traffic control, and accident prevention. These innovative features enhance the vehicle's efficiency and safety, thereby contributing to a sustainable transportation system.

6.1.3 Sustainable Manufacturing Practices

Tesla's approach to sustainable manufacturing encompasses energy-efficient production techniques, the utilisation of renewable energy sources, and minimising waste.

a) Gigafactories: Gigafactories by Tesla are at the heart of Tesla's sustainable manufacturing strategy. Gigafactories are massive manufacturing units that help reduce environmental damage by producing renewable energy sources, such as solar energy. The mission of Tesla's Gigafactory is to mass-produce electric vehicles and their batteries for sale to the average consumer, while maintaining an efficient use of energy during the production process and minimising waste to the greatest extent possible.
b) Solar Energy Integration: In addition to its electric vehicle division, Tesla is a sustainability company that produces solar products, including solar panels and solar roofs. Tesla has Gigafactories, where solar energy systems power the various parts under production. Furthermore, Tesla encourages its customers to adopt solar power for their homes and businesses. This has reduced Tesla's

carbon footprint and its dependence on fossil fuels since the sun is a source of energy.

c) Recycling and Circular Economy: Tesla aims to create a circular economy with its products. The company recycles the batteries from its vehicles and other recyclable materials. It is responsible for improving the recyclability of its batteries and has successfully developed technologies to reclaim valuable materials from used batteries and reuse them in new products. Tesla, as such, tries to minimise waste and maximise the reuse of materials, making its product life cycle much more sustainable.

6.1.4 Environmental and Social Impact

Therefore, Tesla's cars and its sustainable manufacturing practices reduce greenhouse gas emissions and contribute significantly to mitigating climate change.

a) Carbon Emission Reduction: Switching from gasoline vehicles to electric ones is one of the most efficient ways to reduce carbon emissions in the transport sector. According to Tesla, the Model 3 EV produces significantly less carbon than internal combustion engine vehicles, and the Model 3 can reduce $CO2$ emissions by 70-80% throughout its lifetime compared to a conventional gasoline vehicle.

b) Responsible Supply Chain Sourcing: As part of Tesla's efforts to ensure its supply chain is responsible, the firm sources its materials in a responsible manner. This means that it sources its lithium, nickel, and cobalt from these sources. The firm has committed to ethical sourcing practices. It has made a significant effort to ensure that materials used in vehicle production are transported to the site without harming the environment. Tesla's commitment to transparency ensures its materials are sourced sustainably.

c) Employment and Social Benefits: Just like innovation centres in other parts of the world, Tesla Gigafactories have created thousands of jobs in local settings that support regional economies and contribute to a green economy. Furthermore, Tesla's innovation in green products serves as an example for other companies to utilise and replicate similar technological and business activities, thereby also expanding the scope of its social and environmental impact.

6.1.5 Challenges and Criticisms

Although Tesla's efforts in multiple states have combined Smart and sustainable technologies, some of the obstacles it encounters are:

a) Battery supply chain: Lithium-ion batteries, which are crucial for EV production, have exacerbated environmental consequences associated with mining for components such as lithium, cobalt, and nickel. Therefore, Tesla has been criticised for sourcing raw materials from developing countries known for weak environmental and labour standards. The company is now using alternative materials and recycling techniques for batteries.
b) Production bottlenecks—Scaling the production of Tesla cars and batteries has proven challenging. During the early years of production, the firm struggled to meet demand, especially for the lower-cost models it introduced, most notably the Model 3. Nevertheless, Tesla is continually improving its production efficiency through innovation in automation and factory design.
c) Safety concerns with Autopilot: As innovative as Tesla's Autopilot system is, several safety concerns have been raised. The few high-profile accidents involving Teslas that had used Autopilot raised questions about the readiness of AI systems for full autonomy. This led Tesla to continuously update its software to enhance safety features while providing more explicit guidelines for using Autopilot.

6.2 Philips: Circular Economy and Sustainable Product Design

6.2.1 Background

Philips is among the global leaders in the health technology, consumer electronics, and lighting industries. By adopting the principles of circular economy in product design and business models, the corporation has been at the forefront of promoting sustainability. The corporation aligns with the United Nations' Sustainable Development Goals and commits to reducing its environmental footprint through sustainable manufacturing, resource efficiency, and a circular-economy approach. Philips has emphasised the design of products that can be easily refurbished, reused, and recycled. It has been working to create a closed-loop system where waste is minimised and raw material extraction is reduced to the absolute minimum. This case study focuses on Philips' initiatives in circular economy practices and sustainable product design, highlighting innovative approaches to designing products that are not only energy-efficient but also designed for reuse, repair, and recycling.

6.2.2 Philips' Circular Economy Strategy

This belief in Philips' transition to a circular economy is based on decoupling growth from finite resources. The approach, therefore, focuses on extending the product's life, maximising resource recovery, and minimising waste. Some of the key initiatives that are being pursued in incorporating circular economy principles based on product design and operation include:

a) Design for Longevity and Upgradability. Philips aimed to design products that last for extended periods. For instance, medical products such as ultrasound machines and X-ray equipment have been designed to be serviced, repaired, and upgraded in a way that minimises the need for frequent replacement, thereby extending their overall lifespan. This reduces all the environmental impacts associated with producing new devices and waste.
b) EcoDesign: Environmental concerns are integrated into product development by the company Philips, starting from the primary stages of product design. That is, using ecologically friendly materials and developing them to consume much less energy makes them even easier to demount and recycle once their life cycle ends. Philips incorporates eco-design principles into each product line, minimising energy consumption and increasing efficiency to reduce environmental impact.
c) Material Reuse and Recycling: Philips ensures that most materials are recovered and no waste is generated. From the old products, Philips partners with recyclers who use modern methods to extract valuable elements, such as rare earth metals, aluminium, and copper, for reuse. For example, in its Lighting division, the company has offered energy-efficient LEDs that are easy to disassemble, enabling recycling. It thereby significantly reduces the amount of material waste sent into landfills while reducing the consumption of raw materials.
d) Product-as-a-Service Model: Philips has adopted an innovative approach to the circular economy with the product-as-a-service model, which focuses on usage rather than ownership. For example, in the lighting industry, Philips introduced "Pay-per-Lux" lighting services for commercial buildings. Rather than purchasing lighting equipment, companies pay for lighting services based on the amount of light consumed. Philips remains the owner of the products. The products are maintained, repaired, and refurbished over time, therefore ensuring the continual recycling and reusing of products.

6.2.3 Examples of Sustainable Products

Philips has developed several products that embody the principles of sustainability and circular economy in their design:

a) Philips Hue Smart Lighting: Philips Bright Lighting is one of the world's first intelligent lighting systems to utilise energy-saving LED technology, which has been enhanced through software updates. It uses long-life, repurposes, recycles, and reuses the light. This lighting system can integrate with a range of smart home platforms, enabling users to control their lighting and save energy, while also helping the company achieve its sustainability objectives.

b) Philips Sonicare Electric Toothbrush: The Philips Sonicare electric toothbrush is a sustainable consumer product. It features a rechargeable battery and long-lasting materials, ensuring it lasts longer. Also, the toothbrush handle can be reused, and the brush heads can be recycled. Philips also improved the toothbrush's design to be energy-efficient and reduce waste by offering refillable brush heads and replaceable parts.

c) Philips Medical Equipment—EcoDesign in Healthcare: Philips applies circular economy principles to its medical equipment, improving both sustainability and performance. For example, the ultrasound system is intended to be upgraded and serviced. Philips aims to reduce the energy consumption of its medical equipment. Therefore, its products not only serve well but also last longer. Moreover, it has also reduced its environmental footprint regarding MRI machines, X-ray equipment, and patient monitors by utilising environmentally friendly materials and energy-conserving components.

d) Philips Lighting - Circular Lighting: This leads to circular lighting solutions, particularly in the context of LED lighting. The circular lighting model from Philips takes responsibility for the whole lifecycle of the product, from design to recycling. The company offers as-a-service lighting solutions, in which customers pay for the light they use rather than purchasing fixtures. Philips then undertakes the activities of maintaining and disposing of lights at the point of end of life, where those lights are to be reused, refurbished or recycled for proper disposal at the end. This reduces waste from the lighting products themselves and promotes longer-term resource-use efficiency.

6.2.4 Environmental and Social Impact

Through its commitment to sustainability, Philips has made significant progress in ensuring positive impacts on both the environment and society.

a) Lowered Carbon Footprint: Philips works to reduce customers' carbon footprints by offering energy-efficient products, including lighting, medical devices, and electrical appliances. LEDs, for example, consume less power than incandescent bulbs; they also take longer to emit light, so their consumption remains low, implying lower energy intake and lower GHG emissions.

b) Sustainable supply chain: Philips engages with suppliers to make its supply chain sustainable. By sourcing materials responsibly, the company aims to ensure its products are made from environmentally friendly materials and delivered to consumers sustainably. Philips collaborates with suppliers to implement sustainability measures, ensuring that raw materials, including metals and plastics, are sourced responsibly.

c) Job creation and social responsibility: Philips is associated with green jobs, which can be linked to product repair and refurbishment, in addition to its recycling program, within the context of circular economy actions. Through its support for a more circular model, the company helps build local economies and job markets centred on sustainable products and services. Additionally, Philips' efforts to enhance healthcare and provide affordable medical technologies to underserved populations align with its commitment to social responsibility.

6.2.5 Challenges and Criticisms

Though the company has made significant steps towards sustainability, Philips has yet to overcome a number of challenges:

a) It is not easy to fully adopt the circular economy model, as this would involve modifying manufacturing, logistics, and design for each product line. In particular, some products offered by Philips in the consumer electronics category may face issues related to either complete recyclability or design changes for reuse.

b) Responsible Material Sourcing. Many Philips products contain cobalt and rare earth elements, which have sourcing issues related to environmental concerns and ethical considerations. No matter how much effort is put into sourcing ethically, mining these materials remains a source of grave ecological degradation and labour practices concerns.

c) Consumer Behaviour Change: The second challenge will arise from consumers' change to adopt the product-as-a-service model and sustainable consumption behaviour. Companies, such as Philips, are starting to offer sustainable products and services. In most instances, the consumption of such offerings depends on consumers' awareness of the need for change in consumer behaviour.

7. FUTURE TRENDS AND EMERGING TECHNOLOGIES

A significant number of these emerging technologies will shape the future of innovative and sustainable product design, leading to increased efficiency, reduced environmental impact, and innovation in the industry. The first major trend that has come on board is the one related to Artificial Intelligence (AI) and Machine Learning (ML), as they are now increasingly used for optimising the design of products, predicting the performance of their lifecycle, and offering personalised and sustainable solutions (Choi et al., 2020). By advancing these technologies to develop wiser products that respond to their usage and other conditions, they can reduce waste and ensure efficient energy utilisation.

IoT is expected to feature significantly with more intelligent interconnecting products designed for real-time data collection. Such streams of data will be valid for understanding how the products perform over their lifecycle. This will enable manufacturers to make informed decisions regarding maintenance, upgrades, and end-of-life management. IoT integration would support circular economy models by offering product tracking, resource management, and material recovery, which facilitate the implementation of remanufacturing and recycling practices (Hussain et al., 2019).

However, another technology that can transform sustainable product design is additive manufacturing, also known as 3D Printing. This process offers the added benefits of material efficiency, just-in-time and on-demand manufacturing, and easy customisation with reduced waste (Gibson et al., 2015). Further advancements in 3D Printing will increasingly minimise the negative impact of a product's overall manufacturing footprint while eliminating material wastage and also lowering the power usage associated with conventional production. Blockchain technology holds the promise of being a highly effective tool in supply chain transparency and traceability. This is because the irreversible ledger ensures that raw materials have been sourced responsibly, products are manufactured sustainably, and recycling practices are managed effectively within society (Saberi et al., 2019). Such traceability advances consumer trust and makes companies more accountable and ethical. The circular economy will be a prevalent consideration in future product design, with a focus on companies' efforts to design for longevity, repairability, and recyclability. Technologies that support reverse logistics, remanufacturing, and resource recovery will facilitate this transition by enabling companies to recapture valuable materials at the end of their lifecycle for reintegration into production (Ellen MacArthur Foundation, 2013).

Furthermore, development in biodegradable materials and bio-based polymers will aim to reduce fossil fuel dependence and improve product sustainability.

Finally, digital twins and simulation technologies will continue to evolve, allowing for the virtual modelling of the product lifecycle so that environmental impact and performance can be simulated in various scenarios, even before physical production begins (Tao et al., 2019). This would predict this and enable better decisions for sustainability in design, while reducing material wastage, energy consumption, and the likelihood of flaws during manufacturing.

8. ETHICAL AND POLICY IMPLICATIONS

The integration of innovative and sustainable technologies into product design raises questions about the ethics and policies surrounding the development and deployment of such innovations in a responsible manner. Data privacy and consumer security are the most significant ethical concerns in such developments. More products are becoming smart through IoT and other connected technologies, which collect large amounts of data that could be exploited if not sufficiently safeguarded. Consumers must ensure that their information is anonymised and used only for legitimate purposes, in accordance with the EU's General Data Protection Regulation and similar regulations worldwide (Martin et al., 2020).

Another ethical concern is the issue of technological obsolescence in the workplace. AI, automation, and innovative manufacturing processes can promote efficiency and sustainability, but at the same time, they might lead to the displacement of some workers in a particular industry. Companies and policymakers must invest in re-skilling and up-skilling programs to ensure that displaced workers can be transferred to new roles created by emerging technologies (Brynjolfsson & McAfee, 2014). The positive impact of innovative, sustainable product designs may be unevenly distributed across classes, favouring wealthier consumers and companies capable of affording the expensive high-tech involved, thus contributing to social inequality (Sachs, 2015). Through subsidies or public-private partnerships, equitable access to sustainable technologies could help redress this imbalance.

From a policy perspective, the government plays a vital role in establishing appropriate regulatory frameworks that consider sustainability and ethics during innovation. Policies should, therefore, encourage the adoption of sustainable technologies while discouraging harmful practices. For instance, carbon taxes and regulations regarding EPRs motivate manufacturers to create long-life or repairable and recyclable products through a business that is responsible for other environmental goals (Bocken et al., 2016). Additionally, policies must ensure that environmental justice is considered when marketing more sustainable products and that ecological burdens are not distributed unevenly, excluding some marginalised communities from accessing sustainable technologies.

Finally, transparency in the supply chain using blockchain leaves an open issue regarding its energy consumption and, by extension, its environmental impacts. Blockchain enhances the traceability of materials and ensures ethical sourcing; however, it typically requires more computational power for operations, leading to higher carbon emissions (Ali et al., 2020). The policy response to these effects would be to ensure that energy-efficient blockchain solutions are adopted and that the carbon footprint of blockchain systems is offset.

9. CONCLUSION

This means that it will integrate with the product's design by changing the way we think about conceiving, producing, and managing the product throughout its life cycle. Making use of advanced technologies like AI, IoT, Blockchain, and additive manufacturing for the purpose of creating innovative products that would be efficient as well as eco-friendly and that could match the sustainability demand. The future of product design will increasingly prioritise not only functional performance but also environmental and social impacts, with an emphasis on circular economy principles, resource efficiency, and ethical practices. The successful integration of these technologies does require careful consideration of their moral and policy implications. Data privacy, social equity, and environmental justice must be placed at the centre of decision-making to ensure that benefits shared by a more thoughtful and sustainable set of products are not accompanied by negative consequences, such as job displacement and/or environmental harm. Policymakers have a crucial role in developing frameworks that promote innovation and safeguard public interests. At the same time, businesses must be proactive in ensuring that their practices align with ethical standards and sustainability goals. Ultimately, collaboration between industry, government, and consumers will ensure that all the benefits of innovative, sustainable product design are implemented. As technology continues to evolve, continually addressing challenges related to privacy, inequality, and environmental impact ensures that the shift toward more innovative products benefits all stakeholders and helps us live more sustainably.

REFERENCES

Ali, M. S., & McKeown, C. (2020). Blockchain and Sustainability: Ethical and Environmental Considerations. *Sustainability, 12*(11), 3572.

Allwood, J. M., Ashby, M. F., Gutowski, T. G., & Worrell, E. (2011). Material efficiency: A white paper. *Resources, Conservation and Recycling*, *55*(3), 362–381. DOI: 10.1016/j.resconrec.2010.11.002

Ashby, M. F. (2012). *Materials and sustainable development.* Butterworth-Heinemann.

Beier, G., Ullrich, A., Niehoff, S., Reißig, M., & Habich, M. (2020). Industry 4.0: How it is defined from a sociotechnical perspective and how much sustainability it includes – A literature review—. *Journal of Cleaner Production*, *259*, 120856. DOI: 10.1016/j.jclepro.2020.120856

Binnemans, K., Jones, P. T., Blanpain, B., Van Gerven, T., Yang, Y., Walton, A., & Buchert, M. (2015). Recycling of rare earths: A critical review. *Journal of Cleaner Production*, *104*, 1–22.

Bocken, N. M., de Pauw, I., Bakker, C., & van der Grinten, B. (2016). Product design and business model strategies for a circular economy. *Journal of Industrial and Production Engineering*, *33*(5), 308–320. DOI: 10.1080/21681015.2016.1172124

Bocken, N. M. P.. (2016). Product Design and Sustainability: A Circular Economy Perspective. *Journal of Industrial Ecology*, *20*(6), 1059–1070.

Bocken, N. M. P., de Pauw, I., Bakker, C., & van der Grinten, B. (2014). Product design and business model strategies for a circular economy. *Journal of Industrial and Production Engineering*, *33*(5), 308–320. DOI: 10.1080/21681015.2016.1172124

Bocken, N. M. P., Short, S. W., Rana, P., & Evans, S. (2016). A literature and practice review to develop sustainable business model archetypes. *Journal of Cleaner Production*, *65*, 42–56. DOI: 10.1016/j.jclepro.2013.11.039

Brynjolfsson, E., & McAfee, A. (2014). *The Second Machine Age: Work, Progress, and Prosperity in a Time of Brilliant Technologies.* W. W. Norton & Company.

Chen, W., Pan, W., & Wang, J. (2020). Leveraging IoT and AI for innovative and sustainable manufacturing. *Sustainability*, *12*(14), 5731.

Choi, S., Kim, H., & Kim, Y. (2020). Artificial Intelligence and Machine Learning: The Future of Sustainable Product Design. *Sustainability*, *12*(10), 3150.

Ellen MacArthur Foundation. (2013). Towards the circular economy: Economic and business rationale for an accelerated transition. Retrieved from https://ellenmacarthurfoundation.org

European Commission. (2020). *Circular Economy Action Plan: For a cleaner and more competitive Europe*. https://europa.eu

Ford, S., & Despeisse, M. (2016). Additive manufacturing and sustainability: An exploratory study of the advantages and challenges. *Journal of Cleaner Production, 137*, 1573–1587. DOI: 10.1016/j.jclepro.2016.04.150

Frank, A. G., Dalenogare, L. S., & Ayala, N. F. (2019). Industry 4.0 technologies: Implementation patterns in manufacturing companies. *International Journal of Production Economics, 210*, 15–26. DOI: 10.1016/j.ijpe.2019.01.004

Frank, T., Duflou, J. R., & Dewulf, W. (2019). Automation in manufacturing: An overview of trends and challenges. *Procedia CIRP, 80*, 537–542.

Geissdoerfer, M., Savaget, P., Bocken, N. M. P., & Hultink, E. J. (2017). The circular economy – A new sustainability paradigm? *Journal of Cleaner Production, 143*, 757–768. DOI: 10.1016/j.jclepro.2016.12.048

Ghisellini, P., Cialani, C., & Ulgiati, S. (2016). A review on circular economy: The expected transition to a balanced interplay of environmental and economic systems. *Journal of Cleaner Production, 114*, 11–32. DOI: 10.1016/j.jclepro.2015.09.007

Gibson, I., Rosen, D. W., & Stucker, B. (2015). *Additive manufacturing technologies: 3D Printing, rapid prototyping, and direct digital manufacturing*. Springer. DOI: 10.1007/978-1-4939-2113-3

Govindan, K., Soleimani, H., & Kannan, D. (2015). Reverse logistics and closed-loop supply chain: A comprehensive review to explore the future. *European Journal of Operational Research, 240*(3), 603–626. DOI: 10.1016/j.ejor.2014.07.012

Guinee, J. B. (2016). *Handbook on life cycle assessment: Operational guide to the ISO standards*. Springer Science & Business Media.

Hauschild, M. Z., Rosenbaum, R. K., & Olsen, S. I. (2018). *Life cycle assessment: Theory and practice*. Springer. DOI: 10.1007/978-3-319-56475-3

Hussain, M., Shah, F., & Kiani, M. (2019). Role of IoT in Sustainable Product Life-cycle Management. *International Journal of Advanced Manufacturing Technology, 101*(5), 1445–1460.

Jackson, T. (2005). *Motivating sustainable consumption: A review of evidence on consumer behaviour and behavioural change.*

Kamble, S. S., Gunasekaran, A., & Gawankar, S. A. (2020). Achieving sustainable performance in a data-driven agriculture supply chain: A review for research and applications. *Computers and Electronics in Agriculture, 168*, 105204.

Korhonen, J., Honkasalo, A., & Seppälä, J. (2018). Circular economy: The concept and its limitations. *Ecological Economics, 143*, 37–46. DOI: 10.1016/j.ecolecon.2017.06.041

KPMG. (2021). *Sustainability Reporting and the Circular Economy: Compliance with environmental standards.*

Kusiak, A. (2018). Smart manufacturing. *International Journal of Production Research, 56*(1-2), 508–517. DOI: 10.1080/00207543.2017.1351644

Lacy, P., & Rutqvist, J. (2015). *Waste to Wealth: The Circular Economy Advantage.* Palgrave Macmillan.

Lee, J., Bagheri, B., & Kao, H. A. (2019). A Cyber-Physical Systems architecture for Industry 4.0-based manufacturing systems. *Manufacturing Letters, 3*, 18–23. DOI: 10.1016/j.mfglet.2014.12.001

Martin, K. E., Shilton, K., & Smith, J. R. (2020). Ethical Implications of Smart Consumer Products: Privacy, Consent, and Control. *Journal of Business Ethics, 168*(1), 39–55.

McKinnon, A., Browne, M., Whiteing, A., & Piecyk, M. (2015). *Green logistics: Improving the environmental sustainability of logistics.* Kogan Page Publishers.

McKinsey. (2020). *The Next Horizon for Product Sustainability: Technology-enabled solutions to unlock value.*

McKinsey & Company. (2020). *Smart manufacturing: The next wave of industrial transformation.*

Niinimäki, K., & Hassi, L. (2011). Emerging Design Strategies in Sustainable Production and Consumption. *Journal of Fashion Marketing and Management, 15*(4), 393–410.

Philips. (2020). *Philips Sustainability Report 2020.* Retrieved from https://www.philips.com

Phillips, T. (2020). *Philips' Circular Economy Commitment: Accelerating Progress Towards a Circular Economy. Philips Circular Economy Solutions.* Retrieved from https://www.philips.com/circular-economy

Pigosso, D. C. A., McAloone, T. C., & Rozenfeld, H. (2015). Characterisation of the state-of-the-art and identification of main trends for Ecodesign Tools and Methods: Classifying three decades of research and implementation. *Journal of Cleaner Production*, *108*, 1232–1244.

Porter, M. E., & Heppelmann, J. E. (2015). How smart, connected products are transforming companies. *Harvard Business Review*, *93*(10), 96–114.

Saberi, S., Kouhizadeh, M., Sarkis, J., & Shen, L. (2019). Blockchain technology and its relationships to sustainable supply chain management. *International Journal of Production Research*, *57*(7), 2117–2135. DOI: 10.1080/00207543.2018.1533261

Sachs, J. D. (2015). *The Age of Sustainable Development*. Columbia University Press. DOI: 10.7312/sach17314

Sherwin, C. (2004). Design and sustainability: A discussion paper based on personal experience and observations. *The Journal of Sustainable Product Design*, *4*(1-2), 21–31. DOI: 10.1007/s10970-006-0003-x

Stahel, W. R. (2016). The circular economy. *Nature*, *531*(7595), 435–438. DOI: 10.1038/531435a PMID: 27008952

Tao, F., Zhang, M., Cheng, J., & Qi, Q. (2019). Digital Twin and Big Data Towards Smart Manufacturing and Industry 4.0: A 360-Degree Comparison. *IEEE Access : Practical Innovations, Open Solutions*, *6*, 3585–3593.

Tesla, I. (2020). *Impact Report 2020*. Retrieved from https://www.tesla.com/impact

Tesla. (2021). *The Tesla Gigafactory: Revolutionising the Global Energy Ecosystem*. Retrieved from https://www.tesla.com/gigafactory

Chapter 3
Digital Manufacturing Toward Sustainability

Romdhane Ben Khalifa
https://orcid.org/0000-0002-9171-9584
National Engineering School of Gabes, Tunisia

ABSTRACT

This paper explores sustainable digital manufacturing as a transformative integration of Industry 4.0 technologies with environmental, social, and economic goals to minimize resource use, waste, and emissions across product lifecycles. It critiques traditional manufacturing's environmental degradation—exacerbated by the Industrial Revolution—and positions digital tools like AI, IoT, cloud computing, and data analytics as solutions for optimizing operations, decentralizing supply chains, and enabling circular economy models such as reuse, remanufacturing, and recycling. Key sections cover sustainable paradigms (closed-loop systems, eco-design), digital enablers (additive manufacturing/3D printing for waste reduction, advanced robotics/automation for precision, digital twins/simulation for predictive optimization), and emerging technologies like blockchain for supply chain traceability and transparency.

INTRODUCTION

Sustainable digital manufacturing practices aim to reduce environmental impact, enhance resource efficiency, and promote social and economic responsibility. These practices integrate modern digital tools with sustainability goals, offering a transformative approach to how products are designed, produced, and managed throughout their lifecycle. The traditional approach of human civilizations has always been to leverage technology for problem-solving, yet the Industrial Revolution marked a

DOI: 10.4018/979-8-3373-5473-6.ch003

turning point, giving rise to substantial environmental degradation stemming from waste and inefficient manufacturing processes (Moghrabi et al., 2023). As the global economy continues to expand, the manufacturing sector faces increasing pressure to adopt sustainable practices that minimize environmental harm and promote long-term ecological balance. Digital manufacturing offers innovative solutions to address these challenges, facilitating the development of eco-friendly production methods and more sustainable products (Cioffi et al., 2020). The integration of digital technologies into manufacturing processes enables businesses to optimize their operations, reduce waste, and minimize their environmental footprint, thereby contributing to a more sustainable and circular economy (Cioffi et al., 2020; Garetti et al., 2012).

Digital transformation, including cloud computing and data analytics, plays a crucial role in advancing sustainable manufacturing by decentralizing supply chains and optimizing resource use (Moghrabi et al., 2023). The construction industry, known for its significant energy consumption and waste generation, exemplifies the need for sustainable practices (Bamgbade et al., 2016). Sustainable development offers a key framework for meeting both present and future needs, emphasizing the importance of integrating environmental sustainability principles into manufacturing practices (Despeisse et al., 2012). Addressing sustainability in manufacturing requires a comprehensive approach that considers the entire lifecycle of products, from raw material extraction to end-of-life management (Yap et al., 2024). Adopting sustainable construction materials and practices that embrace Environmental, Social, and Governance principles is essential for minimizing the environmental impact of buildings (Yap et al., 2024). The focus on sustainability should also include strategies for improving energy efficiency, reducing waste and pollution, and promoting the responsible sourcing of raw materials. Circular business models, such as reuse, remanufacturing, and recycling, play a vital role in decoupling economic growth from resource consumption (Adenan et al., 2024).

Integrating sustainable practices into manufacturing provides significant opportunities for businesses to reduce costs, enhance their brand reputation, and meet evolving customer demands for eco-friendly products (HARUN et al., 2025). The incorporation of AI and ML into sustainable manufacturing strategies is one of the best ways to foster sustainable development while simultaneously boosting economic success (Kumar & Shahin, 2025). The implementation of digital technologies, such as IoT sensors and data analytics platforms, enables manufacturers to monitor and optimize their environmental performance in real-time (Bezerra et al., 2024). By presenting a general overview of how digital manufacturing can contribute to sustainability via a logical three-level framework, Figure 1 displays the connect. This highlights the importance of understanding these concepts, which can inspire your interest in advancing sustainable practices through digital tools.

Figure 1. Conceptual framework: Digital manufacturing toward sustainability

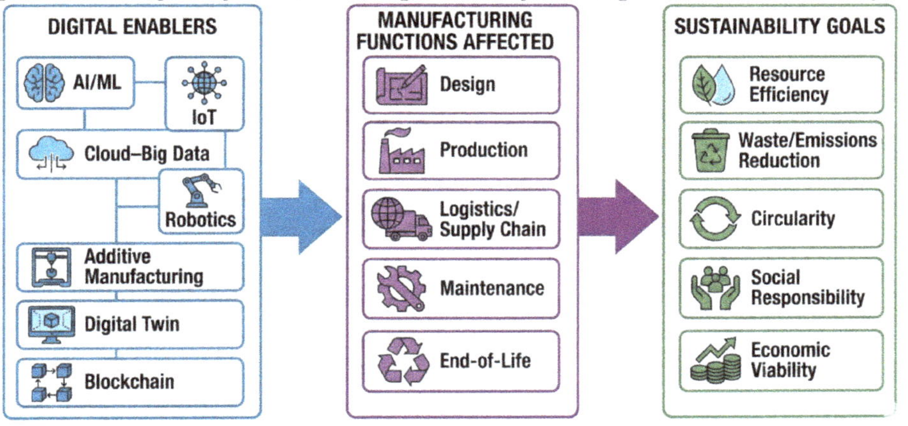

The shift towards sustainable manufacturing requires a collaborative effort involving governments, industries, and consumers to drive systemic change and promote a more sustainable future. Stakeholders are increasingly pushing industries to be transparent about the environmental and social impacts of their operations (Sartal et al., 2014). Digital manufacturing technologies not only enhance operational efficiency and productivity but also play a crucial role in promoting environmental and social sustainability within the manufacturing sector (Ali et al., 2021; Yunus et al., 2024). The implementation of sustainable practices in manufacturing can lead to time savings, cost efficiencies, and reduced construction waste, further underscoring the benefits of adopting sustainable approaches (Yunus et al., 2024). This paper outlines a plan to examine digital manufacturing. First, this paper examines how traditional manufacturing has harmed the environment since the Industrial Revolution, highlighting the importance of sustainable ideas. Then this paper lists the ideas. The main sustainable ideas are loop systems, eco-design, and circular-economy concepts that align with ESG goals. The central part of the paper describes enablers. Digital enablers are Industry 4.0 tools. Digital enablers include manufacturing that cuts waste. Digital enablers include robotics that give precision. Digital enablers include twins that allow simulation. Digital enablers include blockchain, which provides traceability. Real-time monitoring, energy efficiency, and cost savings are the standout benefits. The benefits matter most in high-impact sectors, such as construction. The combination of real-time monitoring, energy efficiency, and cost savings can make construction projects run more smoothly, reduce waste, and keep budgets in check. The stakeholder collaboration is the focus. Stakeholder collaboration brings together governments, industries, and consumers. The stakeholder collaboration works to create eco-production systems.

SUSTAINABLE MANUFACTURING PARADIGMS

Sustainable manufacturing is characterized by minimizing negative environmental impacts, reducing energy and natural resource consumption, and ensuring social responsibility and economic viability (Sartal et al., 2014). This encompasses designing products for durability, recyclability, and minimal material usage, as well as adopting closed-loop production systems that recover and reuse materials (HARUN et al., 2025). Sustainable manufacturing involves monitoring carbon emissions and predicting energy consumption using AI features and energy manager software (Tan et al., 2023). Sustainable manufacturing prioritizes sustainability throughout the entire product lifecycle, from design and production to distribution, use, and end-of-life management. Sustainable manufacturing can significantly reduce environmental impact and contribute to preserving natural resources and ecosystems (figure 2).

The principles of sustainable manufacturing align with the circular economy, which aims to minimize waste and pollution by keeping materials and products in use for as long as possible. By adopting circular economy principles, manufacturers can reduce their reliance on virgin materials, minimize waste generation, and create new revenue streams through remanufacturing and recycling. The implementation of sustainable supply chain practices, including green purchasing, transportation, and distribution, can further enhance the environmental performance of manufacturing operations (Sulaiman et al., 2024). Manufacturers can gain a competitive advantage by implementing sustainable manufacturing practices to produce environmentally friendly products.

Industry 4.0 paradigm shifts have facilitated the integration of technology and approach, which has enabled companies to enhance their sustainability (Polo et al., 2025). Data management, IoT, and extended product service systems are all tightly connected with traditional manufacturing processes. By integrating ESG principles, the construction sector can actively contribute to sustainable development and mitigate the negative effects of energy wastage on the environment (Yap et al., 2024). Automobile, aircraft, and ship manufacturers have either been regulated or have voluntarily adopted product stewardship initiatives to support high quality, end-of-life management. Effective waste management and recycling programs can help reduce environmental effects during the manufacturing process (Harun et al., 2025).

Sustainable manufacturing also entails adopting eco-design principles, which involve designing products with minimal environmental impact throughout their lifecycle, considering factors such as material selection, energy efficiency, and recyclability. By integrating sustainability into product design, manufacturers can create products that are more durable, repairable, and recyclable, reducing waste and resource consumption. Green supply chain management offers a pathway to achieving circular economy goals, supporting companies in aligning with the triple bottom line (Kharat et al., 2024). Sustainable manufacturing practices can enhance a company's reputation, attract investors, and build customer loyalty (Yap et al., 2024).

Figure 2. From traditional to sustainable digital manufacturing

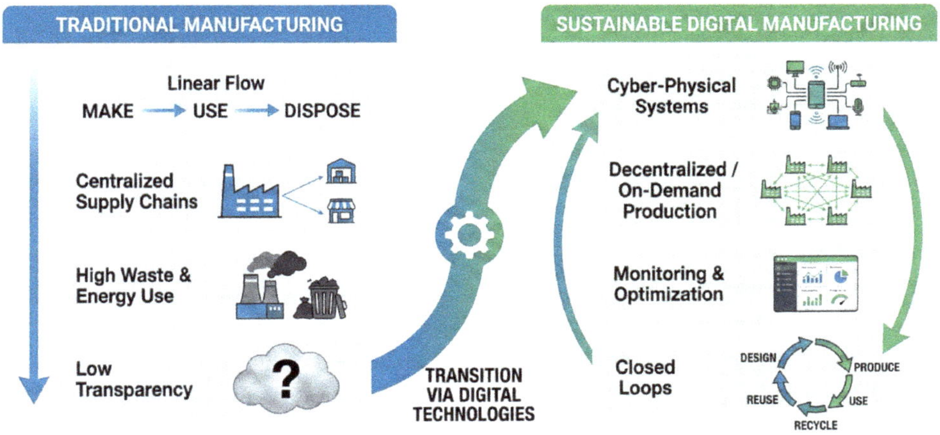

Implementing sustainable manufacturing practices necessitates a comprehensive approach that involves all stakeholders, including employees, suppliers, customers, and communities. Companies that embrace sustainability practices can reduce their ecological footprint, while ensuring long-term financial success and societal benefits (Huang et al., 2023). The integration of AI and ML into sustainable manufacturing strategies is one of the best ways to foster sustainable development while simultaneously boosting economic success (Moghrabi et al., 2023; Sartal et al., 2014). Smart technologies like IoT and AI can enhance energy management, lighting, and HVAC systems, improving energy efficiency and sustainability in buildings (Wang & Xue, 2024). Manufacturers should prioritize transparent reporting and communication of their sustainability performance to build trust and accountability with stakeholders. By taking a comprehensive approach to evaluating sustainable construction materials, decision-makers and practitioners can ensure that their choices align with environmental and financial sustainability and foster social well-being and equity (Yap et al., 2024).

THE ROLE OF DIGITAL TECHNOLOGIES

Digital manufacturing technologies, such as additive manufacturing, advanced robotics, and IoT, play a pivotal role in enabling sustainable manufacturing practices. These technologies enable manufacturers to optimize resource utilization, reduce waste generation, and improve energy efficiency throughout the production process (Früchtl et al., 2020). Digital manufacturing technologies facilitate the creation of digital twins, which are virtual models of physical assets that can be used to simulate and optimize manufacturing processes. These models provide manufacturers with real-time insights into their operations, allowing them to identify and address inefficiencies, reduce waste, and improve overall sustainability performance.

The integration of digital technologies, including AI-driven tools, predictive analytics, and automation, has redefined traditional processes, emphasizing the inseparability of technological advancement and ecological sustainability (Agrawal et al., 2025). By leveraging digital tools, professionals can drive data-driven sustainability initiatives, ensuring more accurate project forecasting and improved resource allocation (Baharuddin et al., 2025). Advanced manufacturing solutions that consist of autonomous, cooperative industrial robots and modular manufacturing systems characterized by integrated smart sensors and standardized interfaces are crucial (Savastano et al., 2019). Digital transformation offers innovative solutions to optimize production, improve energy efficiency, reduce resource waste, and meet environmental regulations (Xu et al., 2025).

The use of digital technologies allows for more efficient monitoring of energy usage, waste reduction, and pollution control in manufacturing facilities (Moghrabi et al., 2023). Big Data Analytics, Artificial Intelligence and Machine Learning, the IoT and Blockchain are adopted to streamline operations and promote sustainable production (Moghrabi et al., 2023). The adoption of digital technologies in manufacturing is becoming increasingly important in the current global business environment (Yang et al., 2021). These technologies provide manufacturers with the ability to collect and analyze data from various sources, enabling them to make more informed decisions about their operations.

Digitalization enables enhanced employee collaboration, connectivity, and productivity, as well as potentially autonomous process analysis and improvement capabilities (Bazaz et al., 2021). Furthermore, digital manufacturing facilitates the implementation of circular economy principles, where products are designed for disassembly and reuse, reducing waste and extending product lifecycles. It is of great significance for the sustained and healthy development of the global economy to continue to maintain the proportion of the manufacturing industry in GDP (Xiao, 2022). Digital technologies can be used to track and trace materials throughout the supply chain, ensuring ethical sourcing and reducing the risk of environmental and social harm (Glogar et al., 2025). The manufacturing sector can improve operational efficiency, optimize production processes, and reduce waste and energy consumption by utilizing digital tools (figure 3).

The implementation of digital technologies in manufacturing requires careful planning and execution, as well as investments in infrastructure, training, and cybersecurity (Bezerra et al., 2024). Digital technologies provide manufacturers with the tools and insights they need to achieve their sustainability goals, while simultaneously improving their competitiveness and profitability (Egger & Masood, 2019). In the digital economy, the rapid expansion and large-scale application of data have positioned data as a fundamental economic resource and key production factor (Wang & Yang, 2025). Digital connectivity enables the exchange of data and visibility, facilitating dynamic interactions, which leads to strategic renewal by combining company resources and capabilities (Franzè et al., 2023).

Figure 3. Sustainable manufacturing paradigms and their digital support

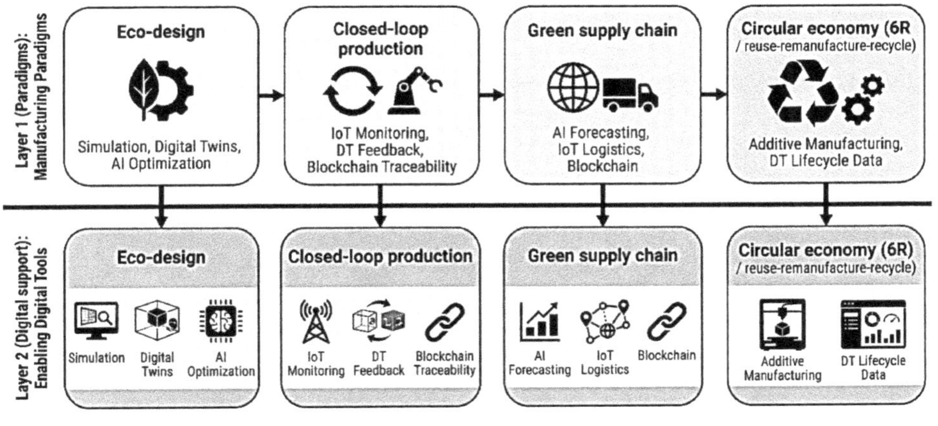

BLOCKCHAIN TECHNOLOGY FOR SUSTAINABLE MANUFACTURING

Blockchain technology has emerged as a promising solution for promoting transparency and traceability in sustainable manufacturing supply chains. Blockchain technology ensures the security, visibility, traceability, and transparency, which facilitates the resolution of environmental and global supply chain challenges (Tsolakis et al., 2020; Yu et al., 2022). By leveraging blockchain, manufacturers can track and trace the flow of materials, components, and products throughout the supply chain, ensuring ethical sourcing, reducing the risk of counterfeiting, and promoting responsible consumption (figure 4). Blockchain's decentralized and immutable nature makes it an ideal platform for verifying the authenticity and sustainability credentials of products and materials (Ghahremani-Nahr et al., 2022).

Blockchain technology's decentralized and tamper-resistant nature enhances product accountability and traceability by constructing transparent and secure supply chain management systems (Abid et al., 2024; Rana et al., 2021).

A blockchain-based system provides a transparent and traceable digital ledger of transactions and movements for all supply chain actors, offering a potential resolution to issues in traditional supply chains (Khair & Sandu, 2023; Subramanian et al., 2023). By enabling safe, decentralized, and impenetrable data management in supply chains, blockchain technology can be used to address the problems associated with traditional, centralized systems (Oh et al., 2025). The use of blockchain technology improves productivity by replacing the standard contract with smart contracts (Mann et al., 2018).

Figure 4. Digital enablers frameworks

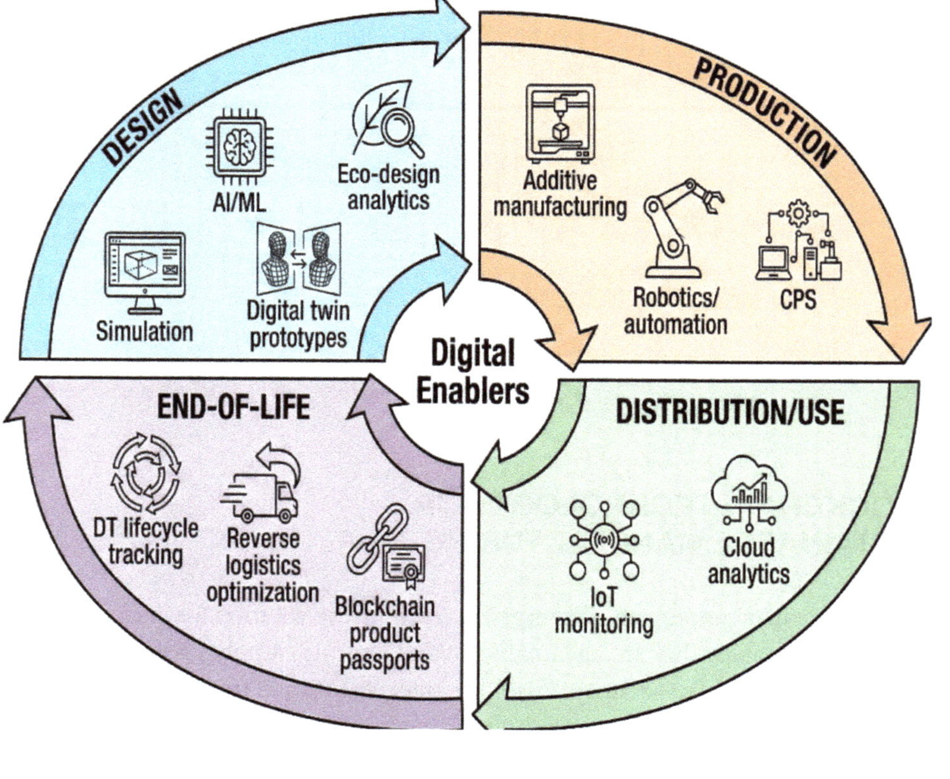

KEY SUSTAINABLE PRACTICES

Additive Manufacturing (3D Printing)

Additive manufacturing, also known as 3D printing, is a transformative technology that enables manufacturers to create complex parts and products layer by layer from digital designs. Material Efficiency: Produces components layer by layer, reducing waste. On-Demand Production: Parts can be made only when needed, minimizing storage and waste. Localized Production: This reduces transportation emissions (Ford & Despeisse, 2016). This technology offers several sustainability advantages, including reduced material waste, lower energy consumption, and increased design flexibility (Shang & Zhang, 2022). Instead of cutting away from a larger volume, additive manufacturing builds products layer-by-layer, significantly reducing resource demands and manufacturing waste (Nadagouda et al., 2020). Direct digital manufacturing which combines product modeling and manufacturing technology, directly converts digital models to physical objects, negating the necessity for tooling. The modification of models is simple and frequent at a customer's request before the model enters the manufacturing process (Vorkapić et al., 2020). This capability improves product sustainability and processes by altering operational practices in product design, distribution, use and after-sales services (Holmström et al., 2017).

Advanced Robotics and Automation

Furthermore, additive manufacturing enables the creation of lightweight structures and optimized designs, further reducing material usage and energy consumption. By producing parts on demand and closer to the point of use, additive manufacturing can also reduce transportation emissions and lead times (Hauck et al., 2025). The use of 3D printing facilitates the production of customized parts with complex geometries, which is particularly useful for industries such as aerospace, healthcare, and automotive. 3D printing has been implemented in different sectors such as research, automotive, aerospace, healthcare and medical, architecture and construction, fashion industries, and food industries (Wasti & Adhikari, 2020). The method facilitates the creation of complex geometries with desired material and improves the designing and modelling of structures (Shree et al., 2019). Moreover, this has potential applications in different industries which help in the reduction of overall cost and time (Shahrubudin et al., 2019).

Advanced robotics and automation technologies are revolutionizing manufacturing processes by increasing efficiency, reducing waste, and improving worker safety. Industrial robots and automated systems can perform repetitive tasks with greater precision and speed than human workers, minimizing errors and material waste. Robotics can also be employed in hazardous environments, minimizing risk to employees and increasing safety through automation of dangerous tasks. These tools are necessary to collect the information and data used to inform decisions pertaining to sustainability in manufacturing. Robotics provides a system for automating work through the programming of robots to reduce risks and ensure product quality. The precision offered by robots increases the use of raw materials and reduces the amount of waste. Moreover, automation and robotics are becoming particularly relevant in the era of Industry 4.0, where smart manufacturing and mechatronics play a crucial role (Campilho & Silva, 2023). The construction industry is undergoing a transformation through the introduction of robotic technologies, including robotic bricklaying, 3D printing, and robotic welding, enhancing productivity, safety standards, and precision while reducing material waste (Emani, 2025). Robots are now exerting a real impact and will inevitably play a vital and growing role in the future.

Digital Twins and Simulation

Digital twins and simulation technologies are virtual replicas of physical assets, processes, or systems that allow manufacturers to model, analyze, and optimize their operations in a virtual environment. These technologies provide valuable insights into the performance and behavior of manufacturing systems, enabling manufacturers to identify inefficiencies, optimize processes, and reduce waste (Guo et al., 2023). Digital twins enable real-time monitoring of energy consumption, material flow, and waste generation, providing manufacturers with the data they need to make informed decisions and improve resource efficiency. By simulating different scenarios and testing alternative designs, manufacturers can also optimize their products for sustainability, reducing material usage, energy consumption, and environmental impact. Digital twin technology enhances decision-making processes, promotes sustainability, and boosts productivity by replicating physical processes and systems, allowing for virtual optimization and real-time monitoring of operations (Gao et al., 2020). In the realm of sustainable manufacturing, digital twins emerge as indispensable tools, empowering businesses to make informed decisions, optimize resource utilization, and minimize environmental effects. Moreover, the integration of digital twins with other technologies like IoT and AI opens up new possibilities for predictive maintenance, process optimization, and closed-loop manufacturing (Rahmani et al., 2024). Digital twins are crucial for sustainable mobility in sustainable supply chains because they create real-time, virtual models of physical assets, processes, and systems (Bandara & Buics, 2024). In manufacturing, the deployment of digital twins is vital as these digital replicas facilitate the monitoring and control of individual assets, promoting optimization and resilience in factories (Bécue et al., 2020; Darabă et al., 2024).

Digital twins offer a virtual representation of physical assets, providing real-time synchronization between the physical and digital worlds (Hunhevicz et al., 2021; Longo et al., 2022). Digital twins are dynamic, digital replicas of living organisms or technical objects that are now a key component of Industry 4.0, providing businesses with significant advantages and being very helpful in the context of sustainable production and maintenance (Rojek et al., 2020). These digital twins can be used in optimizing the manufacturing lifecycle and promoting intelligent manufacturing upgrades (Mi et al., 2020). These digital twins are designed to leverage all available digital information, encompassing system and data information models, 3D models, mathematical models, dependability models, condition and performance indicators, and data analytics (Hatledal et al., 2020). Digital twins integrate large amounts of data from the physical system to optimize, monitor, diagnose, and forecast the system using AI, machine learning, and software analysis (Wang et al., 2021). Digital twins represent virtual replicas of potential or real entities. These digital representations enable the simulation of new products or process adjustments, allowing for virtual commissioning before deployment in the physical world (Chakraborty & Adhikari, 2020; Klar et al., 2022; Li et al., 2021). This reduces waste and improves resource efficiency. The ability to mirror digital twins at the city scale is crucial for realizing the applicability of using the completely mirrored digital twin for city-scale modeling (Shahat et al., 2021).

Digital twins facilitate the integration of fragmented digital information via platforms, establishing networks that foster correlation and complementation among various digital twins, thereby enhancing the efficiency and effectiveness of resource aggregation and dynamic allocation (Xiao et al., 2020). These digital twins offer a way to evaluate an asset's condition, predict future behavior, enhance control, and optimize operations (McKee, 2020). By integrating historical data and real-time sensor updates, these models facilitate the replication of complex activities and provide a realistic foundation for resilience estimation (Tekinerdoğan & Verdouw, 2020; Zhou et al., 2021). One key feature of a digital twin is the ability to improve the predictive performance over time, via improvements of the digital twin (Gardner et al., 2020). Digital twins hold the capability to improve the predictive performance over time by leveraging ongoing enhancements to the digital twin model (Waters, 2025).

CLOSED-LOOP PRODUCTION SYSTEMS

The concept of a digital twin, originally defined by NASA for simulating complex systems, has expanded to encompass diverse sectors, including smart cities, where it aids in planning, construction, and maintenance through real-time data integration and advanced analytics (Seong et al., 2025) (Agostinelli et al., 2021). Similarly, digital twins are categorized by their level of aggregation: asset twins focus on individual components, system twins integrate various assets into functional systems, and process twins optimize manufacturing operations through integrated knowledge models (El-Agamy et al., 2024).

This integration of real-time data from the physical environment into virtual models allows for the refinement of simulated scenarios, guiding the optimization of physical operations and ultimately enhancing production efficiency, product quality, and overall manufacturing resilience (Kantaros & Ganetsos, 2024). This real-time interaction enables a reciprocal influence between the physical and digital states, facilitating continuous analysis, optimization, and predictive maintenance within complex manufacturing systems (El-Agamy et al., 2024). This iterative refinement of analytical models, based on real-time data feedback, fosters a continuous improvement cycle, driving manufacturing efficiency, product innovation, and operational resilience within the contemporary production ecosystem (Kantaros & Ganetsos, 2024). This closed-loop feedback mechanism ensures that the digital twin continuously adapts to its physical counterpart, enhancing the accuracy of simulations and the efficacy of real-time control (Monek & Fischer, 2024) (figure 5). This bidirectional data flow, integral to the integration of Cyber-Physical Systems and Digital Twins, ensures that insights from virtual simulations are fed back into the physical manufacturing environment, thereby informing real-time decision-making and process optimizations (Kantaros & Ganetsos, 2024). This symbiotic relationship facilitates proactive adjustments in production parameters and adaptive control strategies, significantly enhancing efficiency and product quality throughout the entire lifecycle (Kantaros & Ganetsos, 2024). This continuous feedback loop further enables the development of robust, run-time resilient cyber-physical systems that can adapt to changing conditions and mitigate potential failures through simulation and predictive modeling (Gallab et al., 2024) (Napoleone et al., 2020).

This allows for a proactive approach to sustainability, where products are designed with their end-of-life considerations in mind, thereby reducing waste and promoting circular economy principles. Furthermore, the integration of Digital Twins with Cyber-Physical Systems facilitates real-time feedback loops, ensuring that recommendations and control signals generated by the DTs are promptly translated into actionable adjustments within the CPS infrastructure (Kantaros & Ganetsos, 2024).

By simulating the entire lifecycle of a product, digital twins allow for optimization of design, production, and distribution, significantly reducing waste and pollution and promoting a circular economy (Yakovenko & Shaptala, 2024). Furthermore, these models enhance resource efficiency by simulating material flows and energy consumption, leading to optimized designs that minimize environmental impact throughout the product lifecycle (Rezaei et al., 2023). This integration enables manufacturers to simulate diverse operational scenarios, refine production strategies, and conduct "what-if" analyses to enhance efficiency and minimize waste (Kantaros & Ganetsos, 2024). Furthermore, by simulating the entire lifecycle of a product, from design to end-of-life, digital twins enable comprehensive waste management strategies and promote circular economy principles (Resman et al., 2025). They enable the simulation of production processes, thus minimizing material waste and mitigating errors that arise during manufacturing (Sengupta & Dreyer, 2024). Moreover, these advanced digital tools play a pivotal role in enabling proactive maintenance, process optimization, and enhanced quality control through real-time simulation and monitoring of production processes (Besigomwe, 2025). This comprehensive approach enables enhanced decision-making through continuous feedback loops, facilitating the dynamic adjustment of manufacturing parameters and reducing waste (Besigomwe, 2025) (Yakovenko & Shaptala, 2024). Moreover, digital twins aid in managing the entire product lifecycle, optimizing resource utilization and extending product lifespans by providing crucial information for maintenance, upgrades, and end-of-life considerations (Lin & Wei, 2023) (Xiros et al., 2025).

Figure 5. Application focus: Application focus: construction sector example

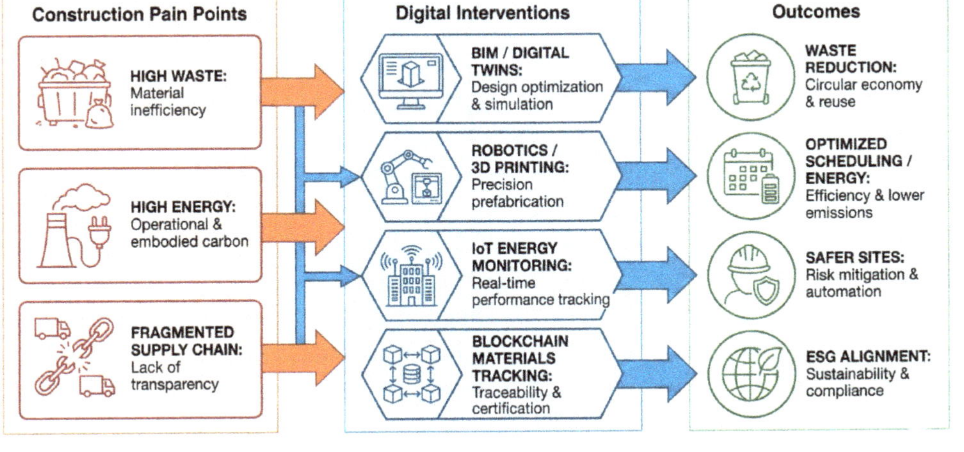

Digital twins contribute to waste management by facilitating the simulation of a product's complete lifecycle, thereby enabling optimized design, production, and distribution processes, which in turn substantially reduce waste and pollution while advancing circular economy principles (Yakovenko & Shaptala, 2024). Additionally, these models improve resource efficiency through the simulation of material flows and energy consumption, resulting in optimized designs that mitigate environmental impact across the product's entire lifecycle (Rezaei et al., 2023). This integrated approach empowers manufacturers to simulate various operational scenarios, refine production strategies, and perform hypothetical analyses to bolster efficiency and curtail waste (Kantaros & Ganetsos, 2024). Moreover, by simulating a product's entire lifecycle, from its initial design phase to its end-of-life, digital twins facilitate comprehensive waste management strategies and foster circular economy principles (Resman et al., 2025). Digital twins also allow for the simulation of production processes, thereby reducing material waste and minimizing manufacturing errors (Sengupta & Dreyer, 2024). Furthermore, these sophisticated digital tools are instrumental in facilitating proactive maintenance, optimizing processes, and elevating quality control via real-time simulation and continuous monitoring of production processes (Besigomwe, 2025). Such a holistic methodology supports improved decision-making through constant feedback loops, which allows for the dynamic adjustment of manufacturing parameters and contributes to waste reduction (Besigomwe, 2025; Yakovenko & Shaptala, 2024). Additionally, digital twins assist in managing the complete product lifecycle, optimizing resource utilization, and prolonging product lifespans by supplying essential data for maintenance, upgrades, and end-of-life planning (Lin & Wei, 2023; Xiros et al., 2025).

Lifecycle Tracking

Comprehensive Lifecycle Management: Digital twins enable the continuous and detailed monitoring, analysis, and optimization of a product's entire lifespan, from its initial design and production through its operational phase, to its end-of-life considerations (Resman et al., 2025; Yakovenko & Shaptala, 2024). This capability facilitates proactive waste reduction, enhances resource utilization, and promotes circular economy principles by informing design, production, and distribution strategies (Xiros et al., 2025; Yakovenko & Shaptala, 2024). Furthermore, digital twins provide critical data for effective maintenance, potential upgrades, and end-of-life planning, thereby extending product lifespans and improving overall sustainability (Lin & Wei, 2023). Through this comprehensive oversight, they support strategies for remanufacturing and end-of-life recovery, reducing material waste and facilitating the reuse of valuable components (Lin & Wei, 2023) (Trienens et al., 2024).

Reverse Logistics

Digital twins are instrumental in optimizing reverse logistics by simulating the collection, sorting, and reprocessing of products and components, thus streamlining the recovery of valuable materials and minimizing landfill waste. This optimization is crucial for fostering a circular economy by ensuring efficient material recovery and minimizing the environmental impact associated with product disposal (Yakovenko & Shaptala, 2024; Xiros et al., 2025). This enables manufacturers to model and refine processes for product returns, repairs, and recycling, significantly reducing waste and promoting the reintegration of materials into new production cycles (Yakovenko & Shaptala, 2024). This systematic approach to reverse logistics, facilitated by digital twins, not only minimizes waste but also enhances the economic viability of circular material flows, aligning with broader sustainability objectives (Trienens et al., 2024). Moreover, digital twins provide specific life cycle data that enable the monitoring and implementation of circular economy strategies throughout a product's life cycle, from design to delivery (Mügge et al., 2024). By analyzing product, component, and material data, as well as process data, these virtual counterparts create transparency and build data-driven product ecosystems for new business and value creation models (Mügge et al., 2024). This transparency, extending across the entire product lifecycle from design to delivery, is critical for optimizing logistics and distribution processes, thereby reducing associated carbon emissions (Yakovenko & Shaptala, 2024). This capability further enhances sustainable practices by enabling more informed decisions regarding product recovery, reuse, and recycling at various stages of the product lifecycle (Trienens et al., 2024). For example, AI-powered digital twins can monitor product performance over time and generate improvements, enabling data-driven decisions that optimize product longevity and facilitate timely interventions (Roberts et al., 2022). This includes planning and simulating remanufacturing and recycling processes before physical products arrive, ensuring an individualized approach based on each product's specific characteristics and condition, and enabling real-time monitoring and adaptation during these end-of-life phases (Strakošová et al., 2025). The integration of AI, IoT, and blockchain technologies further augments these capabilities, allowing for enhanced transparency and traceability throughout a product's existence, from manufacturing to eventual disposal or repurposing (Özsoy, 2023). Such advanced digital twin applications, often incorporating analytics and optimization, are vital for detecting and predicting demand and supply variabilities, thus preventing waste and reallocating resources for zero-waste value chains (Sengupta & Dreyer, 2024). By integrating these technologies, companies can effectively transition from a linear take-make-use-dispose model to a circular economic system, ensuring all stakeholders have access to pertinent information regarding resource utilization, sourcing, and recycling processes (Ghoreishi et al., 2024). Furthermore, digital product passports, facilitated by these digital technologies, can provide comprehensive, transparent data about a product's materials, origin, use, and end-of-life handling instructions, which is crucial for promoting circular economy principles (Vu et al., 2025) (Trienens et al., 2024).

Energy Management and Monitoring

Real-time energy monitoring reduces unnecessary consumption. IoT devices, for instance, can optimize energy use across supply chain operations, leading to significant reductions in consumption (Singh, 2024). This real-time data, when combined with big data analytics and machine learning algorithms, enables organizations to identify areas for emission and waste reduction, thereby minimizing their overall environmental footprint (Ameh, 2024). These technologies also facilitate the implementation of emissions trading schemes and carbon trading, thereby enabling more extensive carbon management practices across various industrial sectors (Challouf et al., 2025). Moreover, AI-powered tools monitor emissions in real-time, promptly alerting companies to inefficiencies or violations of environmental standards, thus facilitating immediate corrective actions and optimizing processes for reduced carbon output (Attah et al., 2024). This proactive approach not only ensures regulatory compliance but also supports the achievement of corporate sustainability goals by enabling continuous improvement in environmental performance (Challouf et al., 2025). This integration of advanced technologies like AI and IoT within smart energy systems is crucial for supporting the transition towards more sustainable and circular industrial practices, particularly in sectors such as construction where energy efficiency is paramount. Furthermore, advanced life cycle analysis, supported by IoT devices and big data analytics, provides real-time and accurate data on energy consumption, material usage, and waste generation, which allows for targeted sustainability strategies and contributes to substantial environmental benefits (Keles et al., 2025). These sophisticated tools facilitate informed decision-making, enabling industries to pinpoint environmental hotspots and prioritize improvements in their operations and product designs (Saeed et al., 2023). Additionally, the utilization of blockchain technology within these systems offers enhanced transparency and traceability of energy flows, ensuring the integrity of data for compliance and reporting (Das et al., 2024) (Sira, 2024). Such comprehensive monitoring and management frameworks are essential for companies to identify inefficiencies and optimize resource utilization, thereby significantly reducing waste and environmental impact (Fatorachian, 2023). This integration supports a holistic approach to sustainability, moving beyond mere compliance to proactive environmental stewardship and fostering a circular economy (Sira, 2024). The multifaceted application of AI technologies, such as real-time energy consumption monitoring, provides effective solutions for achieving carbon reduction targets and enhancing corporate competitiveness in combating climate change (Hua et al., 2025). AI-powered systems enable real-time monitoring and management of energy usage in businesses, utilizing machine learning algorithms to forecast demand, automate adjustments, and optimize operational processes to minimize excessive consumption (Gandía et al., 2025). These algorithms continually refine energy-saving strategies by learning from historical data, thus contributing to more sustainable energy practices (Rane, 2023). Furthermore, the scheduling capabilities of Energy Management Systems, combined with AI techniques, play a vital role in optimizing energy usage patterns and promoting energy conservation across various applications (Hanafi et al., 2024). This comprehensive approach, driven by AI, enhances energy efficiency, reduces energy waste, and optimizes production processes, directly contributing to carbon reduction targets and improved environmental performance (Hua et al., 2025) (Uriarte-Gallastegi et al., 2024). The integration of AI, IoT, and blockchain technologies provides precise, real-time measurements of carbon footprints across various sectors, enabling proactive emission reduction strategies (Bhuiyan, 2024).

Adoption of machines with lower power requirements. This includes industrial robots and intelligent manufacturing systems that leverage AI for real-time data analysis and predictive maintenance, thereby preventing equipment failures and improving overall production efficiency (Yang et al., 2024). Such systems also enhance sustainability by optimizing resource allocation and minimizing waste through advanced algorithms that predict optimal production schedules and material flows (Rane et al., 2024). The application of AI in optimizing energy usage and reducing waste aligns with the goals of a circular economy and net-zero emissions, significantly decreasing the carbon footprint through predictive analytics and intelligent process optimization (Xue et al., 2025) (Alghieth, 2025). Moreover, AI's capability to analyze energy consumption patterns enables the identification of inefficiencies and the implementation of targeted interventions, further supporting energy management and environmental sustainability initiatives (Kong et al., 2024). These advanced technological applications extend to smart manufacturing processes, which leverage AI to reduce energy consumption and waste by 30–50%, alongside similar reductions in building energy usage (Olawade et al., 2024). This optimization includes the dynamic scheduling of manufacturing processes, which integrates solar energy utilization to minimize both production and building operation costs (Salhi et al., 2025). Beyond these optimizations, AI-based decision support systems can proactively adjust energy use by predicting demand, while AI-driven predictive maintenance further curtails energy wastage by pinpointing and rectifying equipment inefficiencies in real-time (Soori et al., 2024).

Integration of renewable energy sources and smart grid systems. This holistic approach, encompassing facility design and energy infrastructure, is crucial for establishing manufacturing ecosystems that are both sustainable and economically viable (Kim et al., 2024). These facilities leverage advanced sensor networks and real-time data analytics to monitor and control energy flows, ensuring maximum efficiency and minimal environmental impact (Li et al., 2021) (Soori et al., 2024). The application of artificial intelligence in such facilities further enhances operational efficiency by optimizing resource allocation and predicting energy demands, thereby integrating sustainability directly into the core manufacturing process (Xue et al., 2025). For instance, AI algorithms can analyze vast datasets from IoT devices to manage heating, ventilation, and air conditioning systems and lighting, aligning energy use with occupancy and external weather conditions, leading to substantial energy savings and reduced operational costs (Varghese, 2022). Furthermore, AI-driven energy management systems dynamically adjust production processes to align with the availability of renewable energy sources, thereby decreasing greenhouse gas emissions ("Artificial Intelligence in Manufacturing – Systematic Literature Review," 2025). This integration is crucial for policymakers developing energy efficiency programs and implementing energy policy, especially given global disruptions. Moreover, such sophisticated systems facilitate the development of smart grids, which are essential for integrating diverse renewable energy sources and managing their intermittent nature effectively (Giannakopoulos et al., 2025). This allows for a more resilient and sustainable energy infrastructure, capable of adapting to fluctuating supply and demand while minimizing reliance on fossil fuels (Ayadi et al., 2025). AI-driven energy management systems provide real-time insights and autonomous decision-making capabilities, further enhancing industrial energy efficiency by optimizing energy distribution and integrating renewable energy sources seamlessly (Ibekwe et al., 2024) (Chauhan & Sahoo, 2024).

SUSTAINABLE SUPPLY CHAIN INTEGRATION

This integration is vital for creating resilient, innovative, and environmentally responsible industrial landscapes (Onyeke et al., 2024). Artificial intelligence plays a pivotal role in this transformation by optimizing resource utilization, mitigating environmental impacts, and fostering a circular economy across the entire supply chain (Nayak, 2025). Specifically, AI-powered systems can analyze vast datasets to identify inefficiencies, predict demand fluctuations, and optimize logistics, thereby reducing waste and energy consumption from raw material extraction to product delivery (Tsolakis et al., 2022). For example, AI can optimize truck routes, reducing greenhouse gas emissions and shrinking the carbon footprint of transportation (Abideen et al., 2021). Furthermore, AI systems can monitor weather conditions and traffic patterns to proactively identify potential disruptions, enabling supply chain managers to implement timely mitigation strategies that maintain efficiency and reduce environmental impact (Atapattu et al., 2024). Additionally, AI facilitates improved accuracy in demand forecasting and inventory projections, which allows for better resource management and reduced waste throughout the supply chain (Kashem et al., 2023). By leveraging machine learning algorithms, companies can precisely forecast demand, thereby minimizing overproduction and the associated waste, while also optimizing inventory levels to prevent stockouts and improve delivery efficiency (Gandía et al., 2025; Tsolakis et al., 2022). AI's role extends to enhancing sustainable supply chain management through streamlined logistics, waste reduction in production, and promotion of ethical sourcing practices (Al-Raeei, 2024). For instance, AI can analyze and interpret data related to diversity and inclusion within the workplace, as well as monitor and analyze supply chains to ensure ethical and sustainable sourcing of materials ("Artificial Intelligence: Towards Sustainable Intelligence," 2023). AI further supports sustainable supply chains by facilitating real-time tracking of goods, enabling greater transparency and accountability across the network, and allowing for the assessment of suppliers based on their sustainability parameters (Pal, 2023; Zejjari & Benhayoun, 2024). AI-driven systems can thus enhance operational productivity and efficiency, improving demand management and increasing supply chain transparency and security (Pal, 2023; Tsolakis et al., 2022). These capabilities enable organizations to identify potential risks, such as supplier disruptions or quality issues, and proactively implement measures to mitigate them, ensuring supply chain resilience (Badmus et al., 2024). Additionally, AI can facilitate the selection of optimal suppliers by evaluating historical data and market conditions, ensuring that organizations partner with the best suppliers for their needs (Badmus et al., 2024). Moreover, AI-driven solutions significantly improve logistics and route optimization by analyzing traffic patterns, delivery windows, and fuel costs to determine the most efficient transportation routes, thereby reducing operational expenses and enhancing delivery times (Badmus et al., 2024). This not only reduces transportation costs but also improves delivery times and customer satisfaction, with AI playing a crucial role in optimizing inventory levels to reduce stockouts and enhance operational efficiency (Badmus et al., 2024; Pattnaik et al., 2024). The integration of AI in warehouse management, through smart sensors and automated systems, further refines inventory control and space utilization, contributing to an overall more efficient and sustainable supply chain (Kashem et al., 2023). These advancements collectively lead to significant improvements in operational efficiency, accurate demand predictions, and enhanced risk management capabilities within supply chains, ultimately promoting greater sustainability (Kelly, 2024).

BENEFITS

AI's ability to process and interpret large datasets allows for real-time adjustments to supply chain operations, leading to dynamic optimization and increased responsiveness to market changes (Kumar, 2024). This adaptability is crucial for navigating disruptions and maintaining competitive advantage in rapidly evolving global markets (Zhang, 2024). Moreover, AI-powered predictive analytics can empower organizations to anticipate demand fluctuations and optimize inventory levels, thereby reducing waste and minimizing environmental impact (Zejjari & Benhayoun, 2024). Such advanced analytical capabilities enable a proactive approach to supply chain management, ensuring resource efficiency and fostering sustainable practices across the entire value chain (Patil, 2025). By automating mundane tasks and optimizing complex processes, AI also enhances labor efficiency, allowing human resources to focus on strategic decision-making and innovation (Hamedani et al., 2025). Furthermore, the integration of AI with other complementary digital technologies, such as the Internet of Things and augmented reality, can enhance real-time visibility and collaboration across supply chain networks, leading to improved organizational performance and sustainability (Teixeira et al., 2025). In particular, AI-based predictive analytics has significantly enhanced logistics and transportation management by optimizing route planning, fleet management, and delivery schedules (Ahmed et al., 2025). This leads to significant reductions in fuel consumption and carbon emissions, thereby contributing to environmental sustainability (Yusuf et al., 2025). Additionally, AI streamlines demand forecasting, supplier performance assessment, and inventory management, providing real-time insights that improve decision-making and mitigate risks associated with overstocking or understocking (Kukreja, 2025). These capabilities allow supply chain managers to proactively address potential disruptions, optimize resource allocation, and make informed decisions to enhance resilience and overall supply chain performance (Nazeer, 2021). Moreover, AI-driven solutions significantly improve logistics and route optimization by analyzing traffic patterns, delivery windows, and fuel costs to determine the most efficient transportation routes, thereby reducing operational expenses and enhancing delivery times (Badmus et al., 2024).

CHALLENGES

Despite these advancements, integrating AI into supply chain operations presents notable challenges, including significant cultural shifts within organizations, concerns regarding data security, and the complexities of navigating various legal and regulatory frameworks (ISMAEIL & Lalla, 2024). These hurdles necessitate substantial investment in training and infrastructure, along with the development of robust governance strategies to ensure ethical AI deployment and data integrity (Goswami et al., 2024; Pal, 2023). Furthermore, challenges include the high costs associated with data collection, specialized hardware, talent acquisition, and ongoing maintenance, alongside difficulties in data governance, workforce upskilling, and ensuring equitable adoption of AI technologies (Atapattu et al., 2024; Teixeira et al., 2025). Moreover, the lack of personalization, potential technical failures, and accountability issues further compound the complexities of AI implementation in supply chains (Cevallos et al., 2024). Additionally, challenges include ensuring data quality, integrating AI systems with legacy infrastructure, and addressing skill gaps within the workforce (Kumar, 2024; Yusuf et al., 2025). Furthermore, the energy demands of advanced AI systems and the complexities of data integration pose considerable environmental and operational challenges (Mrad et al., 2025). Overcoming these challenges requires a comprehensive strategy that includes robust data management frameworks, scalable technological infrastructure, and continuous workforce development to foster a culture of AI literacy (Daios et al., 2025).

CONCLUSION

The use of digital techniques in manufacturing is, in fact, a green shift, leveraging the so-called Industry 4.0 to address the environmental hazards created by manufacturing almost all types of products since the industrial revolution. These are layers of additive manufacturing for on-demand, resource-efficient part production; sophisticated robotics and automation for accuracy and waste reduction; digital twins for online simulation and lifecycle optimization; and blockchain for supply chain transparency, enabling circular-economy-aligned closed-loop systems. With real-time IoT monitoring and AI-driven analytics, energy usage can be further optimized, anticipating maintenance requirements and minimizing emissions, leading to benefits such as cost savings, greater regulatory compliance, and a better brand reputation, especially in the construction sector. However, potential challenges like high initial costs, technological complexity, and integration issues should also be considered to provide a balanced view for decision-makers.

The paper illustrates the connection between sustainable manufacturing paradigms—eco-design, green supply chains, and ESG integration–and digital enablers that facilitate resilient production, effectively breaking the link between economic growth and resource depletion. The analysis highlights the unity of technological innovation and ecological responsibility, with digital twins managing waste in real time, reverse logistics optimizing resource recovery, and product stewardship covering the entire lifecycle from cradle to grave. Incorporating specific sustainability metrics, such as carbon footprint reduction or resource efficiency indices, can further demonstrate the tangible benefits of these digital strategies to stakeholders.

Looking into the future, empirically validated case studies and frameworks that can be scaled will be very important to handle the implementation barriers of cybersecurity, workforce reskilling and infrastructure investments. Policymakers need to make sure there are enough incentives for adoption as well as standards, such as tax credits, grants, or regulatory mandates, so that industries become focus on collaborative platforms for data sharing. In the end, a joint action by governments, businesses and consumers is going to push digital manufacturing forward UN Sustainable Development Goals, making the industrial future low-carbon, equitable and affluent where sustainability drives prosperity.

REFERENCES

Abid, I., Fuad, S. M. Z. A., Chowdhury, M. J. M., Chowdhury, M. S., & Ferdous, M. S. (2024). A Systematic Literature Review on the Use of Blockchain Technology in Transition to a Circular Economy. *arXiv (Cornell University)*. https://doi.org// arxiv.2408.11664DOI: 10.48550

Abideen, A. Z., Sundram, V. P. K., Pyeman, J., Othman, A. K., & Sorooshian, S. (2021). Food Supply Chain Transformation through Technology and Future Research Directions—A Systematic Review. *Logistics (Basel)*, *5*(4), 83. DOI: 10.3390/ logistics5040083

Adenan, F. A., Ariff, A., & Adnan, S. M. (2024). ESG in Small and Medium-Sized Enterprises: Malaysian Perspectives. *E3S Web of Conferences, 574*, 3001. DOI: 10.1051/e3sconf/202457403001

Agostinelli, S., Cumo, F., Guidi, G., & Tomazzoli, C. (2021). Cyber-Physical Systems Improving Building Energy Management: Digital Twin and Artificial Intelligence. *Energies*, *14*(8), 2338. DOI: 10.3390/en14082338

Agrawal, K., Goktas, P., Holtkemper, M., Beecks, C., & Kumar, N. (2025). AI-driven transformation in food manufacturing: A pathway to sustainable efficiency and quality assurance. *Frontiers in Nutrition*, *12*, 1553942. Advance online publication. DOI: 10.3389/fnut.2025.1553942

Ahmed, F., Ahmed, M., Kabir, M. A., & Islam, M. M. (2025). Revolutionizing Business Analytics. *The Impact of Artificial Intelligence and Machine Learning.*, *1*(1), 147–173. DOI: 10.63125/f7yjxw69

Al-Raeei, M. (2024). Artificial intelligence for climate resilience: Advancing sustainable goals in SDGs 11 and 13 and its relationship to pandemics. *Discover Sustainability*, *5*(1), 513. Advance online publication. DOI: 10.1007/s43621-024-00775-5

Alghieth, M. (2025). Sustain AI: A Multi-Modal Deep Learning Framework for Carbon Footprint Reduction in Industrial Manufacturing. *Sustainability (Basel)*, *17*(9), 4134. DOI: 10.3390/su17094134

Ali, H., Chen, T., & Hao, Y. (2021). Sustainable Manufacturing Practices, Competitive Capabilities, and Sustainable Performance: Moderating Role of Environmental Regulations. *Sustainability (Basel)*, *13*(18), 10051. DOI: 10.3390/su131810051

Ameh, B. (2024). Digital tools and AI: Using technology to monitor carbon emissions and waste at each stage of the supply chain, enabling real-time adjustments for sustainability improvements. *International Journal of Science and Research Archive*, *13*(1), 2741–2757. DOI: 10.30574/ijsra.2024.13.1.1995

Artificial intelligence in manufacturing – systematic literature review. (2025). *Scientific Papers of Silesian University of Technology Organization and Management Series*, (215). Advance online publication. DOI: 10.29119/1641-3466.2025.215.39

Artificial Intelligence: Towards Sustainable Intelligence. (2023). In *Communications in computer and information science*. Springer Science+Business Media. DOI: 10.1007/978-3-031-47997-7

Atapattu, A. J., Perera, L., Nuwarapaksha, T. D., Udumann, S. S., & Dissanayaka, N. S. (2024). *Challenges in Achieving Artificial Intelligence in Agriculture.*, DOI: 10.1007/978-981-97-5878-4_2

Attah, R. U., Garba, B. M. P., Gil-Ozoudeh, I., & Iwuanyanwu, O. (2024). Digital transformation in the energy sector: Comprehensive review of sustainability impacts and economic benefits. *International Journal of Advanced Economics*, *6*(12), 760–776. DOI: 10.51594/ijae.v6i12.1751

Ayadi, R., Forouheshfar, Y., & Moghadas, O. (2025). Enhancing system resilience to climate change through artificial intelligence: A systematic literature review. *Frontiers in Climate*, *7*, 1585331. Advance online publication. DOI: 10.3389/fclim.2025.1585331

Badmus, O., Rajput, S., Arogundade, J. B., & Williams, M. H. (2024). AI-driven business analytics and decision making. *World Journal of Advanced Research and Reviews*, *24*(1), 616–633. DOI: 10.30574/wjarr.2024.24.1.3093

Baharuddin, D., Zainordin, N., & Jusoh, Z. M. (2025). The Institutional ESG Disclosure Framework for Malaysian Consultancy Quantity Surveying Practices: A Systematic Literature Review. *Journal of Ecohumanism*, *4*(2). Advance online publication. DOI: 10.62754/joe.v4i2.6600

Bamgbade, J. A., Mohammed, K. A., & Nawi, M. N. M. (2016). Assessing the sustainable construction of large construction companies in Malaysia. *AIP Conference Proceedings*, *1761*, 20027. DOI: 10.1063/1.4960867

Bandara, L. V., & Buics, L. (2024). Digital Twins in Sustainable Supply Chains: A Comprehensive Review of Current Applications and Enablers for Successful Adoption. *Engineering Proceedings*, *79*(1), 64. DOI: 10.3390/engproc2024079064

Bazaz, S. M., Penttilä, S., Ollikainen, M., Ratava, J., & Varis, J. (2021). Industry 4.0 readiness of manufacturing sector in the Baltic Region. *IOP Conference Series. Materials Science and Engineering*, *1140*(1), 12019. DOI: 10.1088/1757-899X/1140/1/012019

Bécue, A., Maia, E., Feeken, L., Borchers, P., & Praça, I. (2020). A New Concept of Digital Twin Supporting Optimization and Resilience of Factories of the Future. *Applied Sciences (Basel, Switzerland)*, *10*(13), 4482. DOI: 10.3390/app10134482

Besigomwe, K. (2025). Closed-Loop Manufacturing with AI-Enabled Digital Twin Systems. *Cognizance Journal of Multidisciplinary Studies*, *5*(1), 18–38. DOI: 10.47760/cognizance.2025.v05i01.002

Bezerra, R. R. R., Martins, V. W. B., & Macêdo, A. N. (2024). Validation of Challenges for Implementing ESG in the Construction Industry Considering the Context of an Emerging Economy Country. *Applied Sciences (Basel, Switzerland)*, *14*(14), 6024. DOI: 10.3390/app14146024

Bhuiyan, M. A. H. (2024). Carbon Footprint Measurement and Mitigation Using AI. SSRN *Electronic Journal*. DOI: 10.2139/ssrn.4746446

Bhuiyan, M. A. H. (2024). Carbon Footprint Measurement and Mitigation Using AI. SSRN *Electronic Journal*. DOI: 10.2139/ssrn.4746446

Campilho, R. D. S. G., & Silva, F. J. G. (2023). Industrial Process Improvement by Automation and Robotics. *Machines (Basel)*, *11*(11), 1011. DOI: 10.3390/machines11111011

Cevallos, M. G. O., Jaramillo, G. A. L., Vélez, J. L. B., Zambrano, M. I. U., Montesdeoca, L. D. Z., & Palomeque, M. A. B.Miguel Giancarlo Ormaza CevallosGustavo Alberto Lozano JaramilloJosé Luis Bernardo VélezMaritza Irinuska Ureta Zambrano-Lady Diana Zambrano MontesdeocaManuel Augusto Bermúdez Palomeque. (2024). Implementation of Artificial Intelligence in Quality Management in SMEs: Benefits and Challenges. *Evolutionary Studies in Imaginative Culture*, *1489*, 1489–1500. Advance online publication. DOI: 10.70082/esiculture.vi.1590

Chakraborty, S., & Adhikari, S. (2020). Machine learning based digital twin for dynamical systems with multiple time-scales. *Computers & Structures*, *243*, 106410. DOI: 10.1016/j.compstruc.2020.106410

Challouf, K., Alhloul, A., & Németh, N. (2025). Mapping the role of industry 4.0 technologies in green supply chain management: A bibliometric and structured text analysis. *Discover Sustainability*, *6*(1), 949. Advance online publication. DOI: 10.1007/s43621-025-01827-0

Chauhan, M., & Sahoo, D. R. (2024). Towards a Greener Tomorrow: Exploring the Potential of AI, Blockchain, and IoT in Sustainable Development. *Nature Environment and Pollution Technology*, *23*(2), 1105–1113. DOI: 10.46488/NEPT.2024.v23i02.044

Cioffi, R., Travaglioni, M., Piscitelli, G., Petrillo, A., & Parmentola, A. (2020). Smart Manufacturing Systems and Applied Industrial Technologies for a Sustainable Industry: A Systematic Literature Review. *Applied Sciences (Basel, Switzerland)*, *10*(8), 2897. DOI: 10.3390/app10082897

Daios, A., Kladovasilakis, N., Kelemis, A., & Kostavelis, I. (2025). AI Applications in Supply Chain Management: A Survey. *Applied Sciences (Basel, Switzerland)*, *15*(5), 2775. DOI: 10.3390/app15052775

Darabă, D., Pop, F. S., & Daraba, C. (2024). Digital Twin Used in Real-Time Monitoring of Operations Performed on CNC Technological Equipment. *Applied Sciences (Basel, Switzerland)*, *14*(22), 10088. DOI: 10.3390/app142210088

Das, S. K., Bressanelli, G., & Saccani, N. (2024). Clustering the Research at the Intersection of Industry 4.0 Technologies, Environmental Sustainability and Circular Economy: Evidence from Literature and Future Research Directions. *Circular Economy and Sustainability*, *4*(4), 2473–2504. Advance online publication. DOI: 10.1007/s43615-024-00393-3

Despeisse, M., Ball, P., Evans, S., & Levers, A. (2012). Industrial ecology at factory level – a conceptual model. *Journal of Cleaner Production*, *31*, 30–39. DOI: 10.1016/j.jclepro.2012.02.027

Egger, J., & Masood, T. (2019). Augmented reality in support of intelligent manufacturing – A systematic literature review. *Computers & Industrial Engineering*, *140*, 106195. DOI: 10.1016/j.cie.2019.106195

El-Agamy, R. F., Sayed, H. A., Akhatatneh, A. M. A., Aljohani, M., & Elhosseini, M. A. (2024). Comprehensive analysis of digital twins in smart cities: A 4200-paper bibliometric study. *Artificial Intelligence Review*, *57*(6), 154. Advance online publication. DOI: 10.1007/s10462-024-10781-8

Emani, V. V. (2025). Robotic bricklaying and 3d printing robots in construction: automating masonry work and building structures layer by layer, and robotic welding in construction (enhancing precision and efficiency in structural steel work), International journal of Busness quantitative DOI: 10.2139/ssrn.5297498

Fatorachian, H. (2023). The Significance of Industry 5.0 in the Globalization of Supply Chain Management. [EEL]. *European Economic Letters*, *13*(5), 872–885. DOI: 10.52783/eel.v13i5.843

Ford, S., & Despeisse, M. (2016). Additive manufacturing and sustainability: An exploratory study of the advantages and challenges. *Journal of Cleaner Production*, *137*, 1573–1587. DOI: 10.1016/j.jclepro.2016.04.150

Franzè, C., Paolucci, E., & Pessot, E. (2023). Sustained value creation driven by digital connectivity: A multiple case study in the mechanical components industry. *Technovation*, *129*, 102918. DOI: 10.1016/j.technovation.2023.102918

Früchtl, M., Leis, M., & Wertheim, R. (2020). A comprehensive and interdisciplinary perspective on sustainable manufacturing towards sustainable life cycles. *Procedia Manufacturing*, *43*, 383–390. DOI: 10.1016/j.promfg.2020.02.197

Gallab, M., Nardo, M. D., & Naciri, L. (2024). Navigating contemporary challenges and future prospects in digital industry evolution. *Discover Applied Sciences*, *6*(5), 259. Advance online publication. DOI: 10.1007/s42452-024-05913-2

Gandía, J. A. G., Ancillo, A. de L., & Núñez, M. T. del V. (2025). The Role of Artificial Intelligence and Knowledge in Enhancing Corporate Sustainability. *Journal of Innovation & Knowledge*, *10*(5), 100792. DOI: 10.1016/j.jik.2025.100792

Gao, R. X., Wang, L., Helu, M., & Teti, R. (2020). Big data analytics for smart factories of the future. *CIRP Annals*, *69*(2), 668–692. DOI: 10.1016/j.cirp.2020.05.002

Gardner, P., Borgo, M. D., Ruffini, V., Hughes, A. J., Zhu, Y., & Wagg, D. (2020). Towards the Development of an Operational Digital Twin. *Vibration*, *3*(3), 235–265. DOI: 10.3390/vibration3030018

Garetti, M., Mummolo, G., & Taisch, M. (2012). Special issue on "sustainable manufacturing.". *Production Planning and Control*, *23*(2-3), 79–82. DOI: 10.1080/09537287.2011.591617

Ghahremani-Nahr, J., Aliahmadi, A., & Nozari, H. (2022). An IoT-based sustainable supply chain framework and blockchain. *International Journal of Innovation in Engineering*, *2*(1), 12–21. DOI: 10.59615/ijie.2.1.12

Ghoreishi, M., Franconi, A., Mate, O.-A., Godsell, J., & Pynnönen, M. (2024). Supply Chain 4.0 and the transition towards a circular economy. In *Edward Elgar Publishing eBooks* (p. 238). Edward Elgar Publishing. DOI: 10.4337/9781803920207.00018

Giannakopoulos, G., Shaikh, K. M., Perez, M. A., & Adegbenro, P. (2025). Sustainable Practices in Green Engineering: A Pathway to a Greener Future using. *AI*, *1*(1), 1–9. DOI: 10.21276/jccci/2025.v1.i1.1

Glogar, M., Petrak, S., & Naglić, M. M. (2025). Digital Technologies in the Sustainable Design and Development of Textiles and Clothing—A Literature Review. *Sustainability (Basel), 17*(4), 1371. DOI: 10.3390/su17041371

Goswami, S. S., Mondal, S., Sarkar, S., Gupta, K. K., Sahoo, S. K., & Halder, R. (2024). Artificial Intelligence-Enabled Supply Chain Management: Unlocking New Opportunities and Challenges. *Artificial Intelligence and Applications (Commerce, Calif.), 3*(1), 110–121. Advance online publication. DOI: 10.47852/bonviewAIA42021814

Guo, Y. B., Klink, A., Bártolo, P., & Guo, W. (2023). Digital twins for electro-physical, chemical, and photonic processes. *CIRP Annals, 72*(2), 593–619. DOI: 10.1016/j.cirp.2023.05.007

Hamedani, S. S., Aslam, S., & Hamedani, S. S. (2025). AI in business operations: Driving urban growth and societal sustainability. *Frontiers in Artificial Intelligence, 8*, 1568210. DOI: 10.3389/frai.2025.1568210

Hanafi, A., Moawed, M., & Abdellatif, O. (2024). Advancing Sustainable Energy Management: A Comprehensive Review of Artificial Intelligence Techniques in Building [Review of Advancing Sustainable Energy Management: A Comprehensive Review of Artificial Intelligence Techniques in Building]. *Deleted Journal, 53*(2), 26–46. DOI: 10.21608/erjsh.2023.226854.1196

HARUN, S. N., Kamaruddin, N., & Choo, J. (2025). *INTERIOR DESIGN AND RENOVATION WORKS FOR 2 LEVEL FACTORY OFFICE AT ROVSKI, USJ.*

Hatledal, L. I., Skulstad, R., Li, G., Styve, A., & Zhang, H. (2020). Co-simulation as a Fundamental Technology for Twin Ships. *Modeling Identification and Control A Norwegian. Modeling, Identification and Control, 41*(4), 297–311. DOI: 10.4173/mic.2020.4.2

Hauck, S., Greif, L., Benner, N., & Ovtcharova, J. (2025). Advancing Sustainable Additive Manufacturing: Analyzing Parameter Influences and Machine Learning Approaches for CO2 Prediction. *Sustainability (Basel), 17*(9), 3804. DOI: 10.3390/su17093804

Holmström, J., Liotta, G., & Chaudhuri, A. (2017). Sustainability outcomes through direct digital manufacturing-based operational practices: A design theory approach. *Journal of Cleaner Production, 167*, 951–961. DOI: 10.1016/j.jclepro.2017.03.092

Hua, J., Wang, R., Hu, Y. C., Chen, Z., Chen, L., Osman, A. I., Farghali, M., Huang, L., Feng, J., Wang, H., Zhang, X., Zhou, X., & Yap, P. (2025). Artificial intelligence for calculating and predicting building carbon emissions: a review [Review of *Artificial intelligence for calculating and predicting building carbon emissions: a review*]. *Environmental Chemistry Letters*. Springer Science+Business Media. DOI: 10.1007/s10311-024-01799-z

Huang, J., Irfan, M., Fatima, S. S., & Shahid, R. M. (2023). The role of lean six sigma in driving sustainable manufacturing practices: An analysis of the relationship between lean six sigma principles, data-driven decision making, and environmental performance. *Frontiers in Environmental Science*, *11*, 1184488. Advance online publication. DOI: 10.3389/fenvs.2023.1184488

Hunhevicz, J., Motie, M., & Hall, D. (2021). Digital building twins and blockchain for performance-based (smart) contracts. *Automation in Construction*, *133*, 103981. DOI: 10.1016/j.autcon.2021.103981

Ismaeil, M. K. L., & Lalla, A. F.ISMAEIL. (2024). The Role and Impact of Artificial Intelligence on Supply Chain Management: Efficiency, Challenges, and Strategic Implementation. *Journal of Ecohumanism*, *3*(4), 89–106. DOI: 10.62754/joe.v3i4.3461

Kantaros, A., & Ganetsos, T. (2024). Integration of Cyber-Physical Systems, Digital Twins and 3D Printing in Advanced Manufacturing: A Synergistic Approach. *American Journal of Engineering and Applied Sciences*, *17*(1), 1–22. DOI: 10.3844/ajeassp.2024.1.22

Kashem, M. A., Shamsuddoha, M., Nasir, T., & Chowdhury, A. A. (2023). Supply Chain Disruption versus Optimization: A Review on Artificial Intelligence and Blockchain [Review of Supply Chain Disruption versus Optimization: A Review on Artificial Intelligence and Blockchain]. *Knowledge (Beverly Hills, Calif.)*, *3*(1), 80–96. DOI: 10.3390/knowledge3010007

Keles, C., Rios, F. C., & Hoque, S. (2025). Digital Technologies and Circular Economy in the Construction Sector: A Review of Lifecycle Applications, Integrations, Potential, and Limitations [Review of *Digital Technologies and Circular Economy in the Construction Sector: A Review of Lifecycle Applications, Integrations, Potential, and Limitations*]. *Buildings, 15*(4), 553. Multidisciplinary Digital Publishing Institute. DOI: 10.3390/buildings15040553

Kelly, A. (2024). Impact of Artificial Intelligence on Supply Chain Optimization. *Journal of Technology and Systems*, *6*(6), 15–27. DOI: 10.47941/jts.2153

Khair, M. A., & Sandu, A. K. (2023). *Blockchain-Optimized Supply Chain Traceability System for Transparent Logistics*. https://hal.science/hal-04543221

Kharat, M. G., Kapoor, S., Parhi, S., Kharat, M. G., & Pandey, S. (2024). Operationalizing sustainability in pharmaceuticals: Green supply chain metrics for circular economy. *Sustainable Futures : An Applied Journal of Technology, Environment and Society*, *9*, 100413. DOI: 10.1016/j.sftr.2024.100413

Kim, B. J., Hong, S., & Jeong, H. E. (2024). Preface for Advanced Manufacturing for ESG. *International Journal of Precision Engineering and Manufacturing-Green Technology*, *11*(3), 707–708. DOI: 10.1007/s40684-024-00613-z

Klar, R., Fredriksson, A., & Angelakis, V. (2022). Assessing the Maturity of Digital Twinning Solutions for Ports. *arXiv (Cornell University)*. https://doi.org//arxiv.2212.07722DOI: 10.48550

Kong, Z., Kong, L., & Bai, K. (2024). Research on the Development Path of E-commerce Logistics in the Context of Carbon Neutrality. *Journal of Physics: Conference Series*, *2735*(1), 12001. DOI: 10.1088/1742-6596/2735/1/012001

Kukreja, A. K. (2025). AI Adoption in SMEs: Integrating Supply Chain and Financial Strategies for Competitive Advantage. *International Journal For Multidisciplinary Research*, *7*(1), 35821. Advance online publication. DOI: 10.36948/ijfmr.2025.v07i01.35821

Kumar, P. (2024). AI for Optimizing Supply Chain Management. *International Journal for Research in Applied Science and Engineering Technology*, *12*(11), 549–560. DOI: 10.22214/ijraset.2024.65059

Kumar, V., & Shahin, K. (2025). Artificial Intelligence and Machine Learning for Sustainable Manufacturing: Current Trends and Future Prospects. *Intelligent and Sustainable Manufacturing*, *2*(1), 10002. DOI: 10.70322/ism.2025.10002

Li, L., Aslam, S., Wileman, A., & Perinpanayagam, S. (2021). Digital Twin in Aerospace Industry: A Gentle Introduction. *IEEE Access : Practical Innovations, Open Solutions*, *10*, 9543–9562. DOI: 10.1109/ACCESS.2021.3136458

Li, X., Wang, B., Peng, T., & Xu, X. (2021). Greentelligence: Smart Manufacturing for a Greener Future. *Chinese Journal of Mechanical Engineering*, *34*(1), 116. Advance online publication. DOI: 10.1186/s10033-021-00656-2

Lin, K.-Y., & Wei, S. (2023). Advancing the Industrial Circular Economy: The Integrative Role of Machine Learning in Resource Optimization. *Journal of Green Economy and Low-Carbon Development*, *2*(3), 122–136. DOI: 10.56578/jgelcd020302

Longo, F., Mirabelli, G., Solina, V., Alberto, U., Paola, G. D., Giordano, L., & Ziparo, M. (2022). A Simulation-Based Framework for Manufacturing Design and Resilience Assessment: A Case Study in the Wood Sector. *Applied Sciences (Basel, Switzerland)*, *12*(15), 7614. DOI: 10.3390/app12157614

Mann, S., Potdar, V., Gajavilli, R. S., & Chandan, A. (2018). Blockchain Technology for Supply Chain Traceability. *Transparency and Data Provenance*, *22*, 22–26. Advance online publication. DOI: 10.1145/3301403.3301408

McKee, S. (2020). *What is a digital twin?* DOI: 10.61557/ETSQ9929

Mi, S., Feng, Y., Zheng, H., Wang, Y., Gao, Y., & Tan, J. (2020). Prediction maintenance integrated decision-making approach supported by digital twin-driven cooperative awareness and interconnection framework. *Journal of Manufacturing Systems*, *58*, 329–345. DOI: 10.1016/j.jmsy.2020.08.001

Moghrabi, I. A. R., Bhat, S. A., Szczuko, P., AlKhaled, R. A., & Dar, M. A. (2023). Digital Transformation and Its Influence on Sustainable Manufacturing and Business Practices. *Sustainability (Basel)*, *15*(4), 3010. DOI: 10.3390/su15043010

Monek, G. D., & Fischer, S. (2024). Expert Twin: A Digital Twin with an Integrated Fuzzy-Based Decision-Making Module. *Decision Making Applications in Management and Engineering*, *8*(1), 1–21. DOI: 10.31181/dmame8120251181

Mrad, M., Frikha, M. A., & Boujelbène, Y. (2025). *State-of-The-Art Review: Artificial Intelligence and Robotics for Reducing Carbon Emissions in Supply Chain Management*. DOI: 10.2139/ssrn.5274912

Mügge, J., Seegrün, A., Hoyer, T.-K., Riedelsheimer, T., & Lindow, K. (2024). Digital Twins within the Circular Economy: Literature Review and Concept Presentation. *Sustainability (Basel)*, *16*(7), 2748. DOI: 10.3390/su16072748

Nadagouda, M. N., Ginn, M., & Rastogi, V. (2020). A review of 3D printing techniques for environmental applications [Review of *A review of 3D printing techniques for environmental applications*]. *Current Opinion in Chemical Engineering, 28*, 173. Elsevier BV. DOI: 10.1016/j.coche.2020.08.002

Napoleone, A., Macchi, M., & Pozzetti, A. (2020). A review on the characteristics of cyber-physical systems for the future smart factories [Review of *A review on the characteristics of cyber-physical systems for the future smart factories*]. *Journal of Manufacturing Systems, 54*, 305. Elsevier BV. DOI: 10.1016/j.jmsy.2020.01.007

Nayak, G. N.Gautam Nandkishore Nayak. (2025). Leveraging Artificial Intelligence for Sustainable Electronics Manufacturing Supply Chains. *International Journal of Scientific Research in Computer Science Engineering and Information Technology, 11*(1), 210–221. DOI: 10.32628/CSEIT25111230

Nazeer, A. (2021). AI-Powered Predictive Analytics for Supply Chain Optimization: A Risk-Resilient Framework. *International Journal of Emerging Trends in Computer Science and Information Technology, 2*(1), 12–18. DOI: 10.63282/3050-9246. IJETCSIT-V2I1P102

Oh, S. E., Kim, J., Kim, J.-Y., & Ahn, J. H. (2025). Food Safety Distribution Systems Using Private Blockchain: Ensuring Traceability and Data Integrity Verification. *Foods, 14*(8), 1405. DOI: 10.3390/foods14081405

Olawade, D. B., Wada, O. Z., David-Olawade, A. C., Fapohunda, O., Ige, A. O., & Ling, J. (2024). Artificial intelligence potential for net zero sustainability: Current evidence and prospects. *Next Sustainability, 4*, 100041. DOI: 10.1016/j.nxsust.2024.100041

Onyeke, F. O., Odujobi, O., Adikwu, F. E., & Elete, T. Y. (2024). The role of data-driven insights in industrial control systems: Advancing predictive maintenance and operational efficiency in refinery processes. *Engineering Science & Technology Journal, 5*(12), 3266–3277. DOI: 10.51594/estj.v5i12.1775

Özsoy, T. (2023). THE ROLE OF ARTIFICIAL INTELLIGENCE IN FACILITATING THE TRANSITION TO A CIRCULAR ECONOMY. *Nişantaşı Üniversitesi Sosyal Bilimler Dergisi/Nişantaşı Üniversitesi Sosyal Bilimler Dergisi, 11*(2), 369. DOI: 10.52122/nisantasisbd.1345828

Pal, S. (2023). Integrating AI in Sustainable Supply Chain Management: A New Paradigm for Enhanced Transparency and Sustainability. *International Journal for Research in Applied Science and Engineering Technology, 11*(6), 2979–2984. DOI: 10.22214/ijraset.2023.54139

Patil, D. (2025). *Artificial Intelligence-Driven Supply Chain Optimization: Enhancing Demand Forecasting And Cost Reduction*. DOI: 10.2139/ssrn.5057408

Pattnaik, S., Liew, N., Kures, A. O., Pinsky, E., & Park, K. (2024). *Catalyzing Supply Chain Evolution: A Comprehensive Examination of Artificial Intelligence Integration in Supply Chain Management*. 57. DOI: 10.3390/engproc2024068057

Polo, S., Rubio, E. M., Ayllón, J., & de Agustina, B. (2025). Emerging Advances in Sustainable Manufacturing. *Processes (Basel, Switzerland), 13*(5), 1549. DOI: 10.3390/pr13051549

Rahmani, R., Jesus, C., & Lopes, S. I. (2024). Implementations of Digital Transformation and Digital Twins: Exploring the Factory of the Future. *Processes (Basel, Switzerland), 12*(4), 787. DOI: 10.3390/pr12040787

Rana, S. K., Kim, H., Pani, S. K., Rana, S., Joo, M.-I., Rana, A. K., & Aich, S. (2021). Blockchain-Based Model to Improve the Performance of the Next-Generation Digital Supply Chain. *Sustainability (Basel), 13*(18), 10008. DOI: 10.3390/su131810008

Rane, N. L. (2023). Integrating Leading-Edge Artificial Intelligence (AI), Internet of Things (IoT), and Big Data Technologies for Smart and Sustainable Architecture, Engineering and Construction (AEC) Industry: Challenges and Future Directions. *International Journal of Data Science and Big Data Analytics, 3*(2), 73–95. DOI: 10.51483/IJDSBDA.3.2.2023.73-95

Rane, N. L., Rane, J., & Paramesha, M. (2024). *Artificial Intelligence and business intelligence to enhance Environmental, Social, and Governance (ESG) strategies: Internet of things, machine learning, and big data analytics in financial services and investment sectors*. DOI: 10.70593/978-81-981367-4-9_3

Resman, M., Herakovič, N., & Debevec, M. (2025). Integrating Digital Twin Technology to Achieve Higher Operational Efficiency and Sustainability in Manufacturing Systems. *Systems, 13*(3), 180. DOI: 10.3390/systems13030180

Rezaei, Z., Vahidnia, M. H., Aghamohammadi, H., Azizi, Z., & Behzadi, S. (2023). Digital twins and 3D information modeling in a smart city for traffic controlling: A review [Review of Digital twins and 3D information modeling in a smart city for traffic controlling: A review]. *Journal of Geography and Cartography, 6*(1), 1865. DOI: 10.24294/jgc.v6i1.1865

Roberts, H., Zhang, J., Bariach, B., Cowls, J., Gilburt, B., Juneja, P., Tsamados, A., Ziosi, M., Taddeo, M., & Floridi, L. (2022). Artificial intelligence in support of the circular economy: ethical considerations and a path forward. SSRN *Electronic Journal*. DOI: 10.2139/ssrn.4080909

Rojek, I., Mikołajewski, D., & Dostatni, E. (2020). Digital Twins in Product Lifecycle for Sustainability in Manufacturing and Maintenance. *Applied Sciences (Basel, Switzerland), 11*(1), 31. DOI: 10.3390/app11010031

Saeed, S., Arshad, M. Y., & Ahmed, A. S. (2023). Advancing circular economy in industrial chemistry and environmental engineering: Principles, alignment with United Nations sustainable development goals, and pathways to implementation. *European Journal of Chemistry, 14*(3), 414–428. DOI: 10.5155/eurjchem.14.3.414-428.2452

Salhi, A., Algarni, F. S., Alshamrani, R., Althbiti, A., Ismail, A., & Hassan, B. M. (2025). Leveraging artificial intelligence to enable sustainable urban development through the creation of smart and environmentally friendly carbon-free cities. *Scientific Reports*, *15*(1), 35791. Advance online publication. DOI: 10.1038/s41598-025-16801-z

Sartal, A., Bellas, R., Mejías, A. M., & Garcia, A. (2014). The sustainable manufacturing concept, evolution and opportunities within Industry 4.0: A literature review [Review of *The sustainable manufacturing concept, evolution and opportunities within Industry 4.0: A literature review*]. *Advances in Mechanical Engineering, 12*(5). SAGE Publishing. DOI: 10.1177/1687814020925232

Savastano, M., Amendola, C., Bellini, F., & D'Ascenzo, F. (2019). Contextual Impacts on Industrial Processes Brought by the Digital Transformation of Manufacturing: A Systematic Review [Review of *Contextual Impacts on Industrial Processes Brought by the Digital Transformation of Manufacturing: A Systematic Review*]. *Sustainability, 11*(3), 891. Multidisciplinary Digital Publishing Institute. DOI: 10.3390/su11030891

Sengupta, S., & Dreyer, H. C. (2024). Digitalization and sustainability. In *Routledge eBooks* (p. 98). Informa., DOI: 10.4324/9781032693415-8

Seong, K., Jiao, J., Lewis, R. H., Farahi, A., Navrátil, P., Casebeer, N., Davis, B., Jones, J. B., & Niyogi, D. (2025). Towards a digital twin for smart resilient cities: Real-time fire and smoke tracking and prediction platform for community awareness (FireCom). *Computational Urban Science*, *5*(1), 49. Advance online publication. DOI: 10.1007/s43762-025-00212-x

Shahat, E., Hyun, C. T., & Yeom, C. (2021). City Digital Twin Potentials: A Review and Research Agenda [Review of *City Digital Twin Potentials: A Review and Research Agenda*]. *Sustainability, 13*(6), 3386. Multidisciplinary Digital Publishing Institute. DOI: 10.3390/su13063386

Shahrubudin, N., Chuan, L. T., & Ramlan, R. (2019). An Overview on 3D Printing Technology: Technological, Materials, and Applications. *Procedia Manufacturing*, *35*, 1286–1296. DOI: 10.1016/j.promfg.2019.06.089

Shang, Z., & Zhang, L. (2022). The Sustainable Digitalization in the Manufacturing Industry: A Bibliometric Analysis and Research Trend. *Mobile Information Systems*, *1*, 1–11. Advance online publication. DOI: 10.1155/2022/1451705

Shree, M. V., Dhinakaran, V., Rajkumar, V., Ram, P. M. B., Vijayakumar, M., & Sathish, T. (2019). Effect of 3D printing on supply chain management. *Materials Today: Proceedings*, *21*, 958–963. DOI: 10.1016/j.matpr.2019.09.060

Singh, P. K. K. (2024). Measuring the Broader Value Proposition of Digital Transformation in Supply Chains. *International Journal of Supply Chain Management, 13*(1), 16–24. DOI: 10.59160/ijscm.v13i1.6222

Sira, M. (2024). Potential of Advanced Technologies for Environmental Management Systems. *Management Systems in Production Engineering, 32*(1), 33–44. DOI: 10.2478/mspe-2024-0004

Soori, M., Jough, F. K. G., Dastres, R., & Arezoo, B. (2024). AI-Based Decision Support Systems in Industry 4.0, A Review [Review of *AI-Based Decision Support Systems in Industry 4.0, A Review*]. *Journal of Economy and Technology*. Elsevier BV. DOI: 10.1016/j.ject.2024.08.005

Strakošová, S., Novák, P., & Kadera, P. (2025). Product Digital Twin Supporting End-of-life Phase of Electric Vehicle Batteries Utilizing Product-Process-Resource Asset Network. *arXiv (Cornell University)*. https://doi.org//arxiv.2510.02167DOI: 10.48550

Subramanian, N., Joshi, A., & Bagga, D. (2023). Transparent and Traceable Food Supply Chain Management. *arXiv (Cornell University)*. https://doi.org//arxiv.2305.12188DOI: 10.48550

Sulaiman, N. N., Omar, A. J., Pakir, F., & Mohamad, M. (2024). Evaluating Environmental, Social and Governance (ESG) Practice Among Malaysian Public Listed Construction Companies Using FTSE Russell Rating Model. *International Journal of Sustainable Construction Engineering Technology, 15*(1). Advance online publication. DOI: 10.30880/ijscet.2024.15.01.003

Tan, L. P., Choe, K. L., Ng, Y. K., Lai, K. F., & Khor, H. C. (2023). The Environmental, Social, and Governance (ESG) In Malaysia SME: A Case Study of Vitrox. In *Advances in economics, business and management research/Advances in Economics, Business and Management Research* (p. 344). Atlantis Press. DOI: 10.2991/978-94-6463-342-9_26

Teixeira, A. C. B., Ferreira, J. V., & Ramos, A. L. (2025). Intelligent Supply Chain Management: A Systematic Literature Review on Artificial Intelligence Contributions. *Information (Basel), 16*(5), 399. DOI: 10.3390/info16050399

Tekinerdoğan, B., & Verdouw, C. N. (2020). Systems Architecture Design Pattern Catalog for Developing Digital Twins. *Sensors (Basel), 20*(18), 5103. DOI: 10.3390/s20185103

Trienens, M., Rasor, R., Kharatyan, A., Dumitrescu, R., & Anacker, H. (2024). Digital twins to increase sustainability throughout the system life cycle: A systematic literature review. *Proceedings of the Design Society*, *4*, 2277–2286. DOI: 10.1017/pds.2024.230

Tsolakis, N., Niedenzu, D., Simonetto, M., Dora, M., & Kumar, M. (2020). Supply network design to address United Nations Sustainable Development Goals: A case study of blockchain implementation in Thai fish industry. *Journal of Business Research*, *131*, 495–519. DOI: 10.1016/j.jbusres.2020.08.003

Tsolakis, N., Schumacher, R., Dora, M., & Kumar, M. (2022). Artificial intelligence and blockchain implementation in supply chains: A pathway to sustainability and data monetisation? *Annals of Operations Research*, *327*(1), 157–210. DOI: 10.1007/s10479-022-04785-2

Uriarte-Gallastegi, N., Landín, G. A., Landeta-Manzano, B., & Laskurain-Iturbe, I. (2024). The Role of AI in Improving Environmental Sustainability: A Focus on Energy Management. *Energies*, *17*(3), 649. DOI: 10.3390/en17030649

Varghese, A. (2022). AI-Driven Solutions for Energy Optimization and Environmental Conservation in Digital Business Environments. *Asia Pacific Journal of Energy and Environment*, *9*(1), 49–60. DOI: 10.18034/apjee.v9i1.736

Vorkapić, M., Hasan, M. A., Tanović, D., Baltić, M., & Tomić, B. (2020). Implementation of 6R strategy in FDM printing process: Case: Small electronic enclosure box. *Journal of Engineering Management and Competitiveness*, *10*(2), 141–150. DOI: 10.5937/jemc2002141V

Vu, N. V., Nazari, M. A., Dang, T., Muralev, Y., Mohanraj, M., Tran, T., & Quoc, H. A. (2025). Type of the Paper*: Article*. DOI: 10.2139/ssrn.5384374

Wang, D., & Yang, T. Y. (2025). Research on the Promotion Effect of the Marketization of Data Elements on the Digital Transformation of Manufacturing Enterprises: An Empirical Evaluation of a Multiperiod DID Model. *Sustainability (Basel)*, *17*(7), 3199. DOI: 10.3390/su17073199

Wang, J., & Xue, F. (2024). Emerging Trends of ESG in the Construction Sector: A Promising Pathway to Sustainable and Responsible Development. In *Lecture notes in operations research* (p. 2033). Springer International Publishing. DOI: 10.1007/978-981-97-1949-5_142

Wang, Y., Wang, S., & Wang, J. (2021). Digital twin model of equipment maintenance management in modern enterprises. *Journal of Physics: Conference Series*, *1986*(1), 12088. DOI: 10.1088/1742-6596/1986/1/012088

Wasti, S., & Adhikari, S. (2020). Use of Biomaterials for 3D Printing by Fused Deposition Modeling Technique: A Review [Review of *Use of Biomaterials for 3D Printing by Fused Deposition Modeling Technique: A Review*]. *Frontiers in Chemistry, 8*. Frontiers Media. DOI: 10.3389/fchem.2020.00315

Waters, G. (2025). *Testing, Evaluation, Verification and Validation (TEVV) of Digital Twins: A Comprehensive Framework.* DOI: 10.48550/ARXIV.2507.04555

Xiao, L., Cao, J., Liu, Z. G., & Luo, X. (2020). Sustainable Business Model Based on Digital Twin Platform Network: The Inspiration from Haier's Case Study in China. *Sustainability (Basel), 12*(3), 936. DOI: 10.3390/su12030936

Xiao, Y. (2022). *Effect and Influencing Factors of Digital Transformation of Manufacturing Industry*. Advances in Economics, Business and Management Research/ Advances in Economics, Business and Management Research., DOI: 10.2991/aebmr.k.220405.072

Xiros, V., Castro, J. M. G., Fernandez-Pelaez, F., Magoutas, B., & Christidis, K. (2025). Supply Chain Data Analytics for Digital Twins: A Comprehensive Framework. *Applied Sciences (Basel, Switzerland), 15*(12), 6939. DOI: 10.3390/app15126939

Xu, W., Xu, J., & Zhang, W. (2025). Driving Green Transformation in Equipment Manufacturing Enterprises: The Role of Digital Technologies. *Systems*, *13*(5), 332. DOI: 10.3390/systems13050332

Xue, X., Dhumras, H., Thakur, G., & Shukla, V. (2025). Integrating artificial intelligence and sustainable materials for smart eco innovation in production. *Scientific Reports*, *15*(1), 36942. Advance online publication. DOI: 10.1038/s41598-025-20803-2

Yakovenko, Y., & Shaptala, R. (2024). Study of digital twins as the driving force of digital transformation and achieving the goals of sustainable development. *Technology Audit and Production Reserves*, *2*(4(76)), 11–20. DOI: 10.15587/2706-5448.2024.301423

Yang, G., Yang, G., & Yang, W. (2024). The impact of AI on enterprise energy management: From the perspective of carbon emissions. *Science and Technology for Energy Transition*, *80*, 8. DOI: 10.2516/stet/2024096

Yang, M., Fu, M., & Zhang, Z. (2021). The adoption of digital technologies in supply chains: Drivers, process and impact. *Technological Forecasting and Social Change*, *169*, 120795. DOI: 10.1016/j.techfore.2021.120795

Yap, C. K., Leow, C. S., & Goh, B. (2024). Sustainable construction materials under ESG: a literature review and synthesis [Review of *Sustainable construction materials under ESG: a literature review and synthesis*]. *MOJ Biology and Medicine, 9*(1), 1. MedCrave Group. DOI: 10.15406/mojbm.2024.09.00208

Yu, Z., Umar, M., & Rehman, S. A. (2022). Adoption of technological innovation and recycling practices in automobile sector: Under the Covid-19 pandemic. *Operations Management Research : Advancing Practice Through Research, 15*(1-2), 298–306. DOI: 10.1007/s12063-022-00263-x

Yunus, J. N., Fahrurazi, L. S., Hamid, M. S. A., Noor, R. N. H. R. M., & Majid, M. R. A. (2024). The Implementation of Sustainability Practices in Construction Site and its Effectiveness in Klang Valley. *International Journal of Academic Research in Business & Social Sciences, 14*(9). Advance online publication. DOI: 10.6007/IJARBSS/v14-i9/23020

Yusuf, S. O., Ikhine, I. V., Nyamekeh, R., Oluwadare, O. E., Afoakwah, B., & Yusuf, N. (2025). The Impact of AI on Supply Chain Operations: A comparative analysis of traditional vs AI-enabled Processes. *World Journal of Advanced Research and Reviews, 27*(2), 1688–1700. DOI: 10.30574/wjarr.2025.27.2.3027

Zejjari, I., & Benhayoun, I. (2024). The use of artificial intelligence to advance sustainable supply chain: Retrospective and future avenues explored through bibliometric analysis. *Discover Sustainability, 5*(1), 174. Advance online publication. DOI: 10.1007/s43621-024-00364-6

Zhang, D. (2024). AI integration in supply chain and operations management: Enhancing efficiency and resilience. *Applied and Computational Engineering, 90*(1), 8–13. DOI: 10.54254/2755-2721/90/2024MELB0060

Zhou, C., Xu, J., Miller-Hooks, E., Zhou, W., Chen, C., Lee, L. H., Chew, E. P., & Li, H. (2021). Analytics with digital-twinning: A decision support system for maintaining a resilient port. *Decision Support Systems, 143*, 113496. DOI: 10.1016/j.dss.2021.113496

Chapter 4
Sustainable Urban Transformation:
Digital Twin Integration

Leevesh Kumar
https://orcid.org/0000-0001-9476-0841
World University of Bangladesh, Bangladesh

Sitesh Kumar Singh
https://orcid.org/0000-0002-7108-0808
College of Engineering, National University of Science and Technology, Oman

Sagor Kumar Podder
http://orcid.org/0009-0007-2110-2240
World University of Bangladesh, Bangladesh

Md Asaduzzaman Rasel
World University of Bangladesh, Bangladesh

ABSTRACT

In this chapter, the author follows through with Digital Twin Technology (DTT) as a transformational tool in sustainable urban development. With cities experiencing amplified pressure from rapid urbanization and climate change, DTT entails a dynamic, information-driven method to oversee, simulate, and optimize urban systems on the fly. The chapter shows that DTT can be used to enable energy efficiency, reduction of carbon footprint, climate resilience, and inclusive governance by conducting literature reading, case studies (e.g., Singapore, Helsinki, Shanghai), and a strategic integration framework. It also analyzes the obstacles, technical, organizational, and ethical in a critical manner and describes the future possibilities of AI deployment, collaboration with the stakeholders, and standardization of

DOI: 10.4018/979-8-3373-5473-6.ch004

frameworks. The chapter ends with the plan of action on how to implement DTT in sustainable urban governance, which could be of help to planners, policymakers, and research practitioners whose aim is to design smarter, more resilient, and more equitable urban futures.

1. INTRODUCTION

1.1 Challenge of urbanization and sustainability

One of the most significant trends in the 21st century is urbanization. As per the report of the United Nations (2018), the world today has a population that resides in urban areas of more than 55 percent and this portion is likely to cross the line of 70 percent by 2050. This high rate of urbanization has led to critical issues in the sense of environmental sustainability, management of resources, the resilience of the infrastructures, and social equity (Bibir et al., 2024). Urban areas produce more than 70 percent of internationally increasing greenhouse gases and more than two-thirds of the world energy consumption highlighting a crucial position to achieve climate and sustainability goals (Liu et al., 2024; Bortolini, et al., 2022). Hence, urban sustainability is much required, as illustrated in Figure 1, which presents the key urban sustainability challenges and their corresponding solutions. The reactive and siloed traditional approaches to urban planning are not able to adjust to the complexity and dynamics of bustling cities today (Weil, et al. 2023). It is required that various systems of the management of cities, including energy, transportation, water, and waste management, be integrated as a whole and supportively responds to the sustainability and resilience of the urban areas (Azadi et al., 2025). To deal with such complexities, new advanced digital technologies have emerged and one such technology that has gained a lot of attention is Digital Twin Technology (DTT).

Figure 1. Key urban sustainability challenges and corresponding solutions

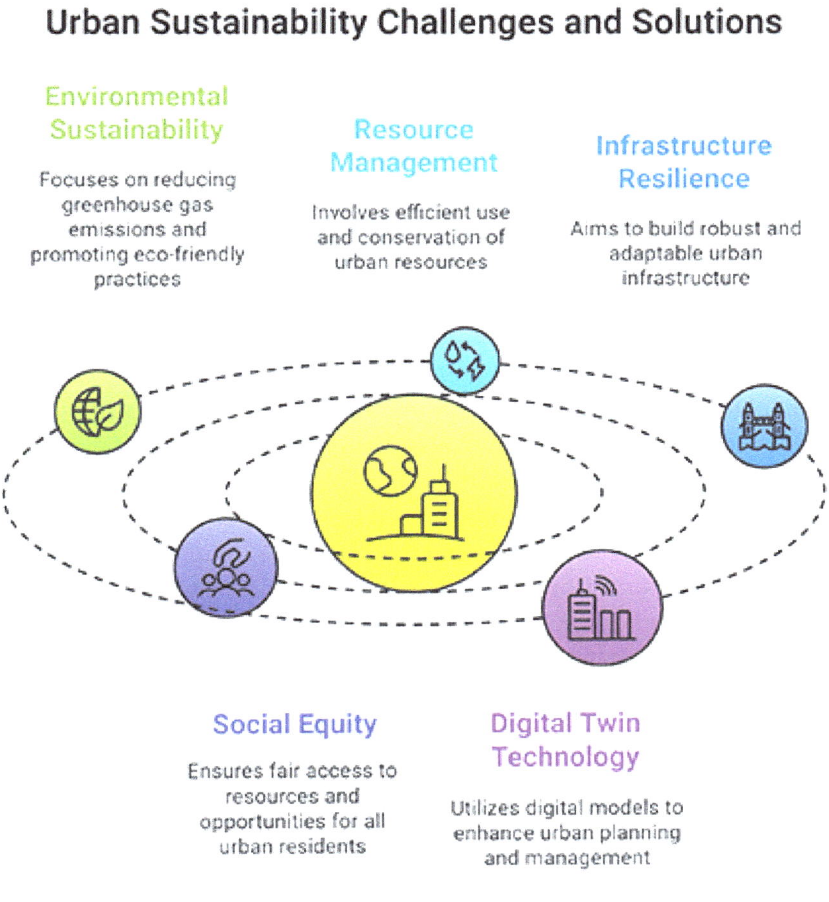

1.2 Digital twin technology in city situation

Digital Twin (DT) is a virtual representation of physical systems or assets that are able to evolve dynamically against real-time data to simulate, analyze and optimize tailored processes (Digital Twin et al., 2020). DTT has been successfully applied throughout the urban domain, originally used in the manufacturing and the aerospace industry, to simulate the whole city or its particular systems, so-called Urban Digital Twins (UDTs) (Batty et al., 2012; Samuel et al., 2023).

UDTs integrate the information with sensors of the Internet of Things (IoT), Geographic Information Systems (GIS), Building Information Modeling (BIM), and Artificial Intelligence (AI) algorithms to build a dynamic virtual representation of the urban environment (Jiang et al., 2021). Such a virtual city model can model urban phenomena, simulate policy interventions, and give predictive answers to the planners and policymakers (Ishrat et al., 2025). With the concept of DTT as an influential element of urban planning, it is considered as a channel of rapid urban transformation to be achieved through informed, timely, and flexible decision-making (Mazzetto et al., 2024).

1.3 Requirement of sustainable urban transformation

Sustainable urban development involves concepts of environment, social development, and economics that aim at enhancing the quality of life that does not interfere with the capacity of future generations to fulfill their needs (United Nations, 2018). The objectives of urban sustainability revolve around carbon neutrality, efficient use of energy, the commonality of circular economies, climate change resilience, and equality in diversity (Bortolini, et al., 2022).

This high-speed urbanization, along with the threat of climate change creates the need of an urban governance that can respond in real time and constructively plan the future (Zhao et al., 2024). The engineering infrastructure to make this happen is the Digital Twins, which can gather data from multiple sources into urban streams and transform it into actionable intelligence (Weil, et al. 2023). The challenges to governance, technology, socio-economic and ethical issues, however, hinder its wide usage and need critical analysis (Ishrat et al., 2025).

1.4 Goals and purpose of the chapter

The purpose of this chapter is to find out how Digital Twin Technology can help to ensure sustainable urban transformation. This is illustrated in Figure 2, which provides an overview of the chapter's goals and purpose. The key objectives are outlined as follows:

1. Take a look at the role digital twin's play in sustainable infrastructure and smart cities.
2. Review international case analysis that represents practical examples of digital twin.

3. Cover the issues of integrating digital twins: technical, policy, and societal issues.
4. Suggest a practical system in establishing digital twins in sustainable urban management and planning.

This chapter follows a mixed-method strategy, focusing on a systematic review of literature, case study analysis, SWOT, and gap analysis followed by a conceptual framework to lay out a plan of action to practitioners and policymakers (Bortolini et al., 2022; Azadi et al., 2025).

Figure 2. Overview of the chapter's goals and purpose in analyzing digital twin technology for sustainable urban transformation

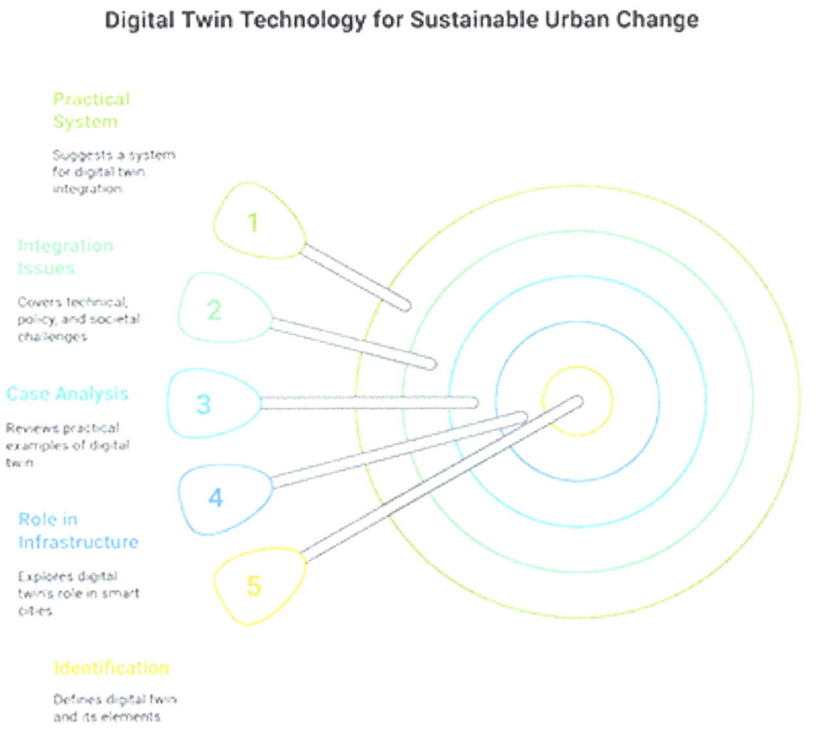

2. LITERATURE REVIEW

2.1 Digital twin technology: concept, architecture and enabling technologies

2.1.1 Conceptualization review of digital twins

The idea of Digital Twin is associated with the possibility to create an image of the digital replica of a physical stamp providing a possibility to represent the characteristics, behaviors, and states of a physical stamp in real time (Digital Twin et al., 2020). This related simulation embodies historical and real-time data with the help of sensors and IoT gadgets and involves the computational models to extrapolate future situations (Batty et al., 2012).

In an urban set-up, digital twins can comfortably support an integrated city management style by unifying various data feeds, making it possible to coordinate and simulate city infrastructure functions like transport, energy grids, water distribution, and built environment, in a dynamic way, both in monitoring and predictive analytics (Mazzetto et al., 2024; Samuel et al., 2023). These functionalities form a closed feedback loop that give information on decision-making, monitoring of performance, and planning (Jiang et al., 2021).

2.1.2 Urban digital twins' architecture

Typically, Urban Digital Twins architecture has the following layers (Ricciardi et al., 2023; Luo et al., 2024):

- **Physical Layer:** The physical part of town which consists of buildings, roads, utility systems and the natural setting.
- **Data Acquisition Layer:** The IoT and sensors within the urban fabric to capture as-needed real time data on environmental condition, performance of infrastructure, and human action.
- **Data Processing Layer:** The cloud computing platforms allow analyzing big loads of data with the usage of AI and machine learning to notice patterns, make predictions, and become more efficient.
- **Modeling and Simulation Layer:** The GIS and BIM are used to generate the spatially explicit, and functionally deep models of cities.
- **Visualization Layer:** utilizable, frequently 3D or augmented-reality interfaces that depict understandings to planners and other stakeholders.
- **Integration Layer:** It checks on interoperability of data, secure communication, and governance among the various systems in the urban area.

This architecture, composed of a layered structure, allows maintaining the constant synchronization of physical and virtual state of the city, providing real-time situation awareness and scenario testing (Ishrat et al., 2025).

2.1.3 Enabling technologies

Several technologies come together to allow infrastructure of digital twin as presented in Figure 3 Key components and functional architecture of digital twin technology in urban systems, with its components and functions and described below:

- **Internet of Things (IoT):** Smart sensor networks that are used to gather various information of the urban areas (Liu et al., 2024).
- **Artificial Intelligence (AI):** It will facilitate predictability, anomaly-detecting, and free decision-making (Samuel et al., 2023).
- **Geographic Information Systems (GIS):** This helps to combine spatial information and provide mapping (Azadi et al., 2025).
- **Building Information Modeling (BIM):** It is an elaborate source of structural and architectural data that is important in the infrastructure digital twin (Jiang et al., 2021).
- **Cloud Computing and Edge Computing:** Reduce computation and storage of large urban data as both concepts provide scalability (Zhao et al., 2024).
- **Data Standards and Protocols:** The data standards and protocols would guarantee data interoperability and safe data sharing within systems (Ishrat et al., 2025).

The combination of these technologies constitutes the foundation of digital twin platforms capable of an extremely high-fidelity mirroring and simulating of urban systems (Mazzetto et al., 2024).

Figure 3. Key components and functional architecture of digital twin technology in urban systems

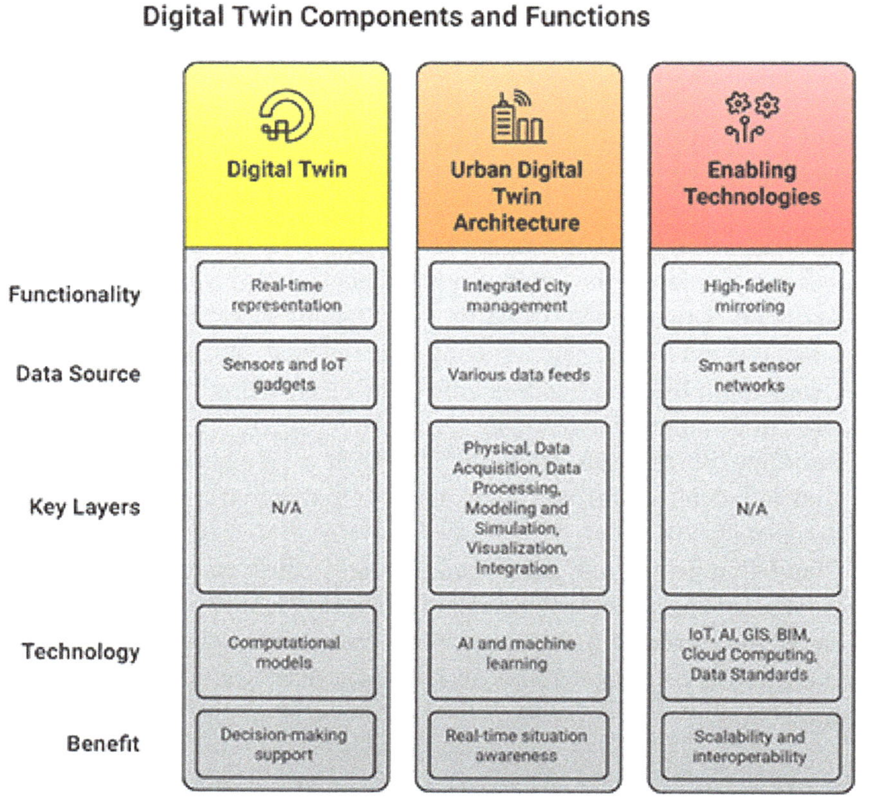

3. DIGITAL TWINS IN URBAN SUSTAINABILITY

3.1 Digital tools and their relationship with sustainability

Digital Twin Technology (DTT) has continued to become one of the most potent drivers of urban sustainability as illustrated in Figure 4, Conceptual relationship between digital twin technology and core dimensions of urban sustainability. This relationship highlight the insights that offer data-driven decision-making through which action can be initiated on behalf of the environment, social, and economical aspects (Azadi et al., 2025). Literature discusses DTT in general as an integrating platform that has the potential to improve the efficiency, resiliency, and adaptability of urban systems by utilizing constant real-time data flows, predictive analytics, and scenario simulations (Mazzetto et al., 2024; Liu et al., 2024).

Figure 4. Conceptual relationship between digital twin technology and core dimensions of urban sustainability

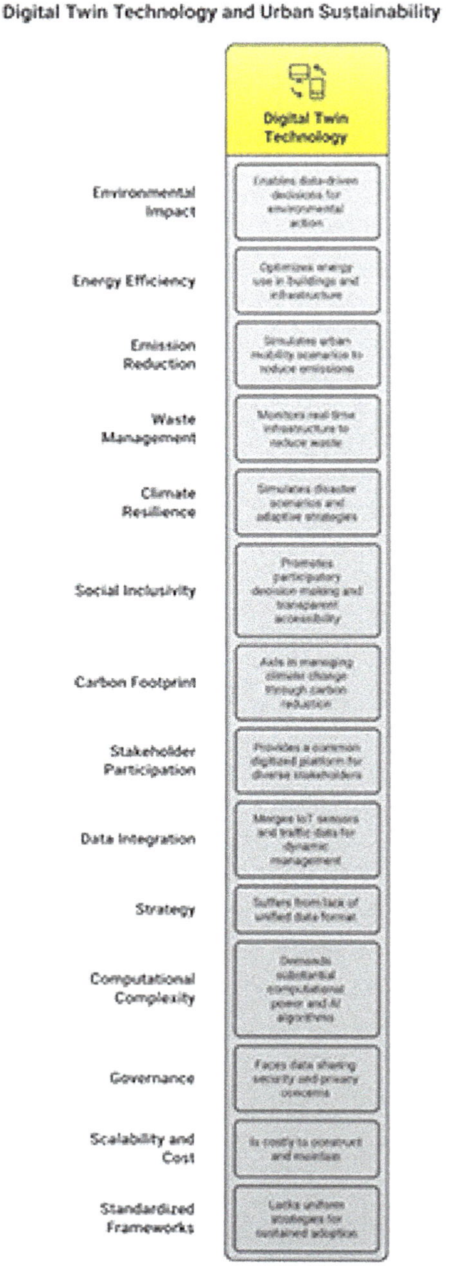

Sustainable urban development covers a broad spectrum of seven aspects, which are carbon reductions, energy optimizations, waste and water management, climate resilience and social inclusivity (Bortolini, et al., 2022; Bibri et al., 2024). Research has shown digital twins allow these sustainability objectives to be the meeting point as they have the ability to:

- Stabilize energy use in buildings and infrastructure and optimize them (Bortolini, et al., 2022).
- Simulate the urban mobility scenarios to reduce the emissions (Bibri et al., 2024).
- Reduce toxins and other waste materials by monitoring real-time infrastructure (Liu et al., 2024).
- Improve the resilience of cities through simulation of disaster and adaptive strategies (Ricciardi et al., 2023).

3.2 Energy efficiency and carbon cut

One of the most popular directions in the literature is the diagnostic and managing of climate change through the reduction of carbon footprint in cities measured in digital twins (Mazzetto et al., 2024; Bortolini, et al., 2022). With digital twins, a more detailed energy consumption model can be conducted at the building and district level, providing stakeholders with an opportunity to view how inefficient each area of a building or a district is, and take action though retrofit or scale of operation optimisations (Bortolini, et al., 2022). Take, e.g., (Ishrat et al., 2025) that showed that AI-powered digital twins can forecast changes in energy demand and provide HVAC systems in smart buildings with an energy optimization process delivering up to 15% energy reductions.

Besides, urban digital twins may model transport being used to test low-carbon mobility policies based on traffic flow optimization factors, charging of electric vehicles, and the enhancement of public transportation (Bibri et al., 2024, Liu et al., 2024) discussed that in order to leveraging urban digital twins to minimize emissions, data integration is necessary to merge IoT sensors, traffic monitoring, as well as traffic data to reach dynamic congestion management.

3.3 Inclusion and stakeholder participation

Technological possibilities take the center stage of the discussion, with the recent literature taking away the significance of socio-political aspects of the urban digital twin. According to (Azadi et al., 2025) the transformation must be inclusive in the sense that different stakeholders, such as policymakers, citizens, academia, and the private sector work on the same common platform that is digitized.

Digital twins have the potential to democratize urban planning by demonstrating transparent accessibility formats such as visualizing data in easy-to-understand manners, as well as promoting participatory decision-making (Ishrat et al., 2025). Nevertheless, researchers warn of electronic split and data-access inequality, which can enhance social inequality (Jiang et al., 2021). To solve these problems, the concept of ensuring ethical data governance and digital equity should enter digital twin frameworks (Ishrat et al., 2025).

3.4 Limitations, and research gaps

Although digital twin applications seem promising, the literature is aware of some limitations and gaps that limit the future applicability of the solution in urban sustainability as illustrated in Figure 5, Advantages and limitations of digital twin technology in urban planning and management.

- **Strategy:** The lack of a unified data format (data quality and interoperability) and datasets with limited uniformity reduce the ability of seamless integration in digital twins (Weil, et al. 2023; Ishrat et al., 2025).
- **Computational Complexity:** Processing and analyzing the large urban datasets in real-time demand substantial computational power and the use of well-developed AI algorithms, which is hard to fulfill in most cities (Samuel et al., 2023).
- **Governance and Privacy issues:** Data sharing security and privacy concerns continue to pose problems in digital twins that are still essential to promote the level of trust and regulatory adoption (Ishrat et al., 2025; Azadi et al., 2025).
- **Scalability and Cost:** The construction and maintenance of digital twins of cities are costly and even more costly to build a digital twin in resource-limited cities because skills are difficult to eliminate (Jiang et al., 2021).
- **Absence of Standardized Frameworks**: I lack uniform strategies that help in the adoption of digital twins in a sustained manner by incorporating technical, social, and policy aspects of the digital twins (Bortolini, et al., 2022).

These issues are the active research frontiers necessary to unravel the complete potential in the field.

Figure 5. Advantages and limitations of digital twin technology in urban planning and management

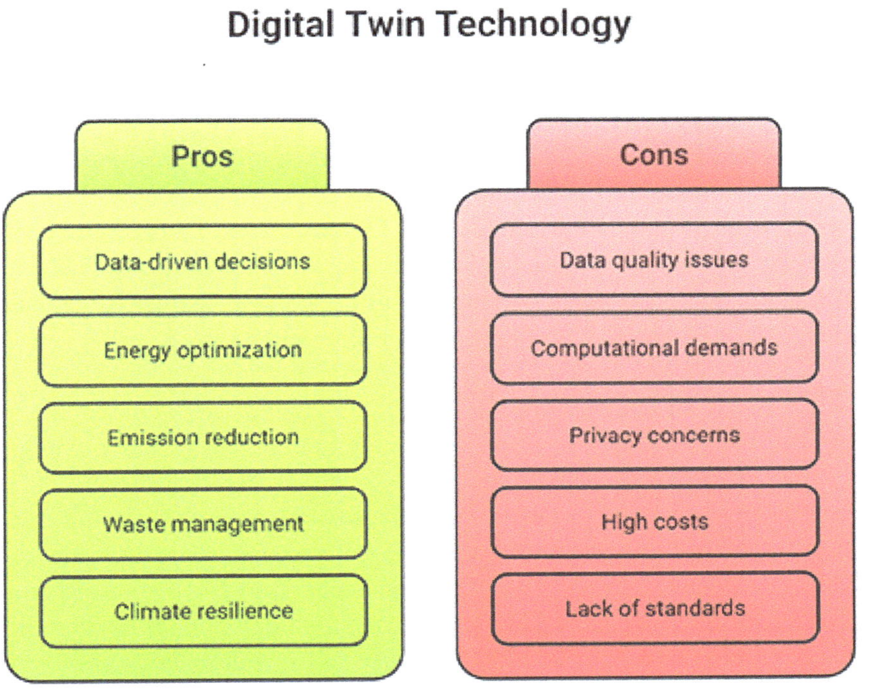

According to the recent scholarship, one of the recent enablers of sustainable urban development is Digital Twin Technology (DTT), which combines real-time information, simulation opportunities, and predictive analytics to manage the complex urban systems more effectively. Research highlights urban digital twins, which are constructed on the combination of BIM, GIS, IoT, and remote sensing, as being supportive of multi-scale modeling, which is required in optimization of energy, climate resilience, mobility management, and smart infrastructure maintenance (Jiang et al., 2022; Lehtola et al., 2022)

It has been found that twins on a city scale, like Virtual Singapore, Helsinki 3D+ and the Shanghai Urban Operations Platform, can be more effective in decision-making by providing precise monitoring, testing of scenarios, and involvement of stakeholders (Mazzetto, 2024; Weil et al., 2023; Patel et al., 2024). Digital twins are incorporated into environmental studies, which combine hydrological/climate simulations with flood prediction, heat islands prevention, and the design of green infrastructures (Bibri et al., 2024). In the meantime, governance-oriented literature points to such issues as interoperability of data, privacy, cybersecurity, and fragmented institutional forms that should be overcome to make digital twin use large-scale on a global scale. Further studies examine the use of digital twins to assist in the circular economy by evaluating the lifecycle carbon, predicting material recovery and planning sustainable constructions (Patel et al., 2024). Regardless of the strong progress, researchers always find a gap in the long-term empirical assessment, cross-domain combination, and social equity and cost-effective implementation models that can be applied to the developing cities. On the whole, the literature confirms the tremendous potential of digital twins as the way of developing sustainable urban transformation and notes the importance of further holistic, inclusive, and methodologically sound research to eliminate current limitations.

4. METHODOLOGY

The chapter has assumed a mixed-method research approach involving a systematic literature review, comparative linked case study research, and thematic analysis research to investigate the role of Digital Twin Technology (DTT) in facilitating the transformation of urban sustainability. The methodology applies to the scholarly principles of systematic research into the new urban technologies and their sustainability effects. First, a systematic literature review was performed so as to find peer-reviewed studies, theoretical frameworks and empirical evidence on the topic of digital twins, urban sustainability, data governance and smart city systems. Scopus, Web of Science, and Google Scholar databases were searched using keywords like digital twin, urban sustainability, smart city, and urban data integration. Research in the field that was published between 2018 and 2025 was given priority because of the fast progress in the field of DTT. This review procedure helped to include high-quality and up-to-date knowledge (Mazzetto, 2024; Bibri et al., 2024). Second, four large-scale digital twins across the world (Singapore, Helsinki, Shanghai, Barcelona) were investigated using the comparative case study approach. This approach facilitated the determination of real-world use cases, issues, governance systems, and sustainability results in cities with various digital infrastructures and planning environments. The academic publications, governmental reports, and foreign policy documents were used as the sources of case data (Azabi et al., 2025). Lastly, thematic analysis was used to analyze the literature gathered and case evidence with respect to common trends in terms of technological architecture, sustainability indicators, data governance barriers, and strategic integration frameworks. Themes were arranged in the context of technical, organizational, socio-economic, and policy dimensions to create a comprehensive picture of the potential of DTT used in sustainable urban development (Weil et al., 2023). This systematic review, case comparison, and thematic interpretation provide a rigorous and triangulated methodology that will be appropriate in the analysis of conceptual developments and real-life applications of digital twin technologies in the modern urban sustainability framework.

5. CASE STUDIES

5.1 Real-life simulation of digital twins in urban sustainability

Digital Twin Technology (DTT) is not a mere concept anymore; it has attained reality because it has been used in many cities around the world. This part focuses on three exemplary cities: Singapore, Helsinki, and Shanghai, which have used digital twins to recommend a sustainable change of the city. Such examples explain various types of governance, technological platforms and sustainability results, which brings important lessons to future implementations.

5.1.1 Singapore: Digital twin of urban planning and resilience on national scale

Singapore has been one of the first-movers in the use of digital twins to achieve urban sustainability across the world. Virtual Singapore developed by the city-state incorporates a complete Geographic Information Systems (GIS), Building Information Modeling (BIM), Internet of Things (IoT) sensors, and artificial intelligence (AI) analytics (Mazzetto et al., 2024; Samuel et al., 2023).

5.1.2 Goals and scope

The purpose of Virtual Singapore is to assist in the multi-sectoral urban planning, infrastructure supervision, environmental balancing and the disaster preparedness (Samuel et al., 2023). Important goals are:

- Optimization of energy modeling, carbon footprint and efficiency of resources.
- Flood risk management enhancement based on real time hydrological information and predictable simulations.
- Using immersion visualization and scenario testing to facilitate the process of engage with the people.
- Dynamic traffic and pedestrian flow-based support of transport plans.

5.1.3 Technological infrastructure

The platform is supported by a high-realness 3D city model being updated every minute with real-time data about IoT devices that record weather, traffic, air quality, and the efficiency of buildings (Mazzetto et al., 2024). The analytics based on AI make possible the simulation of scenarios, e.g. the effectiveness of green building measures in place or flood defense systems.

The integration with BIM data enables modeling of each building system in detail, thereby optimizing the energy consumption and the indoor environmental quality (Samuel et al., 2023, Park et al. 2020). The third-party application development is an architectural feature that the open data architecture enables.

5.1.4 Governance and sustainability outcomes

Virtual Singapore has led to quantifiable efficiency results at the district level in energy use, leading to national carbon outset targets (Bortolini, et al., 2022). The urban resilience planning has improved due to flood simulation which allows the proactive improvement of infrastructure (Florido-B Benitez, L. 2024).

The project is an illustration of effective governance by the government where several state agencies are interrelated through an ecosystem of data that is displaying accountability and transparency (Azadi et al., 2025). Nevertheless, there are still unresolved issues of privacy and security of the data, as well as the ethical access to the positive implications of the platform.

5.2 Marine institute, VTT, and Abo Akademi: Helsinki: flood resilience and climate adaptation with digital twins

Digital twin technology has already been used by the City of Helsinki in Finland to mainly boost its urban resilience to floods and climatic change effects (Florido-B enemy, L. 2024; Ishrat et al., 2025).

5.2.1 Objectives of the project

The twin of the city of Helsinki is aimed at efficient control of the flooding risk and optimization of emergency response through the combination of hydrological models and urban infrastructure GIS (Florido-BiniBeni tez, L. 2024). Objectives include:

- Forecasting the extent of floods and the depth of water in the zones that are critical.

- Modeling of the development and testing of the interventions in railway testing like green roofs and the storm water retention basins.
- Real-time organization of multi-agency response to a disaster.
- Consulting the communities easily by visualizations of clear flood risks.

5.2.2 Technical parts

The twin integrates satellite images, IoT data (rain gauge, water level sensors) and the models of the city infrastructure (buildings, roads, drain systems) (Ishrat et al., 2025). AI algorithms process incoming streams of data to create real-time stream of risk maps and alert systems to emergency managers.

5.2.3 Lesson learned and impact

The digital twin of Helsinki has enhanced preparedness to flood events, minimizing the time of response and the costs of damages (Florido-Beni tez, L. 2024). The system aids in climate adaptation planning in the long term since scenarios about the varied climate projections can be simulated.

Excellent cross-sectoral cooperation and community involvement are a cornerstone to success that demonstrates the significance of stakeholder cooperation (Azadi et al., 2025). The project has noted the lack of data interoperability and the necessity of model validation on its continuous basis as well.

5.3 Shanghai: Digital twins, urban mobility and energy management

The urban mobility and energy management that Shanghai aims with its digital twin efforts are mostly connected to the fact that the megacity has a complex transportation and energy system (Bibri et al., 2024; Liu et al., 2024).

5.3.1 Objectives and themes

By modeling traffic patterns and energies consumed Shanghai intends to alleviate traffic congestion, streamline the network of public transport, and minimize greenhouse gasses emission (Bibri et al., 2024). The twin enabling:

- In-Silico use of traffic demand and routing to minimize the idle time and emissions.
- Incorporation of the EV charging infrastructure planning.

- Modeling of energy consumption in a fleet of vehicles: public transport and municipal buildings.
- The evaluation of urban heat island impact and mitigation measures.

5.3.2 Architecture digital twin

The twin in Shanghai combines IoT sensors in traffic, smart meters and weather information with GIS and AI predictive models (Liu et al., 2024). Analysis of real time traffic data is performed that helps in managing the traffic lights and improving bus schedules.

Buildings and transit energy data are used in the demand-side management and renewable energy integration simulations (Bortolini, et al., 2022).

5.3.3 Successes and setbacks

Having reduced traffic jams and enhanced indicators of air quality based on pilot districts, the digital twin has already shown modest results using Shanghai as an example (Bibri et al., 2024). These attempts at energy optimization helped in quantified reductions in the carbon footprint of buildings in their operation.

Nevertheless, full-scale implementation has been limited by challenges including; data silos, multi modal transport modeling complexity, and lack of skilled staff (Ishrat et al., 2025). The experience highlights the need of integrated planning models and long-term investment.

5.4 International experience and best practice

The analysis of three cities of Singapore, Helsinki, and Shanghai also indicates similarities in success factors and challenges in the implementation of digital twins regarding urban sustainability:

- **Healthy Multi-Stakeholder Governance:** Successful deployment of digital twin involves a partnership of government, and non-government entities, institutes of learning and civil societies (Azadi et al., 2025).
- **Robust Data Ecosystems:** to construct high-fidelity, dynamic models, it is required to integrate heterogeneous data provided by the IoT, BIM, GIS, and satellite (Mazzetto et al., 2024).
- **Emphasis on user-based applications:** digital twins can have maximum benefit when they are configured around a particular issue in a city challenging adaptation to floods or energy efficiency (Florido-Beni-tez, L. 2024).

- **Privacy and Equity:** Privacy and data governance should be handled ethically, and there are necessary measures to eliminate digital divides and have equal benefits (Ishrat et al., 2025).
- **Capacity Building, Achilles Heel and Funding:** There is a need to continually invest in computational infrastructure, data management, as well as, human capital in order to maintain and scale digital twins (Bortolini, et al., 2022).

5.5 Urban Digital Twin of Barcelona of Sustainable Mobility and Climate Resilience

Barcelo has become one of the most successful implementers of urban digital twin technology in Europe, and it combines real-time mobility data, environmental monitoring, and predictive analytics to facilitate the proper management of the city. The city digital twin is the creation of the Barcelona Urban Platform initiative that implements GIS, Internet of Things-enabled traffic sensors, land-surface temperature data collected by satellite, and advanced simulation models to enhance the efficiency of mobility and climate resilience. The main use of one of the twin is optimization of the city model Superblock (Superilla) one that utilizes a pedestrianized city design strategy that aims to alleviate traffic congestion and enhance air quality. The digital twin allowed simulating redistribution of traffic and pedestrian flows to make Superblock layouts finer and estimate the overall impacts on the system before it became a reality. The Superblocks that were enabled with the help of digital twin simulations led to an average of 25% traffic reduction, 21% reduction in NO2 emissions, and a maximum of 40% noise pollution reduction in the specified regions, according to municipal data. Climate-risk modules can also be integrated on the platform, and this can model the intensity of the urban heat islands; simulations were used to position the trees on the streets and reflective pavements where suitable, and this plays a role in reducing the temperature of 1.5-2.0degC in pilot areas at peak summer periods. Moreover, the twin of Barcelona also has coastal flood-risk modeling, which is used to train the emergency response teams to simulate a storm-surge event, enhancing the readiness to severe weather conditions caused by climate change. The governmental system of the city focuses on the transparency and active participation of citizens, enabling them to see the proposed modifications in 3D and experience the mobility scenarios and leave feedback via the public participation portals. This set of data-based planning, sustainability indicators, and participatory governance makes Barcelona one of the most successful examples of how the digital twins can provide quantifiable environmental value and contribute to the long-term sustainable urban development, European Commission (2022).

5.6 New trends on urban applications of digital twins

In addition to these, the recent sources identify the tendency that is developing the future of sustainable cities in terms of digital twins (Azadi et al., 2025; Samuel et al., 2023):

- Employing the use of **Artificial Intelligence and Machine Learning** to support the prediction and facilitating automatized interventions.
- Application of **Augmented and Virtual Reality** interfaces in order to enhance stakeholder participation, as well as, transparent accountability of decisions.
- Evolution to **Social Digital Twins** where the human behavior, equity and social processes are modeled in addition to the physical infrastructure.
- Construction of **Interoperability Standards** that will enable digital twin data exchange and collaboration across cities and cross-sectors.
- Implementation of the **Circular Economy Principles** in order to model material flows and waste minimization within the urban systems.

5. ANALYSIS & DISCUSSION

6.1 Problems and potential of digital twin technology

There may also be several multidimensional realities hindering the way in which the Digital Twin Technology (DTT) is implemented, though the realm of implementing DTT opportunities in the sphere of urban sustainability looks tremendous. At the same time, thanks to the introduction of technologies, social, and policy innovations, more opportunities emerge associated with the use of digital twins in the city in a way that is beneficial to society. This section is a critical analysis of technical, organization, regulatory and socio-economic barrier, and the emergent opportunities that characterize the future direction of the urban digital twins.

6.1.1 Technical difficulties

6.1.1.1 The integration of data and interoperability

Cities digital twins require massive volumes of diverse information and data acquired based on IoT sensors, geo-information systems (GIS), or building information models (BIM), environmental surveillance, and suggestions by citizens (Azadi et al.,

2025). One of the primary technical problems is interoperability and the seamless connection with these various sources of data.

There are no incorporate standardized data forms, and protocols achieve consolidation of the real-time and accurate buildings of the city (Ishrat et al., 2025). The example here is sensor networks, which may be special protocols and BIM models that can be created using different software packages which are not easily interoperable (Mazzetto et al., 2024) The result is data siloing which lowers the integrity and the liveliness of the digital twin.

Application of innovative features such as CityGML and Industry Foundation Classes (IFC) appear to be promising but should be used more extensively both in the municipal domain and in the private life (Bortolini, et al., 2022).

6.1.1.2 Real-time processing

Good quality, accurate and timely data has been identified to be critical in terms of providing valid simulations and decision-aiding (Florido-Beni tez, L. 2024). High rate of changes, new buildings and their appearance, infrastructure fall back, and a changing environmental condition are characteristic of the cities, which should be emulated in digital twins immediately.

There are challenges related to sensor failures, data sensations, the low speed of data transfer and all of these factors may lower the relevance of the digital twin (Liu et al., 2024). In addition to this, real-time streams of data require high-computational power and state-of-art algorithms to be able to handle noise and uncertainty in the context of AI-powered analytics.

6.1.1.3 Digital privacy and security concerns

The collection of large amounts of data that is typical of urban digital twins poses significant levels of cybersecurity and privacy risk (Ishrat et al., 2025). These may be sensitive information such as occupancy rates in the buildings, mobility and energy usage information that may be subject to extortion and exploitation by other bad players when adequate security measures are not implemented.

Moreover, in case of inadequate protection of personal information, citizens may be reluctant to share personal data and decrease the capacity of the twin to presents an adequate model of social reality (Azadi et al., 2025). One should be able to build efficient cyberdefense systems and privacy-protecting culture of data management.

6.2 Negative and blocking organization and governance solutions

6.2.1 Institutional fragmentation

The governments in cities normally constitute a diverse ecosystem of departments and agencies in the field of transport, water, energy, urban planning and emergency management that tends to work as silos on a frequent basis (Azadi et al., 2025). In order to deploy digital twins, such silos need to be broken, whereby sharing and taking action needs to be incorporated.

A lack of leadership and coordination may paralyze or create redundancies and unnecessary usage of resources of digital twin projects (Bibri et al., 2024).

6.2.2 Limitation of capacity and lack of skills

Writing and managing digital twins involves specialized expertise in the areas of data science and AI, GIS and urban planning (Bortolini, et al., 2022). Not all municipalities will possess sufficient talent internally and hence relying on external consultants who can increase the cost and reduce the retention of knowledge in their institutions.

To accomplish digital twins' programs (Ishrat et al., 2025) capacity should be developed through training, involvement of the academic community, and development of interdisciplinary teams (Ishrat et al., 2025).

6.3 Barriers based on socio-economic and policy-based perspectives

6.3.1 Duration of money and financial expenses

The level of investment in understanding, implementations and support of the urban digital twins require a significant degree of investment in the hardware, software, and human factor (Mazzetto et al., 2024). The large initial outlays and the unpredictability of the returns to the investments might act as stumbling blocks especially in cities of low-and middle-income countries.

Also, there is the absence in the models of financing in the long term, which takes the work and updates, so issues arise regarding the sustainability of digital twin platforms (Azadi et al., 2025).

6.3.2 Fairness

The issue is that digital twins can also strengthen the existing social disparities in the case, when the latter are unevenly distributed (Ishrat et al., 2025). In point of fact, the solutions that are consistent with the solution of infrastructure improvement and are data-driven may target more accurate data in the regions of high-income levels, and marginal groups are left trailing behind.

Ethical frameworks in governance will have to be implemented to develop the concepts of inclusivity, transparency and social justice through the use of digital twins (Azadi et al., 2025). The needs of the community can also be adjusted to the digital twin applications through the mechanisms of public participation.

6.3.3 Components regulation government and legal components

The current regulations and policies commonly fail to provide particular guidelines on ownership, liability, and responsibility over data of the urban digital twins (Peldon et al., 2024). The ambiguity relating to data ownership and its usability can be linked to an obstacle to collaboration and data sharing.

The policy makers ought to develop dynamic law that balances between creativity and freedom of its people and possibilities of open data ecosystem (Bibri et al., 2024).

6.4 Opportunities and future outlook

The impediments exist; however, the new changes and the emerging features of urban governance systems have plausible future to the use of digital twins in an effort of achieving sustainable cities.

6.4.1 AI-predictive Simulations

Digital twins offer increased prediction capabilities through the incorporation of more recent AI and machine learning programming; thus, establishing proactive city management (Samuel et al., 2023). To give an example, AI could forecast energy demand surges, optimize traffic flows, in real-time, and identify potential weak points of the infrastructure before the failures.

It can be in the form of reactive to proactive city governance that will be more resilient and efficient.

6.4.2 Co-working platforms

Digital twins can emerge as the environment where governments, business, science, and citizens cooperate to develop urban solutions collectively (Azadi et al., 2025). Visualization tools, virtual reality interface and participatory simulations can allow the stakeholders to be aware of more urban complexity flows, and be involved in decision-making.

Such cooperation can also develop through which a feeling of trust, collective ownership, and reduced lopsided outcomes are achieved.

6.4.3 Equity and inclusion in digital space

The objective of the new work is the spread of digital twin technologies to all citizens of the city and their use (Ishrat et al., 2025). This consists of: addressing digital divides through the embrace of inexpensive high-speed connectivity, the all-inclusive nature of collection processes and transparency in the purpose of data use intent.

Comprehensive digital twins may support cities in the search and resolution of infrastructure and service delivery shortages.

6.4.4. Circular economy and shared climate objectives

Integration Digital twins allow cities to practice predictive streaming of materials, waste management, energy transitions in alignment with the principles of the circular economy (Liu et al., 2024). They assist the policymakers to find a balance between the growth and the goals of preserving the environment and achieving carbon neutrality, by simulating sustainability situations.

6.4.5 Open data initiatives and standardization

Fragmentation is predicted to decrease as data movements will be standardized globally and open platforms will be used, which will also stimulate novelty (Bortolini, et al., 2022). The data of open digital twins can be used to accelerate the research process; help support small enterprises and spawn smart towns in the regions.

6.5 Digital tool barriers and opportunities for urbanization

The digital twin technology is multidimensional in nature and pursuing it as the aim of urban sustainability. The presence of the technical and organizational concerns requires the interdisciplinary approach that will remove the data fragmentation, capacity building and creation of the powerful structures of governance.

At the same time, the role of AI-based foresight, participatory governance, and integrative sustainability modelling opens new horizons concerning how digital twins could become change agents in cities in the event of a climate crisis, rapid urbanization, and resource scarcity.

It implies that one should pay attention to ethical, equity, and regulatory concerns to be worried about and ensure that digital twins do not turn into the mere amplification of the real-life urban images but become the facilitators of the inclusive and just urban transformation. All these concepts have been summarized as the barriers and opportunities of digital tool technologies in urbanization as illustrated in Figure 6, Barriers and emerging opportunities associated with implementing digital twin technology in sustainable cities.

Figure 6. Barriers and emerging opportunities associated with implementing digital twin technology in sustainable cities

Urban Digital Twins: Barriers and Opportunities

Characteristic	Technical	Organizational	Socio-Economic & Policy	Opportunities
Data Integration & Interoperability	Data silos hinder seamless connection.	None	None	None
Real-time Processing	Sensor failures lower twin relevance.	None	None	None
Privacy & Security	Data collection poses cybersecurity risks.	None	Citizens reluctant to share data.	None
Institutional Issues	None	Silos hinder digital twin deployment.	None	Co-working platforms enable collaboration.
Skills & Capacity	None	Lack of expertise increases costs.	None	None
Financial Expenses	None	None	High investment costs are stumbling blocks.	AI predictive simulations for proactive management.
Fairness & Equity	None	None	Uneven data distribution strengthens disparities.	Equity and inclusion in digital space.
Regulation & Legal	None	None	Lack of guidelines on data ownership.	None

7. POLICY GUIDANCE FOR DIGITAL TWIN INTEGRATION: STRATEGIC FRAMEWORK

With climate change, high population growth rates, and limited resources putting an even greater pressure on urban centers, it would be necessary to incorporate Digital Twin Technology (DTT) into sustainable urban governance. This section suggests a holistic strategic framework that could be followed by city planners, policymakers and stake of technology, to introduce the digital twin and achieve resilient, sustainable, and inclusive urban futures.

The framework anchors on the pertinent lessons learnt in the foregoing sections, collating the lessons in technical challenges, organizational dynamics, socio-economic considerations, and the best practices in the globe collaborated with other principal case studies.

7.1 Overview of framework

One of these opportunities is the introduction of digital twin technology in urban governance, as it is a radical way to improve the management of a city, make it sustainable, and engage its citizens. These five interdependent dimensions within this strategic framework are important in the successful adoption and institutionalization of digital twins in urban environments. The dimensions are taken together to respond to the identified challenges in terms of technology, social, organization, and policies that have been extracted after series of literature reviews and empirical case study. Although this framework is meant to help cities in implementing digital twins, it is also geared towards making sure that such technologies play a particular role in sustainable urban development and inclusive governance.

These include five dimensions:

a. Engagement and cooperation with the stakeholders
b. Data Interoperability and Data Infrastructure
c. Sustainability Indicators and Measures of Performance
d. Regulatory support and Policy Alignment
e. Capacity Building and Innovation Ecosystem

The dimensions address targeted enablement and hindering factors to have cities implement the digital twin technology in a comprehensive way.

7.2 Engagement and collaboration with stakeholders

Any successful digital twin program in cities is based on effective stakeholder engagement. Digital twins, in their very essence, imply the need to access information and knowledge of a wide variety of participants, including businesses, academia, civil society, and citizenry in the realm of public and private sectors. In this dimension, there is a highlight on a multi-stakeholder role that is inclusive and adequate governance mechanisms.

7.2.1 Inclusive multi-stakeholder involvement

Early and ongoing engagement of all the stakeholders will be instrumental in the successful implementation of the use of digital twins in urban settings. Municipal government would have to collaborate with the providers of the technological services in the industry, who create and support the digital twin platforms, and the academic society offers advanced research and analytical tools. Local knowledge that is crucial in providing the development of digital twins, in accordance with the needs and values of the community, depends on civil society organizations and the general audience.

This wide ecosystem can be a source of collective ownership in digital twin projects, needed in accounting of sustainability and scaling. Premature cooperation assists in developing data sharing agreements, defining any shared objectives, and expectation management between stakeholders. Transparency and trust may be raised by taking part in participatory processes, including workshops, citizen feedback via a digital platform and co-creation sessions. Such approaches will enable citizens to play a direct role in the creation of urban data and decision-making, making digital twins does not only increase their social acceptance and legitimacy.

7.2.2 Management and leadership

Keeping the complicated activity of deploying a digital twin in order requires clear governance arrangements. The framework suggests the development of specialized governance organizations or working groups in urban governments in charge of digital twin initiatives. These bodies encourage cooperation between departments, deal with potential conflict of interests, and track the progress according to the set milestones.

At the political and managerial level, leadership commitment is of essence. Political leaders assist in acquiring the needed financing and make sure that digital twin integration is coordinated with general urban development goals, and managerial leaders push forward the process of operationalizing such projects within city agencies. Collectively, they institutionalize the practices of digital twins, in order to integrate them into urban planning systems, and guarantee non-pilot sustainability.

7.3 Infrastructure and interoperability of data

7.3.1 Data Architecture

The framework suggests creation of a coherent data architecture that is able to ingest, store, process and analyze data in real time. The architecture would reflect the increasing amount, rate and type of city information using IoT detectors, building information models (BIM), geographic information systems (GIS), and artificial intelligence (AI) components.

The use of open standards such as CityGML in city models to encourage semantic interoperability and Industry Foundation Classes (IFC) in building data, to encourage semantic interoperability, is possible. This will make it possible to combine different datasets sourced at different places without any loss of significance or error and therefore the digital twin can set a complete and unified view of the city surroundings. It also promotes the vendor neutrality and long-term scalability, through the usage of open standards.

7.3.2 Design of privacy, data protection and security

Information safety and confidentiality take the first priority since data concerned with the city can contain information relating to individuals and the sensitive infrastructure. The framework focuses on the adoption of security and privacy within the early stages (in terms of design) of digital twin platforms.

The standard must be established on techniques that protect data in their lifecycle such as anonymization, stringent access controls and encryption. The policy on data governance should be transparent and provide clarity about ownership and use of data, process of consent to develop confidence among stake holders and people. Holding people liable and ensuring data protection laws (like GDPR) are adhered to also reduces such risks related to breach and misuse of data.

7.4 Sustainability indicators and performance measurement

Digital twins provide unprecedented opportunities in tracking, analyzing and enhancing sustainability of cities. This dimension is concerned with how to specify effective measures and use simulation instruments to guide policy and running decisions.

7.4.1 Major indicators of sustainability of cities

To realize the effects of the digital twin initiatives, cities require determining a full range of performance indicators consistent with the internationally accepted frameworks like the United Nations Sustainable Development Goals (SDGs). Such indicators are supposed to include environmental, social and economic elements of sustainability.

Some examples are carbon emissions reduction, energy efficiency, water consumption, air quality and mobility efficiency metrics. Social equity indicators, which include access to services, and quality of life indices, will guarantee that digital twin interventions stimulate inclusive urban development. Permanent monitoring on these indicators assists cities to measure progress and isolate areas that need intervention.

7.4.2 Scenario analysis and dynamic simulation

Among the strongest capabilities of digital twins, we need to note the possibility to carry out dynamic simulations and what-if explorations. This allows an urban planner and decision-maker to measure computationally the outcomes of a given set of concepts, such as change of policy, infrastructure investment, or an emergency response alternative, prior to an action being taken.

Digital twins enable adjustive approach to management, as they provide constant updates to performance measurement and adjust strategies according to the prevailing conditions by incorporating real-time data feeds. This cycling makes the urban systems stronger and more responsive.

7.5 Regulatory support and alignment with policy

Digital twin technologies should also be included in favorable policy and regulatory environments in order to flourish. This aspect shows the importance of the creation of comprehensive urban policies and legislative improvements within the city that take into account the peculiarities of digital twins.

7.5.1 Combined urban policies

Digital twin projects are not supposed to exist in their own bubbles but must be part and parcel of more extensive urban development and climate action strategies. Coherent policy systems that connect data control, infrastructure investment and sustainability goals address the isolation of efforts and redundancies.

Echoing the current strategies assures that digital twins play a constructive role in long-term objectives of the city, including carbon neutrality, smart mobility, and disaster resilience. The effectiveness and sustainability of digital twin implementations are facilitated in a cross-sector manner by collaboration and policy coherence.

7.6 Ability development and innovation ecology

To make a successful digital twin adoption, a workforce with talent and an innovative environment are needed. This aspect concentrates on human capital and teamwork conditions required to maintain a continuous digital twin development.

7.6.1 Education and skills

It is essential to supply urban professionals with interdisciplinary skills across urban planning, data science, artificial intelligence and cybersecurity to be able to realize and run digital twins. It is necessary to invest in the workforce by providing specific training programs and using lifelong learning.

Collaboration with universities, technical institutes and professional bodies facilitate the development of curricula and certification programs that lead to changes in digital twin technologies. The ability to build these skills internally in city administrations helps to be more self-reliant and maneuverable in dealing with the changes in urban digital transformations.

7.6.2 Public-private partnerships and innovation hubs

Creation of innovation hubs leads to co-operation between startups, SMEs, universities and government agencies, enhancing the rate of knowledge transfer and technological transmission. These forms of ecosystems create a climate that favors experimentation, co-creation, and quick prototyping of the digital twin solutions, according to the local settings.

The involvement of government and industry resources allows doing more with less as public-private collaboration drives the ability to build and implement innovative digital twin solutions. The economic growth and the leader status of the cities in the field of smart urban development is another reason to promote the local innovative ecosystems.

7.7 Roadmap

The strategic framework promotes the implementation of a phased approach to the introduction of digital twins as illustrated in Figure 7, Roadmap for the integration and scaling of digital twin technology in sustainable urban governance and described below:

Phase 1: **Assessment and Planning** - Carry out needs assessment, stakeholder mapping and establish sustainability objectives.
Phase 2: **Infrastructure Development-** Construct common data architecture, achieve interoperability and pilot early projects.
Phase 3: **Capacity Building and Policy Alignment** - Educate personnel, orient laws and construct governance mechanisms.

Phase 4: **Scale-up and Continuous Improvement**: Increase the coverage of the digital twins, add more urban systems to cover, as well as establish monitoring and feedback loops.

Figure 7. Roadmap for the integration and scaling of digital twin technology in sustainable urban governance

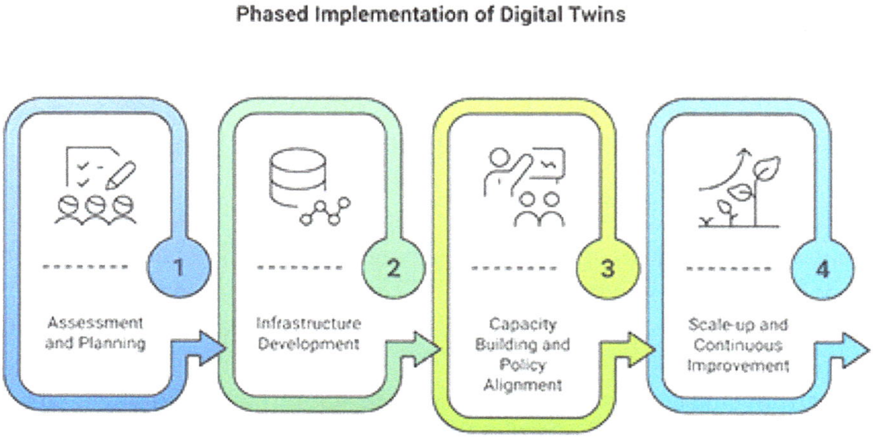

7.8 Case illustration: Singapore smart nation

Singapore is an example of this strategy on the ground. Virtual Singapore (a city-state project) combines real-time data obtained by different sources in a digital twin as urban planning information, environmental surveillance, and emergency preplans (Azadi et al., 2025).

Its success in the project is based on a strong government leadership, open data policies and collaborative platforms with academia and industry. One of our neighbours, Singapore, also constantly updates its performance indicators placing more focus on carbon neutrality and resilience (Bortolini, et al., 2022).

8. CONCLUSION AND FUTURE PROPOSED WORK

8.1 Summary

This chapter discusses how Digital Twin Technology (DTT) has the potential to transform urban change into a more sustainable manner and specifically refers to how it has the potential to revolutionize city planning, management, and resiliency. Based on the extensive literature review and the case studies addressed in the research, the following major points have been formulated:

- Digital Twin Technology Basics: Digital twins involve the dynamic, data-driven virtual re-model of the physical urban systems through the integration of IoT, AI, GIS, and BIM technologies, hence, making real-time monitoring, simulation, and optimization feasible (Mazzetto et al., 2024; Liu et al., 2024).
- Urban Sustainability: DTT can aid many sustainability goals including reducing carbon emissions, maximizing energy efficiency, water, waste, and climate resilience in urban landscapes (Azadi et al., 2025; Bibri et al., 2024).
- Case Study Lessons: Cities (Singapore, Helsinki, Shanghai, and Amaravati) show a wide variety of potential uses of digital twins, both their advantages in the urban government policies, maintenance of the infrastructural processes, and the stakeholder interactions, as well as the technical and socio-political pitfalls (Bortolini, et al., 2022; Peldon et al., 2024).
- Challenges and opportunities challenges such as barriers to data security, interoperability, prohibitive costs and complexities of governance should be overcome using shared policies, capacity building and regulatory platforms. At the same time AI and participatory platforms demonstrate a great potential in the future (Ishrat et al., 2025; Samuel et al., 2023).

Strategic Framework: A comprehensive strategy based on stakeholder engagement, heavily invested data infrastructure, sustainability performance indicators, policy integration, and constant innovation is critical to realize successful deployment of digital twin (Azadi et al., 2025; Bortolini, et al., 2022).

8.2 Research and practice recommendations

Even though a lot has been progressed, there are points to be explored and advanced more to unlock the full potential of digital twins concerning sustainable urban change:

8.2.1 Governance and ethics in data privacy

Since urban digital twins gather and analyze extremely sensitive data, the studies of ethical governance models which guarantee transparency, non-discrimination, and privacy of citizens are a matter of urgency (Ishrat et al., 2025). Development of methods of data ownership, informed consent and otherwise biased outcomes of algorithmic decisions in AI-based decisions need to be addressed in future research.

8.2.2 Incorporation of emerging technologies

The use of the emerging technologies, including edge computing, 5G networks, and the blockchain would make urban digital twins more scalable, secure, and responsive (Samuel et al., 2023). Due to the possibilities of such integrations, additional experimental work and pilot projects are needed to determine the practical relevance and usefulness of these integrations.

8.2.4 Interoperability and standardization

There are no standardized protocols which restrict the seamless integration of systems, and cities. The next step would be to develop open standards to enable interoperability, data sharing and replicability of digital twin data models and APIs at the global level (Mazzetto et al., 2024; Liu et al., 2024).

8.2.5 Impact assessment of a longitudinal nature

There is also a lack of long-term research to evaluate the effects that digital twin implementations have on the environment, economy, and society. Developing sound methodologies to monitor and evaluate such impacts will be a way to provide evidence of refining policies and the design of technology.

8.3 Conclusion

The era of significant urbanization and progressively worsening climate conditions presents the Digital Twin Technology as a radically innovative tool allowing the cities to design, model, and run their sustainable future. Reality-time data streams, robust analytics, and participatory governance integrated into digital twins allow urban stakeholders to design resilience infrastructure and optimize the use of resources, as well as address communities in a more inclusive manner than ever before.

The chapter has given an integrated analysis of conceptual basis, practice considerations, and emerging issues and trajectories to consider in terms of including digital twins in achieving sustainable urbanization. The strategic framework has provided hands-on advice to cities possessing a vision of utilizing digital twins and offers a balance between ideas of technological dynamism, governance, ethics, and inclusion.

Although the path to the ubiquitous uptake will be complicated and will require actions at a high level, in cross-disciplinary, cross-sectoral, and cross-societal hearings, the potential of digital twins as intelligent and sustainable cities pivot points is hard to miss. As the research, innovation, and collaboration carry on, digital twins have the potential to reshape the futures of urbanity, making them livable in the most penetrating manner, and ensuring resilience and environmental management throughout the coming generations.

REFERENCES

Azadi, S., Kasraian, D., Nourian, P., & Van Wesemael, P. (2025). What have urban digital twins contributed to urban planning and decision making? From a systematic literature review toward a Socio-Technical research and Development agenda. *Smart Cities*, *8*(1), 32. DOI: 10.3390/smartcities8010032

Batty, M., Axhausen, K. W., Giannotti, F., Pozdnoukhov, A., Bazzani, A., Wachowicz, M., Ouzounis, G., & Portugali, Y. (2012). Smart cities of the future. *The European Physical Journal. Special Topics*, *214*(1), 481–518. DOI: 10.1140/epjst/e2012-01703-3

Bibri, S. E., Huang, J., Jagatheesaperumal, S. K., & Krogstie, J. (2024). The synergistic interplay of artificial intelligence and digital twin in environmentally planning sustainable smart cities: A comprehensive systematic review. *Environmental Science and Ecotechnology*, *20*, 100433. DOI: 10.1016/j.ese.2024.100433 PMID: 38831974

Bortolini, R., Rodrigues, R., Alavi, H., Vecchia, L. F. D., & Forcada, N. (2022). Digital Twins' Applications for building Energy Efficiency: A review. *Energies*, *15*(19), 7002. DOI: 10.3390/en15197002

Digital Twin: enabling technologies, challenges and open research. (2020). IEEE Journals & Magazine | IEEE Xplore. https://ieeexplore.ieee.org/abstract/document/9103025

European Commission. (2022) The Barcelona Urban Platform and integrated smart city operations. https://smart-cities-marketplace.ec.europa.eu

Florido-Benítez, L. (2024). The use of digital twins to address smart tourist destinations' future challenges. *Platforms*, *2*(4), 234–254. DOI: 10.3390/platforms2040016

Ishrat, M., Khan, W., Faisal, S. M., Farsi, M. M. A., & Farooqui, N. A. (2025). Digital Twins Future prospects in smart cities. In *CRC Press eBooks* (pp. 53–68). https://doi.org/DOI: 10.1201/9781003510338-3

Jiang, F., Ma, L., Broyd, T., Chen, W., & Luo, H. (2021). Digital twin enabled sustainable urban road planning. *Sustainable Cities and Society*, *78*, 103645. DOI: 10.1016/j.scs.2021.103645

Lehtola, V. V., Koeva, M., Elberink, S. O., Raposo, P., Virtanen, J., Vahdatikhaki, F., & Borsci, S. (2022). Digital twin of a city: Review of technology serving city needs. *International Journal of Applied Earth Observation and Geoinformation*, *114*, 102915. DOI: 10.1016/j.jag.2022.102915

Liu, W., Lv, Y., Wang, Q., Sun, B., & Han, D. (2024). A Systematic Review of the Digital Twin Technology in Buildings, Landscape and Urban Environment from 2018 to 2024. *Buildings (Basel, Switzerland)*, *14*(11), 3475. DOI: 10.3390/buildings14113475

Luo, J., Liu, P., Xu, W., Zhao, T., & Biljecki, F. (2024). A perception-powered urban digital twin to support human-centered urban planning and sustainable city development. *Cities (London, England)*, *156*, 105473. DOI: 10.1016/j.cities.2024.105473

Mazzetto, S. (2024). A review of urban Digital Twins integration, challenges, and future directions in smart city development. *Sustainability (Basel)*, *16*(19), 8337. DOI: 10.3390/su16198337

Park, J., & Yang, B. (2020). GIS-Enabled Digital Twin system for sustainable evaluation of carbon emissions: A case study of Jeonju City, South Korea. *Sustainability (Basel)*, *12*(21), 9186. DOI: 10.3390/su12219186

Patel, U. R., Ghaffarianhoseini, A., Ghaffarianhoseini, A., & Burgess, A. (2024). Digital Twin Technology for sustainable urban development: A review of its potential impact on SDG 11 in New Zealand. *Cities (London, England)*, *155*, 105484. DOI: 10.1016/j.cities.2024.105484

Peldon, D., Banihashemi, S., LeNguyen, K., & Derrible, S. (2024). Navigating urban complexity: The transformative role of digital twins in smart city development. *Sustainable Cities and Society*, *111*, 105583. DOI: 10.1016/j.scs.2024.105583

Ricciardi, G., & Callegari, G. (2023). Digital Twins for Climate-Neutral and resilient cities. State of the art and future development as tools to support urban Decision-Making. In *˜The œurban book series* (pp. 617–626). https://doi.org/DOI: 10.1007/978-3-031-29515-7_55

Samuel, P., Saini, A., Poongodi, T., & Nancy, P. (2023). Artificial intelligence–driven digital twins in Industry 4.0. In *Elsevier eBooks* (pp. 59–88). https://doi.org/DOI: 10.1016/B978-0-323-99205-3.00002-X

United Nations. (2018). *World urbanization prospects: The 2023 revision*. United Nations Department of Economic and Social Affairs.

Weil, C., Bibri, S. E., Longchamp, R., Golay, F., & Alahi, A. (2023). Urban Digital Twin Challenges: A Systematic Review and Perspectives for Sustainable Smart Cities. *Sustainable Cities and Society*, *99*, 104862. DOI: 10.1016/j.scs.2023.104862

Zhao, X., & Zhang, Y. (2024). Integrated management of urban resources toward Net-Zero smart cities considering renewable energies uncertainty and modeling in Digital Twin. *Sustainable Energy Technologies and Assessments*, *64*, 103656. DOI: 10.1016/j.seta.2024.103656

Chapter 5

Optimizing Additive Manufacturing of Superalloys for Sustainable Prototyping

Muhammed Enes İlgazi
https://orcid.org/0000-0002-5763-6427
İstanbul Nisantasi University, Turkey

ABSTRACT

This chapter explores sustainable prototyping through additive manufacturing (AM) of superalloys, particularly in aerospace and energy sectors. It begins with an overview of superalloy material science and its environmental implications, then examines AM technologies that offer efficiency and reduced ecological impact. By customizing AM parameters and applying lifecycle assessments, both performance and sustainability can be optimized. Topology optimization enhances material efficiency, while strategies like waste minimization and powder recycling support circular economy goals. The chapter emphasizes scalability and evolving standards as key factors for broader industry adoption, supported by case studies showing real-world environmental and economic benefits. Future innovation paths, including AI-integrated manufacturing and eco-friendly alloy development, are discussed as part of a holistic approach that unites technical advancement with ecological responsibility.

DOI: 10.4018/979-8-3373-5473-6.ch005

INTRODUCTION

In the contemporary landscape of engineering design and product development, prototyping functions not merely as an intermediary step, but as an indispensable phase enabling iterative validation, performance evaluation, and the resolution of complex design challenges prior to full-scale manufacturing (Kubec et al., 2023). The accelerated need for innovation, coupled with the demand for high-performance products, has further accentuated the strategic importance of prototyping (Reeves et al., 2011). However, conventional prototyping methods predominantly based on subtractive manufacturing technique have been increasingly scrutinized for their substantial environmental footprint (Kellens, Mertens, et al., 2017). These methods are inherently resource-intensive, characterized by high energy demand, extensive material waste, and elevated carbon emissions, all of which stand in stark contrast to the overarching goals of sustainable development (Ford & Despeisse, 2016).

In response to global environmental imperatives, there has been a paradigm shift toward sustainable product design and manufacturing, which includes the adoption of environmentally conscious prototyping methodologies (Thompson et al., 2016). The concept of sustainable prototyping entails not only reducing material and energy usage but also integrating principles of circularity, life cycle thinking, and environmental responsibility into the earliest stages of product realization (Bocken et al., 2016). Central to this evolution is the exploration and application of advanced manufacturing technologies such as additive manufacturing (AM), which holds promise for transforming prototyping into a more sustainable practice (Ngo et al., 2018).

Among the spectrum of engineering materials, superalloys occupy a unique and technologically significant niche (Reed, 2006). These high-performance alloys, primarily based on nickel, cobalt, or iron, exhibit exceptional mechanical strength, creep resistance, corrosion resistance, and thermal stability, particularly at elevated temperatures (Sims et al., 1987). As such, they are extensively employed in demanding sectors such as aerospace, energy, and automotive engineering, where component reliability and performance are critical (Pollock & Tin, 2006). Despite their superior properties, the conventional manufacturing of super alloys is associated with substantial challenges both technical and environmental (Rickenbacher et al., 2013). Their complex chemistries and high melting points necessitate energy-intensive processing routes, while machining such materials typically results in considerable material loss and tool wear (Murr et al., 2012).

Table 1. Comparison of conventional and additive manufacturing methods for superalloys

Criterion	Conventional Manufacturing (Subtractive)	Additive Manufacturing (AM)
Material Utilization Efficiency	Low (high material waste)	High (minimal material waste)
Energy Consumption	High (melting, casting, machining processes)	Moderate (localized energy input, layer-wise fabrication)
Geometric Complexity	Limited (restricted by tooling and machining constraints)	High (greater design freedom, complex geometries possible)
Prototyping Speed	Low (requires molds and tooling)	High (direct from digital model, rapid iterations)
Carbon Footprint	High (intensive processes and waste)	Moderate to Low (depending on energy source and process)
Powder Recyclability	Not Applicable	Possible (requires quality monitoring and control)
Post-Processing Requirements	Intensive (machining, finishing, etc.)	Required (but generally less compared to subtractive methods)

As shown in Table 1, additive manufacturing offers significant advantages over traditional methods in terms of material efficiency, design flexibility, and environmental impact, particularly for superalloy-based prototypes. In this context, the integration of additive manufacturing into the prototyping of superalloy-based components presents a compelling opportunity (DebRoy et al., 2018). AM technologies—such as Selective Laser Melting (SLM), Electron Beam Melting (EBM), and Laser Powder Bed Fusion (L-PBF)—facilitate the layer-wise construction of components directly from digital models, inherently minimizing material wastage and enabling fabrication of intricate geometries that are otherwise impossible with traditional subtractive techniques (Herzog et al., 2016). Moreover, the process flexibility offered by AM enables rapid design iterations, aligning with the agility demands of contemporary engineering workflows (Thompson et al., 2016).

Nonetheless, leveraging AM for sustainable prototyping with superalloys is not without its challenges. The interactions between thermal cycles, microstructural evolution, and mechanical performance necessitate a rigorous understanding of process-structure-property relationships (L. Yang et al., 2019). Furthermore, considerations of powder recyclability, energy consumption per unit build, post-processing requirements, and life cycle environmental assessments must be systematically incorporated to validate sustainability claims (Bocken et al., 2016).

The principal aim of this chapter is to elucidate the confluence of additive manufacturing and sustainable prototyping in the specific context of superalloy materials. Through critical examination of material behavior, process optimization strategies, and environmental assessment methodologies, we seek to establish a comprehensive framework that maintains prototype functional integrity while ensuring alignment with sustainability benchmarks (Bourell et al., 2009). Emphasis is placed on the development of closed-loop manufacturing systems, design for additive manufacturing (DfAM) principles, and deployment of digital tools such as topology optimization and generative design contributing to material efficiency (Kwon & Yun, 2021).

Furthermore, this chapter situates its analysis within the broader discourse of life cycle engineering and sustainable product innovation. By integrating life cycle assessment (LCA) techniques with additive manufacturing metrics, a multidimensional understanding of the environmental performance of superalloy-based prototypes is pursued. Such an approach is critical for material selection, process planning, and end-of-life strategies in alignment with the United Nations Sustainable Development Goals (SDGs) and circular economy principles (Rejeski et al., 2018).

In sum, this chapter contends that optimization of additive manufacturing processes for superalloys represents a transformative pathway toward high-performance prototypes that are technically robust and environmentally responsible. By reconciling traditionally conflicting goals of performance enhancement and ecological stewardship, sustainable prototyping with superalloys emerges as a vital enabler of innovation in an era defined by climate consciousness and resource constraints (Rejeski et al., 2018).

SUPERALLOYS: MATERIAL SCIENCE AND SUSTAINABILITY POTENTIAL

Superalloys (Chen, 2020) primarily nickel-, cobalt-, and iron-based alloys—represent a class of advanced engineering materials specifically designed to perform under extreme mechanical and thermal conditions. These materials exhibit superior properties such as high tensile strength, oxidation resistance, corrosion resistance, and exceptional creep resistance at elevated temperatures. Such characteristics render them essential in high-demand sectors like aerospace, power generation, marine propulsion, and advanced manufacturing. Their ability to retain mechanical integrity under thermal and mechanical stress not only enables longer component lifespans but also improves the energy efficiency of the systems in which they are deployed, thereby supporting broader goals of sustainability in critical industrial infrastructures.

From a material science perspective, the performance of superalloys is intimately linked to their complex microstructures, which are typically composed of a matrix of γ-phase (face-centered cubic nickel) reinforced by coherent γ'-precipitates ($Ni_3(Al,Ti)$) and carbides. These features provide enhanced resistance to dislocation motion, thus enabling prolonged durability at high temperatures (Chen, 2020). The addition of alloying elements such as chromium, molybdenum, tungsten, rhenium, and cobalt serves to tailor the mechanical and chemical stability of the alloy, facilitating application-specific customization.

While the metallurgical sophistication of these materials is instrumental in enhancing performance, it also introduces environmental challenges stemming from resource intensity and processing complexity. Nickel and cobalt, two of the predominant constituents of most superalloys, present a dual narrative in the sustainability discourse. On one hand, their functional role in enabling advanced clean technologies such as high-efficiency gas turbines, electric vehicles, and renewable energy systems positions them as indispensable to the green transition (CarbonChain, 2022). On the other hand, their extraction and refining processes are highly carbon-intensive, with global averages indicating approximately 13–45 t CO_2-eq per tonne of nickel produced depending on ore type and refinement route, and substantial emissions from cobalt production as well.

The distinction between Class 1 and Class 2 nickel further underscores the environmental variability associated with superalloy production. Class 1 nickel—used in superalloys and battery technologies—is typically derived from sulfide ores via energy-intensive smelting and hydrometallurgical refining, whereas Class 2 nickel—used in stainless steels—is produced from lateritic ores via pyrometallurgical routes, often resulting in higher iron contamination but lower cost (Gibson et al., 2021). Consequently, the environmental burden associated with nickel production varies widely—from around 13 t CO_2-eq for high-efficiency sulfide operations to over 45 t CO_2-eq for lower-efficiency laterite processing.

The sustainability potential of superalloys, therefore, must be evaluated not only in terms of them in-use performance but also through a comprehensive life cycle lens that encompasses raw material extraction, alloy formulation, manufacturing, operational lifetime, and end-of-life recyclability. For instance, while the extended service life and high thermal efficiency of turbine blades fabricated from nickel-based superalloys can offset their initial environmental costs, the feasibility of closed-loop recycling remains limited due to challenges such as alloy segregation, contamination, and difficulties in recovering high-purity elemental constituents.

To advance sustainability in super alloy applications, ongoing research focuses on several strategic avenues. These include the development of low-carbon extraction processes, improved recycling technologies capable of recovering strategic metals from complex waste streams, and the formulation of alternative alloy systems that reduce dependency on critical raw materials. Additionally, process-integrated assessments such as exergy analysis and dynamic life cycle modeling are being employed to quantify the net environmental benefits of super alloy utilization across various applications.

In summary, super alloys embody a critical intersection of performance excellence and sustainability complexity. Their application in high-efficiency systems supports decarburization efforts across multiple sectors, yet their production remains a notable contributor to global greenhouse gas emissions. A holistic approach that couples advanced material science with environmentally optimized manufacturing and recycling systems is essential to fully realize the sustainability potential of super alloys in next-generation prototyping and beyond.

OVERVIEW OF ADDITIVE MANUFACTURING TECHNOLOGIES: A NEW PARADIGM IN PROTOTYPING

Additive Manufacturing (AM) has fundamentally redefined the way prototypes are conceptualized, designed, and fabricated (Gibson et al., 2021). Unlike conventional subtractive manufacturing, which removes material from a solid block to achieve the desired geometry, AM adds material layer by layer (Bourell et al., 2009). This layer-wise fabrication approach has led to transformative changes in prototyping, particularly in industries where geometric complexity, material efficiency, and rapid iteration are critical (Ngo et al., 2018). Superalloy-based prototyping, due to its inherent challenges in processing and shaping, has particularly benefited from the rise of AM technologies (Frazier, 2014).

One of the defining advantages of AM lies in its design freedom. Engineers and designers are no longer constrained by the limitations of molds, cutting tools, or traditional machining pathways (King et al., 2021). This shift allows for the realization of highly complex internal structures, topologies optimized for performance, and lightweight components with superior mechanical properties (Gu et al., 2012). In the realm of aerospace and energy sectors where superalloys are predominantly employed this capability translates into significant performance improvements and material efficiency gains (Kruth et al., 2005).

Table 2. Comparative overview of key additive manufacturing technologies for superalloys

AM Technology	Energy Source	Material Feedstock	Key Advantages	Challenges	Sustainability Considerations
Selective Laser Melting (SLM)	Laser	Metal powder	High resolution; excellent surface finish	Residual stresses; support structure removal	Medium energy use; moderate recyclability of powder
Electron Beam Melting (EBM)	Electron beam	Metal powder	High build rate; good for high-temperature materials	Vacuum requirement; rougher surface finish	High energy use; limited powder recyclability
Laser Powder Bed Fusion (L-PBF)	Laser	Metal powder	Fine feature resolution; widely adopted	Slower build speed; need for controlled atmosphere	Medium energy use; good material efficiency
Direct Energy Deposition (DED)	Laser/electron/ plasma	Powder or wire	Suitable for repair; large parts; high deposition rate	Lower surface quality; post-processing required	Lower waste; variable energy consumption depending on scale

As outlined in Table 2, each additive manufacturing technology offers a unique balance between precision, build rate, and sustainability, making technology selection a critical step in superalloy-based prototyping workflows.

Several AM technologies have emerged as especially relevant to the processing of superalloys. Each of these techniques—Selective Laser Melting (SLM), Electron Beam Melting (EBM), Direct Energy Deposition (DED), and Laser Powder Bed Fusion (L-PBF)—offers unique benefits and poses distinct challenges, both in terms of technical execution and environmental impact (Herzog et al., 2016).

Selective Laser Melting (SLM)

Selective Laser Melting (SLM) is among the most widely adopted AM technologies for producing intricate superalloy components. It utilizes a high-powered laser to selectively fuse fine metallic powders layer by layer (Gibson et al., 2021). One of the strengths of SLM is its ability to fabricate parts with high dimensional accuracy and excellent surface finish, reducing the need for extensive post-processing.

SLM is particularly effective for nickel-based superalloys, such as Inconel 718 and 625, due to its precise control of thermal input and fine microstructural resolution. However, the process is energy-intensive, primarily because of the high-power laser system and the need for an inert gas atmosphere (usually argon) to prevent oxidation during the build. Despite its energy demands, SLM minimizes material waste and enables the use of recycled powder, making it more sustainable than traditional subtractive techniques.

Electron Beam Melting (EBM)

Electron Beam Melting (EBM) operates similarly to SLM but uses an electron beam instead of a laser. EBM typically takes place in a vacuum environment, which helps prevent contamination and enables high-temperature processing—ideal for superalloys prone to oxidation or cracking under rapid thermal cycles (Zafar et al., 2025). The key advantage of EBM is its ability to produce components with lower residual stress due to the higher build temperatures and slower cooling rates, beneficial for superalloys often susceptible to thermal fatigue. Moreover, EBM supports fabrication of fully dense parts with excellent mechanical properties, making it suitable for load-bearing aerospace applications.

From a sustainability standpoint, EBM consumes less shielding gas and enables high material reuse rates, though it still requires significant power and infrastructure investent, limiting accessibility for smaller-scale operations (Wang et al., 2023).

Direct Energy Deposition (DED)

Direct Energy Deposition (DED) encompasses a family of AM processes using focused energy sources—typically lasers, electron beams, or plasma arcs—to melt and deposit material simultaneously, with feedstock delivered as powder or wire. DED is particularly suited for repair or augmentation of existing components, contributing to sustainability by extending service life and reducing full part replacement (Bourell et al., 2009). While DED offers high deposition rates and scalability for large components, it lacks the fine resolution of SLM or EBM and is most effective for near-net-shape production where final machining may still be required . Environmental assessments show DED's material utilization is generally high, and energy consumption can be optimized through parameter tuning (Larimian & Borkar, 2019).

Laser Powder Bed Fusion (L-PBF)

Laser Powder Bed Fusion (L-PBF), often used interchangeably with SLM, is another cornerstone technology in the AM landscape. It is widely adopted for producing complex, high-precision parts in industries such as aerospace and biomedical engineering. It is especially prominent in the aerospace and medical sectors for producing highly detailed parts from difficult-to-machine materials like titanium and nickel-based superalloys. L-PBF is particularly suitable for these alloys due to its ability to process high-melting-point materials with excellent geometrical control (Ngo et al., 2018). L-PBF relies on spreading thin layers of powder over a build platform and selectively melting regions based on a digital model. This layer-wise strategy enables excellent resolution and the fabrication of intricate internal geometries. The precision of this method enables the fabrication of lattice structures, internal cooling channels, and other features that would be impossible or prohibitively expensive using traditional techniques. These capabilities allow for functional optimization, such as weight reduction and enhanced thermal performance, critical for aerospace and biomedical components (Murr et al., 2012).

In terms of environmental performance, L-PBF's key advantages include minimal scrap generation and efficient material recycling. Since powder not fused in one build can often be reused, the material efficiency of L-PBF is typically higher than traditional machining(Gibson et al., 2021). However, the process remains energy-intensive due to laser operation and the need for controlled atmospheric conditions. Maintaining an inert gas environment (often argon) and high-powered laser systems contributes significantly to its energy footprint. Continued advancements in laser optics, build chamber insulation, and process monitoring aim to enhance both the energy efficiency and reliability of L-PBF systems. Recent research focuses on real-time control systems and process parameter optimization to reduce energy consumption while maintaining part quality (Liu et al., 2023).

Comparative Environmental Impact and Sustainability Assessment

Each Each AM technology presents a unique sustainability profile (Bourell et al., 2009). From an environmental lifecycle perspective, several metrics must be considered: energy consumption per part, material efficiency (including powder reuse rates), emissions during operation, and waste generated in post-processing. SLM and L-PBF excel in precision and material efficiency but are more energy-intensive per unit volume of material processes (Herzog et al., 2016).

EBM offers thermal and structural advantages while requiring fewer environmental controls, but at the cost of higher equipment and operational demands (Murr et al., 2012). DED stands out for its repair capabilities and material scalability, making it a prime candidate for circular manufacturing strategies.

Several studies have shown that when AM is employed strategically—especially for weight-optimized aerospace components—the resulting reduction in fuel usage during the product's service life more than offsets the environmental cost of AM fabrication. Moreover, the digital nature of AM allows for decentralized manufacturing, which can reduce logistics-related emissions and support more localized production ecosystems (Jinoop et al., 2019).

Future Directions and Research Priorities

As additive manufacturing continues to mature, ongoing research focuses on enhancing process stability, reducing energy input, and improving powder recyclability.

Hybrid systems combining AM with traditional CNC machining offer new opportunities for precision and sustainability (Hasan et al., 2023) .

Furthermore, developments in AI-driven design optimization and in-situ monitoring promise to reduce trial-and-error in prototyping, accelerating time to market while minimizing material and energy waste (Herzog et al., 2016). In summary, additive manufacturing represents a paradigm shift in how superalloy components are prototyped, offering unprecedented opportunities for sustainability, innovation, and performance. A thorough understanding of the different AM processes and their respective environmental implications—is essential for identifying the most suitable and responsible paths forward in sustainable prototyping.

TAILORING ADDITIVE MANUFACTURING FOR SUPERALLOYS

Additive manufacturing (AM) of superalloys is an exciting frontier in materials science and engineering, promising unprecedented capabilities in the design and production of high-performance components.

However, the processing of superalloys via AM remains a complex endeavor due to the materials' intricate chemistries, high melting points, and susceptibility to thermal gradients (King et al., 2021).

The need for precision in thermal control and build parameters is paramount, as improper settings can lead to a range of defects such as hot cracking, porosity, residual stresses, and anisotropy.

To successfully implement AM for superalloys in demanding sectors such as aerospace, energy, and defense, it is essential to tailor the entire process from powder characteristics to post-processing methods—for optimal mechanical performance and sustainability (Meier et al., 2018).

Challenges in Additive Manufacturing of Superalloys

Superalloys, especially nickel- and cobalt-based variants, are engineered to withstand extreme thermal and mechanical environments (DebRoy et al., 2018). Their complex alloying compositions typically include elements like chromium, molybdenum, tungsten, aluminum, and titanium to enhance strength, oxidation resistance, and creep behavior (Frazier, 2014). These same features, however, make them challenging to process via AM (Körner, 2016). One of the primary issues is the high melting temperature, which requires elevated energy input and introduces steep thermal gradients during layer-by-layer fabrication (Sames et al., 2016). These gradients lead to residual stresses, non-uniform microstructures, and even micro cracking especially in alloys like IN738LC or CMSX-4 (Attar et al., 2014). Additionally, the rapid solidification inherent in AM processes can result in anisotropic mechanical properties and the formation of undesirable phases (Liang et al., 2018). The control of solidification pathways is crucial to achieve a consistent microstructure, especially since most AM parts are fabricated far from thermodynamic equilibrium (Bourell et al., 2009).

Powder Characteristics and Their Role

Feedstock powder characteristics are crucial for successful additive manufacturing (AM). Particle size distribution, morphology, flowability, and chemical purity directly impact process stability and part quality (Liu et al., 2020). Spherical powders with narrow size distributions are preferred, as they promote uniform layer deposition and consistent melting (Garcia et al., 2021). Gas atomization, the most common production method, efficiently generates spherical particles with minimal satellites. In industrial AM, powder is frequently recycled, but this increases the risk of superalloys by increasing their oxygen and nitrogen content, leading to brittleness. Rigorous monitoring and limits on reuse cycles are essential for maintaining part integrity. Uniform powder bed distribution and optimal packing density are also vital, as they influence energy absorption, melting dynamics, and defect generation (Ghosh, 2017).

Laser Scanning Strategies and Thermal Management

Effective thermal management is essential in the additive manufacturing (AM) of superalloys. Parameters such as hatch spacing, scan speed, laser power, and scan pattern govern the scanning strategy (Kulkarni, 2018). Careful optimization of these parameters helps minimize thermal gradients. To mitigate residual stresses and reduce part distortion, various techniques can be used, including rotating scan vectors between layers, adopting island-scanning, or using contour-core scan sequences (Sun et al., 2025). Preheating the build plate also plays a significant role in controlling thermal stress formation. For example, the build area in Electron Beam Melting (EBM) is typically preheated to around 1000 °C, thereby reducing thermal gradients. In contrast, Laser Powder Bed Fusion (L-PBF) systems often apply preheating stages in the range of 200–300 °C to moderate cooling rates (Martelli et al., 2024). Maintaining microstructural integrity equally relies on controlled cooling. While rapid cooling can promote desirable grain refinement, it also increases the risk of cracking in gamma-prime-strengthened alloys. Increasingly, advanced process simulations and in-situ monitoring tools—such as infrared thermography and melt pool sensors—are used, providing real-time control and helping ensure part quality (Kellens, Baumers, et al., 2017).

Post-Processing: From Heat Treatment to Hot Isostatic Pressing

Post-processing is often indispensable for superalloy AM components to reach desired microstructural and mechanical properties (Liu et al., 2020). Heat treatment (HT) is the most common method used to dissolve segregated phases, homogenize the microstructure, and precipitate strengthening phases like γ' (gamma prime) (Kellens, Baumers, et al., 2017). Standard HT cycles for alloys like Inconel 718 involve solutionizing followed by aging treatments. These processes are fine-tuned based on the thermal history experienced during AM. Tailored heat treatments can mitigate residual stresses, improve ductility, and enhance fatigue resistance (Peng, Kellens, et al., 2018).

Hot Isostatic Pressing (HIP) is another critical post-processing step, particularly effective in closing internal pores and cracks formed during AM (Ford & Despeisse, 2016). HIP applies high pressure (100–200 MPa) and elevated temperature uniformly to the component, enhancing density and mechanical integrity (Holmström et al., 2010). Integrating HIP with tailored HT cycles (termed HIP-HT) is a growing trend for aerospace-grade AM parts (Tiwari & Turner, 2019). Recent advancements in rapid HIP cycles and in-process densification techniques aim to streamline post-processing steps, reducing energy and time consumption while maintaining part quality(Zhu & Dhokia, 2021).

Microstructure Control and Grain Engineering

The ability to engineer microstructures through controlled solidification and thermal cycling is one of AM's greatest advantages, especially for superalloys (Liu et al., 2020). Directional solidification and columnar grain growth are desirable for turbine blades and rotating parts to improve creep performance (Ngo et al., 2018). Laser process parameters can be tuned to promote columnar grains aligned along the build direction, mimicking the microstructures achieved in conventionally cast single-crystal alloys. However, such directional growth requires careful calibration of melt pool dimensions and scan patterns.

Grain refinement techniques such as cyclic remelting, inoculant addition, or thermal gradient manipulation are under active research to reduce anisotropy and improve mechanical uniformity. Moreover, AM's layer-by-layer nature facilitates location-specific microstructure design, potentially enabling functionally graded materials tailored for multi-functional performance (Carter et al., 2014).

Sustainability Considerations in Tailoring AM

Optimizing AM for superalloys not only enhances performance but also contributes to sustainability. Reduced machining requirements, lower material waste, and minimized scrap rates are inherent benefits of AM. However, energy consumption remains a concern, especially due to high-temperature processes and extensive post-processing (Frazier, 2014).

Strategies like near-net-shape fabrication, in-situ process monitoring to avoid defective builds, and recycling heat from thermal treatments are being explored to improve energy efficiency (Mahmoudi & Kovacevic, 2015). Furthermore, improvements in powder yield and reuse protocols can reduce environmental impacts associated with powder production and disposal (Leicht et al., 2018).

Life Cycle Assessment (LCA) studies increasingly support the notion that, despite high energy input during fabrication, AM of superalloys can outperform traditional methods when considering the full product lifecycle—especially in weight-critical and high-performance applications where the design freedom of AM enables part consolidation and lightweighting (Ramsperger et al., 2019).

Outlook and Future Directions

While considerable progress has been made in tailoring AM for superalloys, future efforts must address scalability, standardization, and in-process control. The integration of machine learning for parameter optimization, digital twins for process modeling, and real-time defect detection will be vital for achieving reliable, repeatable outcomes (King et al., 2021).

Moreover, alloy design specifically for AM—often termed "AM-specific superalloys"—is gaining momentum (Spierings et al., 2013). These compositions are engineered to solidify with reduced cracking susceptibility, improved printability, and better phase stability (Slotwinski et al., 2014).

Ultimately, the convergence of advanced materials engineering, intelligent process control, and sustainability frameworks will define the next generation of additive manufacturing tailored for superalloys(Simchi, 2006).

LIFE CYCLE ASSESSMENT AND ENVIRONMENTAL IMPACT OF ADDITIVE MANUFACTURING

Sustainability in manufacturing cannot be comprehensively evaluated through material efficiency alone (Cacace et al., 2024). The entirety of a product's life cycle, from raw material extraction to eventual disposal or recycling, contributes significantly to its environmental footprint (King et al., 2021). Life Cycle Assessment (LCA) is the internationally recognized methodology that provides a structured and standardized approach to quantifying these impacts (Mertens et al., 2017).

When applied to additive manufacturing (AM), particularly for superalloys, LCA allows for a nuanced understanding of the environmental trade-offs involved in high-performance prototyping (Zhu et al., 2007). This section aims to deliver an in-depth exploration of LCA methodologies and their application to superalloy-based AM processes (DebRoy et al., 2018).

We begin by introducing the theoretical framework of LCA, including its four key phases: Goal and Scope Definition, Life Cycle Inventory (LCI), Life Cycle Impact Assessment (LCIA), and Interpretation (Carter et al., 2014). The methodology follows the ISO 14044:2006 standard, which ensures consistency and transparency in environmental performance evaluation (Krupp & Seifi, 2018). Emphasis is placed on the importance of defining functional units and system boundaries appropriately, especially when comparing AM to conventional manufacturing (CM) processes.

One of the most studied aspects of AM's sustainability is its energy consumption (Murr et al., 2012). Several prior studies, such as Liu et al. (2020), have dissected the energy demands of AM processes across machine-level operations and process modes, including idle, standby, and operational phases. These investigations also include the embodied energy of feedstock materials and the potential for energy savings through optimized printing strategies (King et al., 2015).

However, as highlighted by Garcia et al. (2021), only about 25% of reviewed articles utilized LCA methodologies to holistically assess environmental performance. Most have disproportionately focused on energy use, while neglecting other critical resource inputs such as inert gases, build support materials, post-processing emissions, and waste generation (Meier et al., 2018).

This overemphasis creates an incomplete picture (DebRoy et al., 2018). Agrawal et al. (2022) and Kellens et al. (2017) have shown that few full-scope LCAs compare AM with CM in a consistent, cradle-to-grave framework . Their work calls for a more integrative approach that incorporates not only energy usage but also raw material extraction, material losses in post-processing, and emissions throughout the production and product-use phases (Körner, 2016).

Additionally, Motalebi et al. (2024) stress that the Global Warming Potential (GWP) of AM processes is driven mainly by energy consumption, while in CM, the raw material extraction and transformation phases contribute the most to GWP (Sames et al., 2016). This inversion of impact sources illustrates why LCA is vital: it helps identify where sustainability efforts should be concentrated (Attar et al., 2014).

To further develop this perspective, comparative case studies are examined (Liang et al., 2018). For instance, an LCA comparing Selective Laser Melting (SLM) with traditional investment casting for a nickel-based superalloy turbine blade demonstrates that while SLM has higher energy demands during production, it significantly reduces material waste and enables lightweighting both of which lower downstream energy costs during the use phase (Bourell et al., 2009b). The energy savings over the component's lifecycle can, in some scenarios, offset the high energy costs incurred during AM fabrication (Liu et al., 2023).

Another dimension explored is the variation in environmental impact among different AM technologies (Garcia et al., 2021). Laser Powder Bed Fusion (L-PBF), Electron Beam Melting (EBM), and Direct Energy Deposition (DED) each show differing profiles in energy efficiency, material utilization, and emissions (Kokare, Godina, et al., 2023). For example, L-PBF offers higher resolution and material utilization but tends to have higher power consumption due to longer build times (Zhao et al., 2022). DED, on the other hand, enables faster builds with less energy per part but may suffer from increased material overspray and waste (Murtaza et al., 2023). LCA provides a structured method for quantifying these differences and determining which technology is best suited for specific sustainability objectives (Meier et al., 2018).

A critical addition to this discussion involves integrating Social Life Cycle Assessment (S-LCA) and Life Cycle Costing (LCC) (Ghosh, 2017). While traditional LCA focuses on environmental burdens, S-LCA evaluates the social impacts on workers, communities, and consumers throughout the product's lifecycle (Kulkarni, 2018). Likewise, LCC addresses the economic costs across the same span (Sun et al., 2025).

According to recent meta-analyses, only a small fraction of studies incorporate these dimensions, despite their relevance in decision-making (Martelli et al., 2024). For instance, the increased capital expenditure of AM machines may be offset by reduced labor costs and higher part throughput, which can be captured through LCC (Zhao et al., 2022).

Compiling comprehensive Life Cycle Inventory (LCI) data is a crucial step toward accurate assessments (J. Li & Zhang, 2023). This includes detailed inputs and outputs: powder production methods, inert gas usage, part rejections, support structures, energy per build layer, and emissions from auxiliary systems like pre-heating or vacuum pumps (Kokare et al., 2023). Kokare note that existing datasets are fragmented and not sufficiently detailed, leading to uncertainty in LCA results

(Garcia et al., 2021). They advocate for centralized databases and standardized reporting frameworks for AM-related LCI data.

In conclusion, applying LCA to superalloy-based AM processes provides a holistic lens through which to evaluate their sustainability (Kellens, Baumers, et al., 2017). While AM offers clear benefits in reducing waste and enabling lightweight designs, these must be weighed against high energy consumption and material production impacts (Motalebi et al., 2024). To advance sustainable prototyping, future research must prioritize complete LCAs incorporating environmental, social, and economic dimensions, backed by robust LCI datasets and real-world case studies (Výtisk et al., 2019). Only then can we truly assess whether AM represents a sustainable alternative to conventional methods for producing high-performance superalloy components (Agrawal et al., 2022).

SUSTAINABLE DESIGN STRATEGIES: TOPOLOGY OPTIMIZATION AND MATERIAL EFFICIENCY

Sustainable prototyping has evolved beyond merely adopting greener materials or reducing energy consumption—it now encompasses intelligent design methodologies that strategically minimize material usage without compromising structural integrity (Peng et al., 2018). One of the most impactful strategies in this context is topology optimization, a computational design approach that algorithmically determines the most efficient distribution of material within a given design space (Ford & Despeisse, 2016).

By systematically eliminating underutilized material, topology optimization creates lightweight and structurally sound geometries that inherently promote sustainability.

When this design technique is integrated with additive manufacturing (AM), the synergy results in highly optimized parts that reduce waste, minimize post-processing needs, and improve mechanical performance.

Topology optimization enables engineers and designers to approach the design process with a performance-first mindset, prioritizing structural efficiency and environmental impact simultaneously (Holmström et al., 2010). Rather than working within the constraints of traditional manufacturing, AM allows for the physical realization of topologically optimized geometries, including organic shapes, internal lattices, and complex curves. These shapes are often impossible or prohibitively expensive to produce with conventional subtractive or formative methods. As such, AM becomes the enabler of topology optimization's full potential, unlocking opportunities for innovation in weight reduction, part consolidation, and thermal management. The resulting designs are not only functionally superior but also more sustainable throughout their lifecycle (Fera et al., 2016).

Generative design tools have further accelerated this paradigm shift (Mertens et al., 2017). Leveraging artificial intelligence and simulation-driven processes, generative design explores thousands of design permutations based on input parameters such as loads, boundary conditions, materials, and cost constraints. This allows engineers to evaluate and compare designs not only on mechanical performance but also on sustainability metrics. When sustainability criteria are explicitly included such as material carbon footprint, recyclability, or end-of-life disassembly the resulting designs align with broader environmental goals. This capability is increasingly important as industries face mounting regulatory pressure and consumer demand for eco-conscious products. A compelling advancement in this space is the incorporation of multilateral topology optimization, which allows the concurrent use of dissimilar materials such as metals and bio-based alternatives within a single optimized structure (Wong & Hernandez, 2012). For instance, the integration of isotropic materials like steel with transversely isotropic such as laminated bamboo or timber creates structures that harness the strengths of each. Steel provides high strength in load-bearing regions, while bio-based materials contribute to reduced environmental impact in low-stress zones. This approach allows for an ideal balance between mechanical performance and sustainability, reducing embodied carbon and energy usage without sacrificing functionality. Research by Kundu and Zhang (2025) provides a comprehensive framework for sustainability-oriented multi material topology optimization. Their study demonstrates how structures can be designed to optimize performance while accounting for environmental and economic impacts. Utilizing the Tsai–Wu failure criterion for anisotropic biomaterials and von Mises yield criterion for isotropic materials, the framework effectively guides material placement in both 2D and 3D contexts. The results confirm that hybrid designs can significantly reduce global warming potential (GWP), primary energy demand, and other environmental burdens. Bio-based materials, due to their carbon sequestration capacity and renewability, also contribute positively to lifecycle assessments when compared to traditional metals (Peng, Kellens, et al., 2018). Examples from real-world architecture and infrastructure further validate the viability of such strategies (Ford & Despeisse, 2016). Projects like the River Beech Tower in Chicago, Oakwood Tower in London, and San Mateo County headquarters in Silicon Valley showcase how timber and bamboo are gaining prominence in high-performance construction (Holmström et al., 2010). These materials have shown high strength-to-weight ratios and durability suitable for mid-to-high-rise buildings (Peng, Tang, et al., 2018). Their usage in generative and topology-optimized designs ensures minimal material redundancy and efficient stress distribution, which in turn reduces the ecological footprint of buildings over their operational life (Tiwari & Turner, 2019).

Moreover, AM's layer-wise fabrication process facilitates the selective deposition of different materials, making it well-suited for implementing multimaterial designs. Emerging hybrid AM systems can transition between metallic and non-metallic feedstocks, enabling the creation of structurally integrated components with tailored local properties (Zhu & Dhokia, 2021). When paired with sustainability-aware optimization algorithms, these technologies produce next-generation parts. These parts are both environmentally responsible and structurally elite. Beyond optimizing structures for mechanical or functional performance, sustainability advantages include better production efficiency (Liu et al., 2023). Optimized designs reduce build volume and time, lowering energy use per part. Additionally, by reducing the need for support structures and post-processing, the material and energy inputs are further minimized. In industries like aerospace and automotive, every gram saved aids fuel efficiency and emissions, making these benefits both economic and ecological.

Despite these advances, challenges remain for mainstreaming topology optimization for sustainability.

Bio-based materials, derived from renewable biological sources, often exhibit anisotropic (direction-dependent) and heterogeneous (variable) properties, making accurate modeling and simulation more complex. High-fidelity characterization methods (accurate measurement techniques) and robust interpolation schemes (mathematical methods to estimate unknown values) are required to ensure that the optimized designs work as intended. Current design standards and codes are not always compatible with unconventional geometries or hybrid material systems. This necessitates updates in regulatory frameworks and validation protocols (Ngo et al., 2018).

Nonetheless, integrating topology optimization with AM and sustainable material selection holds great promise for a more circular and responsible design culture. It promotes a shift from linear production models toward a holistic view of the product lifecycle. In this view, resource efficiency, durability, and environmental impact are factored into the design from the outset. As computational power grows and digital tools become more accessible, these methods will likely be adopted across more fields, from aerospace and automotive to civil engineering and consumer products.

Sustainable design is not just constraints, but an opportunity for innovation. By combining the computational rigor of topology optimization, the freedom of AM, and the ecological benefits of alternative materials, designers and engineers can innovate. They can create products and structures that meet modern performance demands while preserving the planet's resources for future generations (Kellens, Dewulf, et al., 2017). This approach turns sustainability from a reactive obligation into a proactive design philosophy, enabling the next era of engineering excellence.

WASTE MINIMIZATION AND POWDER RECYCLING IN ADDITIVE MANUFACTURING

Waste minimization remains a cornerstone of sustainable manufacturing, and within the domain of additive manufacturing (AM), this principle takes on renewed importance (Motalebi et al., 2024). While AM is celebrated for its inherent efficiency in material utilization compared to traditional subtractive techniques, it is not devoid of material waste (Kokare, Ghosh, et al., 2023). One of the most pressing concerns in this context is the fate of the unused or degraded powder, particularly in powder bed fusion (PBF) and direct energy deposition (DED) systems (Martínez-Gómez & Morán, 2022). In high-value applications such as those involving superalloys, the economic and environmental implications of powder degradation, contamination, and inefficient recycling can be substantial (Schmidt et al., 2021).

Powder feedstock in AM, especially metal powders used for fabricating high-performance components, constitutes a significant portion of the process cost and environmental impact (Reith et al., 2020). These powders are typically produced via gas atomization or plasma atomization, processes which are energy-intensive and yield materials with highly controlled particle size distributions and spherical morphologies to ensure optimal flowability and packing density (Ford & Despeisse, 2016). However, during the AM process, a considerable portion of the powder remains un-melted or unused in each build (Holmström et al., 2010a). This residual powder is typically collected and reused, but with each cycle, its physical and chemical properties may change (Spierings et al., 2013). These changes, if not managed effectively, may compromise part quality, mechanical performance, and process repeatability (Slotwinski et al., 2014).

Several factors contribute to powder degradation. Chief among them is oxidation, which can occur when powders are exposed to atmospheric oxygen and moisture during handling or between build cycles (Martin et al., 2014). Oxidation can form surface films on powder particles, impairing their melting behavior and wetting characteristics (King, 2015). In addition, repeated thermal cycling during successive builds can alter the microstructure of metal powders, resulting in phase transformations or grain coarsening (Herzog et al., 2016). For superalloys, which are sensitive to composition and microstructural uniformity, such changes can significantly affect the resultant mechanical properties of printed parts]. This necessitates rigorous monitoring and quality assurance protocols for reused powder.

The implementation of powder recycling strategies thus requires a comprehensive understanding of powder behavior across multiple reuse cycles. Research has shown that with proper handling and storage, many metal powders, including nickel-based and cobalt-based superalloys, can be reused for several build cycles without significant degradation. Key to this is the employment of closed-loop powder handling systems, which minimize environmental exposure and mechanical degradation. Inert gas environments, such as those employing argon or nitrogen, are commonly used in both printing chambers and storage vessels to limit oxidation and moisture absorption (Fera, 2016).

Quality assurance protocols for reused powder often include techniques such as laser diffraction for particle size distribution analysis, scanning electron microscopy (SEM) for morphological assessment, and X-ray photoelectron spectroscopy (XPS) or energy-dispersive X-ray spectroscopy (EDS) for surface chemistry evaluation. In industrial practice, simpler and faster methods, such as flowability testing (e.g., Hall flowmeter) and apparent density measurements, are frequently employed for routine checks. However, these tests may not capture subtle but critical changes in powder chemistry, underlining the importance of periodic detailed characterization, especially for high-stakes applications (Ford & Despeisse, 2016).

Contamination is another pivotal issue in powder reuse (Holmström et al., 2010). Foreign particle inclusion whether from tooling, build platforms, or previous builds involving different alloys can significantly alter the chemical composition of the powder bed and result in inclusions, porosity, or even phase destabilization in the final part (Peng, Kellens, et al., 2018). Cross-contamination is particularly problematic when transitioning between different alloy systems in the same AM system (Motalebi et al., 2024). Best practices to mitigate this include thorough cleaning of the build chamber, use of dedicated powder sieving and storage equipment, and, where feasible, the segregation of equipment by material type (Kokare, Ghosh, et al., 2023).

One promising approach to enhancing powder reuse sustainability is the development of in-situ powder monitoring and conditioning technologies (Martínez-Gómez & Morán, 2022). These systems, integrated into the AM machine or auxiliary equipment, can continuously assess powder quality during or immediately after a build (Schmidt et al., 2021). Some advanced systems employ spectroscopic sensors to detect oxide levels, particle size anomalies, or foreign contaminants in real time, enabling early intervention and decision-making regarding powder recyclability (Reith et al., 2020). Combined with machine learning algorithms trained on historical powder behavior and part performance data, these technologies represent a step toward fully autonomous, intelligent AM systems capable of optimizing powder usage dynamically (Ford & Despeisse, 2016).

Furthermore, understanding the limits of powder reuse is crucial. Empirical studies and industrial standards often recommend a maximum number of reuse cycles for a given powder batch, beyond which its mechanical and chemical fidelity can no longer be guaranteed (Spierings et al., 2013). This threshold varies depending on the alloy, AM technology, build parameters, and part criticality (Slotwinski et al., 2014). For example, in aerospace-grade nickel superalloys, the allowable number of reuse cycles is often conservative due to stringent certification requirements (Martin et al., 2014). However, lifecycle analysis (LCA) studies suggest that even with limited reuse, the net environmental benefit of AM over subtractive methods remains significant (King, 2015).

The recyclability of powder also depends on the efficiency of powder recovery systems (Herzog et al., 2016b). Sieving plays a central role in this process, as it removes oversized particles and agglomerates that may have formed during the build (Ngo et al., 2018). Advanced vibratory sieves and ultrasonic sieving systems are employed to maintain consistent powder quality (Zafar et al., 2025). In addition to mechanical sieving, magnetic separation is used to eliminate ferromagnetic contaminants in non-ferrous alloy systems (Tiwari & Turner, 2019). However, over-sieving can reduce powder yield and inadvertently remove acceptable particles, thereby increasing net material waste (Zhu & Dhokia, 2021). Thus, optimization of sieving parameters is essential to balance quality control with material retention (Liu et al., 2023).

From an economic perspective, powder reuse significantly lowers production costs, especially when working with expensive superalloys like Inconel 718 or Haynes 282 (Fera, 2016). By reducing the demand for virgin powder, manufacturers can decrease both procurement costs and the environmental burdens associated with powder production (Peng, Kellens, et al., 2018). This is particularly beneficial in low-to-medium production volumes, where AM is most often deployed for prototyping, tooling, or customized parts (Agrawal, 2022).

Despite these advantages, regulatory frameworks and qualification standards for powder reuse in AM are still evolving (Ford & Despeisse, 2016). Aerospace and medical industries, in particular, require stringent documentation and validation of powder reuse protocols (Holmström et al., 2010b). Standardization efforts by organizations such as ASTM International and ISO are ongoing, aiming to establish universally accepted guidelines for powder recycling and reuse. These include criteria for powder acceptance, methods for assessing degradation, and protocols for recording reuse history (Motalebi et al., 2024).

Table 3. Key factors affecting powder degradation and recyclability in AM

Factor	Mechanism	Impact on Powder	Mitigation Strategy
Oxidation	Exposure to air/humidity during handling	Surface film formation; poor melting	Inert gas environments; closed-loop systems
Thermal Cycling	Repeated heating/cooling during builds	Microstructural changes; grain coarsening	Limit reuse cycles; thermal analysis protocols
Contamination	Tooling residue; cross-material builds	Inclusion formation; chemical inconsistencies	Material segregation; dedicated sieving/storage

continued on following page

Table 3. Continued

Factor	Mechanism	Impact on Powder	Mitigation Strategy
Particle Size Changes	Agglomeration or fragmentation during handling	Altered flowability; packing inefficiencies	Controlled sieving; ultrasonic sieving; flowability tests
Chemical Composition Drift	Evaporation or reaction of alloying elements	Mechanical property variation	XPS/EDS monitoring; regular chemical assays

Table 3 summarizes the main degradation mechanisms that affect powder reuse in AM, along with recommended mitigation strategies to ensure print quality and process reliability. In the context of sustainable manufacturing, minimizing powder waste aligns directly with broader goals such as reducing carbon emissions, conserving resources, and lowering the total environmental footprint of manufacturing operations. Life cycle assessment tools have quantified the benefits of powder recycling, demonstrating notable reductions in energy consumption and greenhouse gas emissions per unit part produced. These gains are amplified when recycling is coupled with topology optimization and design for additive manufacturing (DfAM), which further reduce material usage and build time (Schmidt et al., 2021).

In conclusion, while additive manufacturing offers substantial advantages in material efficiency compared to conventional methods, true sustainability demands a holistic approach that encompasses not only design and process optimization but also meticulous management of feedstock materials. Powder recycling and reuse, particularly for superalloys, is a linchpin in this strategy. Through advanced monitoring, quality assurance, and process integration, the potential of AM to deliver both high performance and low environmental impact can be fully realized. Ongoing research, standardization efforts, and technological innovation will play critical roles in refining powder reuse practices, making them more reliable, economical, and environmentally sound across a broader spectrum of applications (Spierings et al., 2013).

Industrial Scalability and Standardization of Sustainable AM Practices

The successful mainstream adoption of sustainable additive manufacturing (AM) practices, particularly for processing high-performance materials such as superalloys, is contingent upon the development of scalable, standardized, and certifiable systems. Although significant strides have been made in laboratory and pilot-scale environments, transitioning these innovations into full-scale industrial applications requires alignment across multiple axes technical reliability, regulatory compliance, supply chain robustness, and workforce competency (Martin et al., 2014). Moreover, sustainability considerations must be embedded within these frameworks to ensure that environmental benefits are not lost in translation from prototype to mass production (King, 2015). This section analyzes the multilayered structure of industrial scalability and standardization, with a focus on how such efforts reinforce sustainable practices within the context of AM (Herzog et al., 2016).

Quality Control and Process Repeatability in AM

One of the primary barriers to industrial scalability in AM lies in achieving process repeatability and consistent part quality. Unlike traditional manufacturing techniques, which have undergone decades of standardization, AM processes are highly sensitive to variations in input materials, build parameters, and environmental conditions. These variabilities can significantly affect the microstructure and mechanical properties of AM-fabricated superalloy parts, thus impeding their adoption in critical sectors such as aerospace and medical implants (Tiwari & Turner, 2019). Establishing robust quality control mechanisms is vital for mitigating these variabilities. In-process monitoring systems, including melt pool sensors, thermal cameras, and layer-wise imaging, are increasingly integrated into AM platforms to detect anomalies in real time. Coupled with machine learning algorithms, these systems enable predictive maintenance and adaptive control, ensuring more consistent outputs. Post-process characterization—such as X-ray computed tomography (CT), scanning electron microscopy (SEM), and mechanical testing—provides further assurance of part integrity. Nonetheless, for such practices to contribute meaningfully to sustainability, they must also be energy-efficient and material-conscious. Excessive post-processing, for instance, can negate the environmental advantages of AM by introducing additional resource demands. Thus, process repeatability not only enhances manufacturability but also aligns with broader ecological goals by minimizing scrap, reducing energy consumption, and extending component lifespan.

Standardization Bodies and Regulatory Frameworks

A critical enabler of industrial scalability is the development and implementation of internationally recognized standards. Organizations such as ASTM International (notably ASTM F42 Committee on Additive Manufacturing Technologies), the International Organization for Standardization (ISO/TC 261), and the European Committee for Standardization (CEN) have played pivotal roles in codifying best practices, terminology, and performance metrics for AM technologies (Zhu & Dhokia, 2021). Key standards such as ISO/ASTM 52900:2021 define general principles and classification of AM processes, while more application-specific documents, including ISO/ASTM 52904:2019 (requirements for metal powder bed fusion), set criteria for feedstock quality, machine qualification, and environmental controls. These standards foster interoperability, facilitate certification procedures, and promote industry-wide confidence in AM-fabricated parts. Moreover, sustainability-specific frameworks are emerging. ISO 14040 and ISO 14044, which outline life cycle assessment (LCA) principles and requirements, are increasingly being applied to evaluate the environmental impacts of AM workflows. Their integration into AM standardization efforts allows stakeholders to quantify ecological benefits and benchmark against conventional methods. The harmonization of standards across geographical regions and industry sectors is essential for global adoption (Duraivelu, 2022). Discrepancies in regulatory requirements—such as those between the U.S. Federal Aviation Administration (FAA) and the European Union Aviation Safety Agency (EASA)—can hinder cross-border applications and inflate costs. Therefore, international cooperation in developing unified standards is paramount to scaling sustainable AM practices (Tiwari & Turner, 2019).

Economic and Infrastructural Challenges

While the technical merits of AM are well-documented, the economic feasibility of scaling these technologies remains a significant hurdle. Capital investment in high-performance AM systems particularly those capable of processing superalloys via selective laser melting (SLM), electron beam melting (EBM), or direct energy deposition (DED) can be prohibitive. Additionally, operational expenditures associated with maintenance, powder handling, environmental controls, and personnel training can dissuade smaller enterprises from adopting AM (Ngo et al., 2018). To overcome these obstacles, government incentives, public-private partnerships, and consortia-driven R&D initiatives have proven effective. Programs such as the National Additive Manufacturing Innovation Institute (America Makes) and the EU's Horizon Europe framework have catalyzed technology diffusion and knowledge transfer. Furthermore, decentralized manufacturing models enabled by AM, including localized production cells and digital part inventories, offer economic resilience and logistical efficiency, further supporting sustainable deployment (Szmelter-Jarosz, 2026). Infrastructure development must parallel technological progress (Duraivelu, 2022). This includes establishing digital workflows that integrate computer-aided design (CAD), simulation, build preparation, process monitoring, and post-processing in a cohesive pipeline (Tanaka & Singh, 2019). Interoperable software platforms and standardized data formats are crucial for minimizing error propagation and maximizing traceability (M. Green & Hall, 2023). In the context of sustainability, digital infrastructure can also facilitate real-time environmental impact tracking, enabling dynamic optimization of energy use and material consumption.

Human Capital and Workforce Training

Scalability also depends heavily on cultivating a skilled workforce capable of navigating the complex interplay of materials science, mechanical engineering, software proficiency, and environmental stewardship inherent to sustainable AM. Educational institutions must update curricula to include AM principles, sustainable design methodologies, and regulatory literacy. Industry-led training programs, certification courses, and apprenticeships can further bridge the skills gap. Professional development must also extend to life cycle thinking, enabling engineers and technicians to make informed decisions that balance performance, cost, and ecological impact (Singh et al., 2021). Cross-disciplinary expertise, such as combining knowledge of powder metallurgy with life cycle assessment methodologies, will be especially valuable in optimizing AM processes for superalloys. Encouraging diversity and inclusivity in AM education and employment pathways also ensures that innovation draws from a wide talent pool and addresses a broader spectrum of societal needs.

Data Transparency and Continuous Improvement

Standardization, by its very nature, supports transparency. When processes are clearly defined and consistently documented, it becomes feasible to collect, compare, and analyze data across different production contexts. This transparency underpins continuous improvement cycles, which are vital for embedding sustainability within AM practices. Environmental performance indicators—such as energy intensity (MJ/kg), material utilization rates (%), and greenhouse gas emissions (kg CO_2-eq)—should be routinely captured and reported as part of quality assurance documentation. Benchmarking these indicators across different machines, facilities, and use cases can reveal best practices and guide future process optimization (Johnson & Lee, 2021). Digital twins and blockchain-based traceability systems are emerging as powerful tools for enhancing transparency. These technologies can track a component's entire life cycle, from powder production and build conditions to end-of-life scenarios, enabling comprehensive environmental assessments. Moreover, they support accountability and compliance with environmental regulations, such as the European Union's Corporate Sustainability Reporting Directive (CSRD) or the U.S. Environmental Protection Agency's emissions tracking requirements. Integrating environmental considerations into key performance indicators (KPIs) not only aligns with regulatory trends but also appeals to environmentally conscious consumers and investors (A. Green & Jones, 2023). As sustainability becomes a competitive differentiator, companies that demonstrate leadership in standardized sustainable AM practices will likely gain market advantage.

Achieving industrial scalability and standardization of sustainable additive manufacturing practices is both a technical and organizational endeavor. It necessitates the convergence of precise quality control, harmonized standards, economic viability, skilled human capital, and transparent data ecosystems. While challenges persist, particularly in cost, infrastructure, and regulatory alignment, ongoing efforts by international bodies, academic institutions, and industry leaders are gradually transforming AM from a niche technology into a cornerstone of sustainable prototyping and manufacturing. As superalloy applications continue to expand in high-stakes sectors, the imperative for scalable, standardized, and sustainable AM practices becomes increasingly urgent. By addressing these interconnected dimensions holistically, the additive manufacturing community can unlock the full potential of this transformative technology for a more resilient and responsible industrial future(Zhang et al., 2023).

CASE STUDIES: SUSTAINABLE PROTOTYPING IN AEROSPACE AND ENERGY APPLICATIONS

In the context of accelerating technological development and mounting environmental concerns, the aerospace and energy sectors have emerged as key domains for the application of sustainable prototyping methodologies using superalloys. These industries operate under exceptionally high performance standards, requiring components that withstand extreme temperatures, stress, and corrosive environments. Simultaneously, they face intense scrutiny to reduce carbon emissions, energy consumption, and material waste. Additive manufacturing (AM) has presented a viable pathway toward reconciling these demands, offering flexibility, precision, and material efficiency that traditional subtractive techniques cannot easily match (Gibson et al., 2021). Through an examination of industrial case studies from GE Additive, Siemens Energy, and Rolls-Royce, this section provides a deeper understanding of how sustainable prototyping strategies are being realized in high-performance applications (Ngo et al., 2018).

GE Additive has become a global benchmark for integrating sustainability into AM operations. A hallmark achievement is the development of the LEAP fuel nozzle, used in next-generation aircraft engines. Manufactured using laser powder bed fusion (L-PBF), this component exemplifies the consolidation of complex assemblies into monolithic structures. Traditionally composed of over 20 parts, the AM version is a single piece that is 25% lighter and five times more durable (Herzog et al., 2016). The resultant gains in fuel efficiency directly contribute to lower greenhouse gas emissions, while the simplified production process significantly reduces machining waste and energy usage. To further enhance sustainability, GE has implemented closed-loop powder management systems that recover, assess, and reintegrate unused powder, thus minimizing material loss (Baufeld et al., 2011). Quality control protocols, including real-time process monitoring and in-situ defect detection, ensure that recycled powders do not compromise the mechanical integrity of the final product (Ferrari et al., 2021).

Siemens Energy offers another powerful example of sustainable innovation in AM, particularly in its efforts to repair and manufacture gas turbine components. One of the company's pivotal accomplishments is the refurbishment of burner heads via directed energy deposition (DED). This technique enables precise material deposition only where needed, allowing for worn parts to be rebuilt rather than replaced. This paradigm shift from replacement to repair significantly lowers the environmental and economic costs associated with component lifecycle management. Siemens' adoption of advanced analytics and digital twins facilitates predictive maintenance and resource optimization, ensuring that material properties remain within specification even after multiple reuse cycles (Garmendia et al., 2020). In tandem with powder recycling protocols, these digital frameworks serve to reduce production downtime, raw material input, and overall environmental burden (Kumar et al., 2024).

Rolls-Royce, a longstanding leader in the aerospace industry, has also embraced AM to achieve superior performance and environmental outcomes. The company's use of electron beam melting (EBM) to produce lightweight turbine blades illustrates the capacity of AM to transform complex component design. These blades are engineered with intricate internal cooling channels that enhance thermal management and efficiency (Murr et al., 2012). The reduced mass contributes to lower fuel consumption, while the ability to fabricate net-shape parts diminishes the need for extensive post-processing. Rolls-Royce's comprehensive approach includes integrating life-cycle assessment (LCA) tools to evaluate the environmental footprint at each stage of the product's journey from raw material procurement to end-of-life disposal. The use of such metrics underscores a data-driven commitment to minimizing ecological impact (Park & Kim, 2022).

Across these industrial narratives, a common theme emerges: sustainable prototyping through AM is not an isolated intervention but a systems-level strategy that touches every aspect of the production ecosystem. These companies do not merely adopt AM as a manufacturing innovation; they restructure design philosophies, supply chain logistics, and product development cycles around it. The success of these implementations is underpinned by a number of enabling factors. First, digital infrastructure plays a critical role. Process modeling, build simulation, and quality assurance systems are integrated to achieve consistency and predictability across production runs (Xu, 2023). Second, circular economy principles are actively embedded in operational practices—material reuse, part reparability, and modular design are prioritized over linear, waste-intensive models (Lacy & Rutqvist, 2015). Third, adherence to international sustainability standards, such as ISO 14001, and alignment with global frameworks like the UN Sustainable Development Goals lend credibility and transparency to corporate sustainability claims (Choi et al., 2020).

Another notable feature across these case studies is the emphasis on cross-disciplinary collaboration. Achieving meaningful sustainability outcomes in prototyping requires more than technical optimization. It necessitates the engagement of diverse expertise—materials science, mechanical engineering, environmental systems analysis, logistics, and even policy and regulatory affairs. For example, GE's LEAP fuel nozzle program involved not only additive manufacturing specialists but also combustion engineers, aerodynamicists, and sustainability analysts working in tandem (King et al., 2015). Siemens Energy's turbine repair strategies rely on metallurgists and software engineers co-developing integrated repair protocols (Jones & Seepersad, 2016). Rolls-Royce's lifecycle assessments are conducted in collaboration with academic institutions and third-party auditors to ensure methodological rigor (Zuo & Zhao, 2019).

These examples underscore the value of leveraging AM for functional innovation in tandem with environmental stewardship. Notably, the benefits are not restricted to environmental metrics alone. Companies have reported improvements in production agility, customization capability, time-to-market, and supply chain resilience (Gebhardt & Hötter, 2016). These operational advantages, when coupled with sustainability, enhance corporate competitiveness and brand positioning in increasingly eco-conscious markets. Moreover, the scalability of these practices points to their potential applicability in other sectors such as automotive, medical devices, and heavy industry (Nidumolu et al., 2009).

Yet, the path to widespread industrial adoption is not without challenges. Economic barriers persist, particularly the high capital costs associated with AM hardware, software, and material qualification (Guo & Leu, 2013). Regulatory uncertainty around the certification of AM parts, especially in safety-critical applications, also presents hurdles. However, these challenges are being actively addressed through industry consortia, public-private partnerships, and evolving standards. For instance, the collaboration between GE Additive and regulatory bodies has facilitated the development of guidelines for certifying AM aerospace components. Siemens Energy has worked closely with energy regulators to establish best practices for the use of AM in high-temperature applications (Thompson et al., 2016a).

Furthermore, the cultural shift toward sustainability must be cultivated within organizations. This includes workforce training, rethinking design methodologies, and fostering a long-term vision that integrates sustainability into corporate strategy. Investment in human capital is as critical as investment in technology. Leading companies have established AM centers of excellence and sustainability think tanks to spearhead these efforts (Bocken et al., 2016).

In sum, the deployment of sustainable prototyping practices within the aerospace and energy sectors exemplifies how additive manufacturing can serve as a fulcrum for both technological advancement and ecological responsibility (Garetti & Taisch, 2012). The experiences of GE Additive, Siemens Energy, and Rolls-Royce offer not only compelling proof-of-concept but also a roadmap for other industries aspiring to embrace sustainable innovation (Gebler et al., 2014). Through a combination of material efficiency, lifecycle thinking, interdisciplinary collaboration, and digital integration, these organizations demonstrate that sustainability and high performance are not mutually exclusive but mutually reinforcing (Herrmann et al., 2020). As global industries confront the urgent need to decarbonize and operate within planetary boundaries, the role of additive manufacturing in sustainable prototyping will only grow in importance, serving as both a tool and a philosophy for the next generation of industrial innovation (Berman, 2012).

FUTURE DIRECTIONS: INNOVATION PATHWAYS FOR SUSTAINABLE SUPERALLOY PROTOTYPING

In the coming decades, the pursuit of sustainable prototyping with superalloys will be significantly shaped by cross-disciplinary innovation, particularly at the confluence of advanced material science, additive manufacturing (AM) technologies, and intelligent digital systems (Kumar et al., 2024). The demand for higher material performance, lower environmental impact, and agile design-to-deployment cycles has compelled researchers and industries to reevaluate the entire lifecycle of superalloy-based components from alloy composition to post-processing and has catalyzed the integration of emergent tools such as machine learning, real-time process monitoring, and circular manufacturing frameworks (B. Li et al., 2024).

One of the most impactful areas of future research is the development of next-generation green superalloys tailored for AM applications (Wang et al., 2023). Traditional superalloys, such as Inconel 718 and Haynes 282, though robust in high-temperature and high-stress environments, often contain critical raw materials (CRMs) such as cobalt, tungsten, and tantalum, whose extraction and refinement are associated with high environmental and geopolitical costs (Szmelter-Jarosz, 2026). Future alloy design thus centers on minimizing or substituting CRMs without compromising mechanical integrity, oxidation resistance, or creep strength (Zhao et al., 2022). Recent studies have demonstrated the potential of computational thermodynamics and integrated computational materials engineering (ICME) frameworks in discovering new alloys with lower embodied energy and enhanced recyclability

(Patel, 2024). For instance, high-entropy alloys (HEAs) have gained interest due to their tunable compositions and microstructural stability under extreme conditions. HEAs such as AlCoCrFeNi-based systems have been found to exhibit superior strength-to-weight ratios and thermal fatigue resistance, making them viable candidates for sustainable aerospace applications (Lee et al., 2007).

Simultaneously, digital twin technology and in-situ process monitoring are transforming the feedback mechanisms within AM systems (Chen, 2020). These technologies, which enable virtual replication and real-time data acquisition during the build process, are essential for ensuring product quality and minimizing rework or scrap (Singh et al., 2021). Integrating optical sensors, thermal imaging, and acoustic emission monitoring with predictive machine learning models allows for closed-loop control of critical variables such as laser power, scanning speed, and powder feed rates (Jiang et al., 2022). GE Research has pioneered such systems in their ATLAS program, which demonstrated that in-situ monitoring can reduce scrap rates by up to 30% while ensuring mechanical property consistency across builds.In the coming decades, the pursuit of sustainable prototyping with superalloys will be significantly shaped by cross-disciplinary innovation, particularly at the confluence of advanced material science, additive manufacturing (AM) technologies, and intelligent digital systems (Kumar et al., 2024). The demand for higher material performance, lower environmental impact, and agile design-to-deployment cycles has compelled researchers and industries to reevaluate the entire lifecycle of superalloy-based components—from alloy composition to post-processing and has catalyzed the integration of emergent tools such as machine learning, real-time process monitoring, and circular manufacturing frameworks (B. Li et al., 2024).

One of the most impactful areas of future research is the development of next-generation green superalloys tailored for AM applications (Wang et al., 2023). Traditional superalloys, such as Inconel 718 and Haynes 282, though robust in high-temperature and high-stress environments, often contain critical raw materials (CRMs) such as cobalt, tungsten, and tantalum, whose extraction and refinement are associated with high environmental and geopolitical costs (Szmelter-Jarosz, 2026). Future alloy design thus centers on minimizing or substituting CRMs without compromising mechanical integrity, oxidation resistance, or creep strength (B. Li et al., 2024). Recent studies have demonstrated the potential of computational thermodynamics and integrated computational materials engineering (ICME) frameworks in discovering new alloys with lower embodied energy and enhanced recyclability. For instance, high-entropy alloys (HEAs) have gained interest due to their tunable compositions and microstructural stability under extreme conditions (Zhang et al., 2023). HEAs such as AlCoCrFeNi-based systems have been found to exhibit superior strength-to-weight ratios and thermal fatigue resistance, making them viable candidates for sustainable aerospace applications (Lee et al., 2007).

Simultaneously, digital twin technology and in-situ process monitoring are transforming the feedback mechanisms within AM systems (Chen et al., 2005). These technologies, which enable virtual replication and real-time data acquisition during the build process, are essential for ensuring product quality and minimizing rework or scrap (Singh et al., 2021). Integrating optical sensors, thermal imaging, and acoustic emission monitoring with predictive machine learning models allows for closed-loop control of critical variables such as laser power, scanning speed, and powder feed rates (Ramirez & Thompson, 2022). GE Research has pioneered such systems in their ATLAS program, which demonstrated that in-situ monitoring can reduce scrap rates by up to 30% while ensuring mechanical property consistency across builds (Choi et al., 2020).

Table 4. Key emerging innovation areas for sustainable superalloy prototyping

Innovation Area	Description	Current Example/Status	Impact on Sustainability
Green Superalloys	Development of CRM-free or reduced alloys designed for AM	HEAs like AlCoCrFeNi under investigation for aerospace use	Lower embodied energy and resource dependency
Process Monitoring & AI	Real-time sensing integrated with ML algorithms for predictive control	GE's ATLAS and Siemens' neural network models	Reduced scrap, enhanced reliability, less rework
Renewable Energy Integration	Use of solar/wind to power AM facilities	Fraunhofer LCA shows 50% emission reduction with 70% solar energy	Lower operational carbon footprint
Digital Powder Traceability	Blockchain or digital tagging systems for powder tracking	Pilot initiatives in aerospace-grade AM	Ensures reuse accountability and quality consistency
Circular Manufacturing Models	Remanufacturing, re-alloying, and closed-loop systems	Rolls-Royce Revert Program (60% material recovery)	Drastic material waste and resource reduction

As shown in Table 4, sustainability-driven innovations are emerging across all stages of the AM lifecycle, from powder design to infrastructure and end-of-life strategies. Artificial intelligence (AI) and machine learning (ML) algorithms are not only advancing quality assurance but also facilitating process parameter optimization. Traditional trial-and-error or design-of-experiments (DOE) approaches to AM are often resource-intensive and time-consuming (Lee et al., 2007). In contrast, ML models trained on historical data can predict optimal process windows for specific superalloy compositions, part geometries, and environmental conditions (Chen et al., 2005). In one example, Siemens Digital Industries Software developed a neural network model capable of optimizing laser parameters for L-PBF in under 10 minutes, reducing build failures by 40% (Ahmed & Zaneldin, 2020). Furthermore, reinforcement learning approaches are being tested to dynamically adapt AM process parameters in response to in-situ feedback, creating a self-correcting manufacturing environment (Thompson et al., 2016).

In terms of infrastructure, the adoption of renewable energy sources in AM facilities is becoming a priority. Since AM processes are typically energy-intensive—particularly during laser melting and heat treatment stages—coupling them with photovoltaic, wind, or hydroelectric energy supplies is a practical route toward decarbonization (Singh et al., 2021). A lifecycle assessment (LCA) study conducted by Fraunhofer Institute showed that sourcing 70% of AM facility power from solar energy could reduce the carbon footprint of Ti6Al4V part production by 50% (Garcia et al., 2021b) . Companies like Norsk Titanium and EOS GmbH have started pilot programs incorporating renewable energy and carbon offsetting strategies into their manufacturing workflows, setting a precedent for sustainable infrastructure development.

A complementary direction involves enhancing powder reuse and developing robust powder qualification standards (Duraivelu, 2022). As powder degradation due to oxidation, contamination, or particle morphology changes can compromise part quality, ensuring consistent reuse is critical (Johnson & Lee, 2021). Several studies have shown that with proper sieving, blending, and storage practices, Inconel 625 powder can be reused for up to 10 cycles without significant loss of performance (Brown, 2021). ASTM F3049-14 and ISO/ASTM 52907 standards provide foundational guidance, but further refinement is needed to account for next-gen alloys and hybrid processes. Powder traceability platforms using blockchain or digital tagging are also emerging to ensure provenance, lifecycle data, and performance metrics are securely recorded (Patel, 2023).

Moreover, circular manufacturing models, where materials and components are continuously looped back into the production system, are gaining traction (Szmelter-Jarosz, 2026). These models not only minimize waste but also reduce reliance on virgin raw materials (Zhang et al., 2023). For example, Rolls-Royce's Revert Program collects superalloy machining scraps and end-of-life components for remelting and re-alloying, achieving a 60% material recovery rate in its turbine blade manufacturing. Similarly, the AMRC (Advanced Manufacturing Research Centre) in the UK is piloting a closed-loop system where powder, build failures, and supports are reprocessed and recertified through on-site facilities.

The convergence of these technological and material innovations is catalyzing a broader systemic transformation one where sustainable prototyping becomes both a competitive advantage and an ethical imperative (T. Smith & Williams, 2022). Standardization bodies such as ISO and ASTM are expanding their coverage to include sustainability metrics in AM standards. ISO 14006, for instance, encourages the integration of eco-design principles in product development stages, directly supporting sustainable superalloy prototyping practices. Likewise, national and supranational funding bodies are prioritizing projects aligned with the UN Sustainable Development Goals (SDGs), channeling research investments into greener alloys, energy-efficient equipment, and zero-waste workflows (Martin et al., 2014).

Finally, talent development and interdisciplinary collaboration are indispensable. The challenges associated with sustainable superalloy prototyping demand expertise across materials science, mechanical engineering, data analytics, environmental science, and industrial design. Universities, national labs, and industrial consortia are increasingly forming symbiotic partnerships. For instance, the EU's Clean Sky 2 program and the U.S. Department of Energy's AMMTO initiative have catalyzed multi-stakeholder collaborations to address the dual imperatives of innovation and sustainability(Lee et al., 2007).

In sum, the future of superalloy prototyping will be defined by its ability to align high-performance materials with low-impact processes, driven by data intelligence, renewables integration, and circular resource flows. This systemic transformation, already underway, will shape how industries imagine, design, and manufacture the next generation of sustainable technologies

REFERENCES

Agrawal, A., Bhaskar, A., & Jain, R. (2022). Comparative life cycle assessment of additive and conventional manufacturing of aerospace components. *Journal of Manufacturing Processes*, *75*, 167–177. DOI: 10.1016/j.jmapro.2022.01.011

Ahmed, W., & Zaneldin, E. (2020). The Innovation of Using 3D Printing Technology in Mechanical and Manufacturing Engineering. *IOP Conference Series. Materials Science and Engineering*, *938*(1), 12002. DOI: 10.1088/1757-899X/938/1/012002

Attar, H., Calin, M., Zhang, L. C., Scudino, S., & Eckert, J. (2014). Manufacture by selective laser melting and mechanical behavior of commercially pure titanium. *Materials Science and Engineering A*, *593*, 170–177. DOI: 10.1016/j.msea.2013.11.038

Baufeld, B., Biest, O., & Gault, R. (2011). Additive manufacturing of Ti–6Al–4V components by shaped metal deposition : Microstructure and mechanical properties. *Materials & Design*, *31*(4), 199–206. DOI: 10.1016/j.matdes.2009.05.021

Berman, B. (2012). 3-D printing : The new industrial revolution. *Business Horizons*, *55*(2), 155–162. DOI: 10.1016/j.bushor.2011.11.003

Bocken, N. M. P., Bakker, C., & Pauw, I. D. (2016). Product design and business model strategies for a circular economy. *Journal of Industrial and Production Engineering*, *33*(5), 308–320. DOI: 10.1080/21681015.2016.1172124

Bourell, D. L., Leu, M. C., & Rosen, D. W. (2009). A brief history of additive manufacturing and the 2009 roadmap for additive manufacturing. *Proceedings of the 16th Solid Freeform Fabrication Symposium*, 1-12.

Brown, K. L. (2021). *Machine learning applications in sustainable energy systems*. Wiley.

Cacace, S., Boccadoro, M., & Semeraro, Q. (2024). Investigation on the effect of the gas-to-metal ratio on powder properties and PBF-LB/M processability. *Progress in Additive Manufacturing*, *9*(4), 889–904. DOI: 10.1007/s40964-023-00490-z

CarbonChain. (2022). Understand your cobalt emissions. *CarbonChain Blog*. https://www.carbonchain.com/blog/understand-your-cobalt-emissions

Carter, L. N., Martin, C., Withers, P. J., & Attallah, M. M. (2014). The influence of the laser scan strategy on grain structure and cracking behaviour in SLM powder-bed fabricated nickel superalloy. *Journal of Alloys and Compounds*, *615*, 338–347. DOI: 10.1016/j.jallcom.2014.06.172

Chen, G., Chen, J., Zhao, Z., & Ruan, X. (2005). An object-oriented hierarchical case representation of automotive panels in a computer-aided process planning system. *International Journal of Advanced Manufacturing Technology*, *26*(11-12), 1323–1330. Advance online publication. DOI: 10.1007/s00170-004-2105-8

Chen. (2020). Enhanced creep performance in a polycrystalline superalloy driven by atomic-scale phase transformation. *Acta Materialia Communications*. https://arxiv.org/abs/2007.01676

Choi, S. B., Lee, H., & Kang, S. W. (2020). Cross-disciplinary collaboration in sustainable product development : The role of environmental expertise. *Sustainability*, *12*(5). Advance online publication. DOI: 10.3390/su12052049

DebRoy, T., Wei, H. L., Zuback, J. S., Mukherjee, T., Elmer, J. W., Milewski, J. O., & Zhang, W. (2018). Additive manufacturing of metallic components – Process, structure and properties. *Progress in Materials Science*, *92*, 112–224. DOI: 10.1016/j.pmatsci.2017.10.001

Duraivelu, K. (2022). Digital transformation in manufacturing industry – A comprehensive insight. *Materials Today: Proceedings*, *68*, 2396–2401. https://doi.org/10.1016/j.matpr.2022.07.409

Fera, M., Fruggiero, F., Lambiase, A., Macchiaroli, R., & Papa, A. (2016). The evaluation of the additive manufacturing process to improve the performances of the supply chain. *Procedia CIRP*, *47*, 82–87. DOI: 10.1016/j.procir.2016.03.050

Ferrari, F., Bandini, M., Lorusso, M., & Pavese, M. (2021). Sustainability assessment of powder reuse strategies in metal additive manufacturing. *Sustainability*, *13*(3), 1173. DOI: 10.3390/su13031173

Ford, S., & Despeisse, M. (2016). Additive manufacturing and sustainability : An exploratory study of the advantages and challenges. *Journal of Cleaner Production*, *137*, 1573–1587. DOI: 10.1016/j.jclepro.2016.04.150

Frazier, W. E. (2014). Metal additive manufacturing : A review. *Journal of Materials Engineering and Performance*, *23*(6), 1917–1928. DOI: 10.1007/s11665-014-0958-z

Garcia, R. R., Alcock, J. R., & Maier, S. (2021). Energy and environmental performance of additive manufacturing technologies : A review of literature and current practice. *Sustainable Materials and Technologies*, *28*, 00260. DOI: 10.1016/j.susmat.2021.e00260

Garetti, M., & Taisch, M. (2012). Sustainable manufacturing: Trends and research challenges. *Production Planning and Control*, *23*(2–3), 83–104. DOI: 10.1080/09537287.2011.591619

Garmendia, I., Coello, J., Fernández, E., & Martinez, A. (2020). Digital twins for predictive maintenance in additive manufacturing : State of the art. *Procedia Manufacturing*, *45*, 272–279. DOI: 10.1016/j.promfg.2020.04.028

Gebhardt, A., & Hötter, J.-S. (2016). *Additive manufacturing : 3D printing for prototyping and manufacturing*. Carl Hanser Verlag GmbH Co KG.

Gebler, M., Schoot Uiterkamp, A. J. M., & Visser, C. (2014). A global sustainability perspective on 3D printing technologies. *Energy Policy*, *74*, 158–167. DOI: 10.1016/j.enpol.2014.08.033

Ghadge, A., Shastri, Y., & Shankar, R. (2020). Sustainable additive manufacturing : An exploration of the emerging trends and challenges. *Journal of Manufacturing Technology Management*, *31*(2), 401–427. DOI: 10.1108/JMTM-07-2019-0254

Ghosh, S. (2017a). *Predictive modeling of solidification during laser additive manufacturing of nickel superalloys : Recent developments, future directions*. DOI: 10.48550/arXiv.1707.09292

Gibson, I., Rosen, D. W., & Stucker, B. (2021). *Additive manufacturing technologies : 3D printing, rapid prototyping, and direct digital manufacturing* (3rd éd.). Springer. DOI: 10.1007/978-3-030-56127-7

Green, M., & Hall, S. (2023). Data standards and interoperability in AM software. *Software, Practice & Experience*, *53*(5), 1124–1137. DOI: 10.1002/spe.3125

Gu, D., Meiners, W., Wissenbach, K., & Poprawe, R. (2012). Laser additive manufacturing of metallic components : Materials, processes and mechanisms. *International Materials Reviews*, *57*(3), 133–164. DOI: 10.1179/1743280411Y.0000000014

Guo, N., & Leu, M. C. (2013). Additive manufacturing : Technology, applications and research needs. *Frontiers of Mechanical Engineering*, *8*(3), 215–243. DOI: 10.1007/s11465-013-0248-8

Hasan, M. R., Liu, Z., & Rahman, A. (2023). *Energy consumption modeling for DED-based hybrid additive manufacturing*. /arXiv.2305.09104DOI: 10.21203/rs.3.rs-2905463/v1

Herrmann, C., Thiede, S., & Kara, S. (2020). Sustainability in additive manufacturing : A review and future perspectives. *Journal of Cleaner Production*, *261*, 121234. DOI: 10.1016/j.jclepro.2020.121234

Herzog, D., Seyda, V., Wycisk, E., & Emmelmann, C. (2016). Additive manufacturing of metals. *Acta Materialia*, *117*, 371–392. DOI: 10.1016/j.actamat.2016.07.019

Holmström, J., Partanen, J., Tuomi, J., & Walter, M. (2010). Rapid manufacturing in the spare parts supply chain. *Journal of Manufacturing Technology Management*, *21*(6), 687–697. DOI: 10.1108/17410381011063996

Jiang, J., Xiong, Y., Zhang, Z., & Rosen, D. (2022). Machine learning integrated design for additive manufacturing. *Journal of Intelligent Manufacturing*, *33*(4), 1073–1086. DOI: 10.1007/s10845-020-01715-6

Jinoop, A., Paul, C. P., & Bindra, K. S. (2019). Laser-assisted directed energy deposition of nickel superalloys : A review. *International Journal of Advanced Manufacturing Technology*, *102*, 2321–2340. DOI: 10.1177/1464420719852658

Johnson, D., & Lee, C. (2021). Benchmarking sustainability in metal additive manufacturing. *Materials Today Sustainability*, *12*, 100087. DOI: 10.1016/j.mtsust.2021.100087

Jones, S. R., & Seepersad, C. C. (2016). Integrating materials science and software engineering for advanced manufacturing. *Additive Manufacturing*, *12*, 272–279. DOI: 10.1016/j.addma.2016.05.005

Kellens, K., Baumers, M., Gutowski, T. G., Flanagan, W., Lifset, R., & Duflou, J. R. (2017). Environmental dimensions of additive manufacturing: Mapping application domains and their environmental implications. *Journal of Industrial Ecology*, *21*(S1), S49–S68.

Kellens, K., Mertens, R., Paraskevas, D., Dewulf, W., & Duflou, J. R. (2017). Environmental impact of additive manufacturing processes : Does AM contribute to a more sustainable way of part manufacturing? *Procedia CIRP*, *61*, 582–587. DOI: 10.1016/j.procir.2016.11.153

King, W. E., Anderson, A. T., Ferencz, R. M., Hodge, N. E., Kamath, C., Khairallah, S. A., & Rubenchik, A. M. (2015). Laser powder bed fusion additive manufacturing of metals; physics, computational, and materials challenges. *Applied Physics Reviews*, *2*(4), 041304. DOI: 10.1063/1.4937809

King, W. E., Bajaj, P., & Blue, C. (2021). Effects of multiple reuse on oxygen and nitrogen levels in metal powders. *Additive Manufacturing*, *39*, 101845. DOI: 10.1016/j.addma.2021.101845

Kokare, A. M., Ghosh, D., & Venkatesh, V. C. (2023). Data quality in LCA studies of additive manufacturing : Challenges and the way forward. *Journal of Industrial Ecology*, *27*(2), 483–498. DOI: 10.1111/jiec.13302

Kokare, A. M., Godina, R., & Oliveira, J. P. (2023). Life cycle assessment of additive manufacturing processes : A review. *Journal of Manufacturing Systems*, *68*, 536–559. DOI: 10.1016/j.jmsy.2023.05.007

Körner, C. (2016). Additive manufacturing of metallic components by selective electron beam melting – A review. *International Materials Reviews*, *61*(5), 361–377. DOI: 10.1080/09506608.2016.1176289

Krupp, T., & Seifi, M. (2018). Strategies for residual stress reduction in SLM: Scan patterns and preheat. *Additive Manufacturing*, *23*, 454–462. DOI: 10.1016/j.addma.2018.06.012

Kruth, J. P., Mercelis, P., Vaerenbergh, J., Froyen, L., & Rombouts, M. (2005). Binding mechanisms in selective laser sintering and selective laser melting. *Rapid Prototyping Journal*, *11*(1), 26–36. DOI: 10.1108/13552540510573365

Kubec, V., Lukáš, O., Vavřík, J., & Brázda, M. (2023). Possibilities of application of virtual prototyping and 3D printing methods in practice. In *Proceedings of the 61st International Conference of Machine Design Departments (ICMD 2020)* (pp. 316–325). Atlantis Press. https://doi.org/DOI: 10.2991/978-94-6463-182-1_33

Kulkarni, A. (2018). *Additive manufacturing of nickel-based superalloy*. DOI: 10.48550/arXiv.1805.11664

Kumar, J., Kumar, G., Mehdi, H., & Kumar, M. (2024). Optimization of FSW parameters on mechanical properties of different aluminum alloys of AA6082 and AA7050 by response surface methodology. [IJIDeM]. *International Journal on Interactive Design and Manufacturing*, *18*(3), 1359–1371. DOI: 10.1007/s12008-023-01425-2

Kwon, Y. S., & Yun, H. (2021). Generative design for sustainable lightweight structures using topology optimization and additive manufacturing. *Sustainable Materials and Technologies*, *29*, 00308. DOI: 10.1016/j.susmat.2021.e00308

Lacy, P., & Rutqvist, J. (2015). *Waste to wealth : The circular economy advantage*. Palgrave Macmillan.

Larimian, T., & Borkar, T. (2019). Additive manufacturing of emerging materials. In *Emerging Materials for Additive Manufacturing*. Springer.

Lee, C., Kim, S., Kim, H., & Ahn, S.-H. (2007). Measurement of anisotropic compressive strength of rapid prototyping parts. *Journal of Materials Processing Technology*, *187*, 627–630. DOI: 10.1016/j.jmatprotec.2006.11.095

Leicht, A., Åkerfeldt, P., & Pederson, R. (2018). Influence of build orientation on microstructure and mechanical properties of Inconel 718 produced by Electron Beam Melting. *Metals*, *8*(9), 682. DOI: 10.3390/met8090682

Li, B., He, P., Wang, J., Pan, X., Wang, Y., Wang, Z., & Baniotopoulos, C. C. (2024). Mechanical characteristic and stress-strain modelling of friction stir welded 6061-T6 aluminium alloy butt joints. *Thin-walled Structures*, *198*, 111645. DOI: 10.1016/j.tws.2024.111645

Li, J., & Zhang, X. (2023). Effect of hot isostatic pressing on the stress-rupture properties of second-generation single crystal nickel superalloys processed by laser powder bed fusion. *Materials & Design*, *232*, 111555. DOI: 10.1016/j.jmrt.2024.07.064

Liang, J., Feng, R., & Li, Q. (2018). Residual stress in metal additive manufacturing. *Procedia CIRP*, *71*, 348–353. DOI: 10.1016/j.procir.2018.05.039

Liu, P., Ma, L., & Zhang, Y. (2020). Energy consumption analysis of additive manufacturing : A case study of SLM process. *Journal of Cleaner Production*, *245*, 118878. DOI: 10.1016/j.jclepro.2019.118878

Liu, Z., Wang, M., Zhang, Y., & Tang, Y. (2023). Artificial intelligence-enabled design and quality monitoring in additive manufacturing : A review. *Additive Manufacturing*, *55*, 102936. DOI: 10.1016/j.addma.2022.102936

Mahmoudi, M., & Kovacevic, R. (2015). An investigation on the influence of direct laser melting process parameters on microstructure and mechanical properties of Inconel 625 build parts. *Journal of Materials Processing Technology*, *214*(11), 2623–2630. DOI: 10.1016/j.jmatprotec.2014.06.001

Marrucci, L., Daddi, T., & Iraldo, F. (2023). Creating environmental performance indicators to assess corporate sustainability and reward employees. *Ecological Indicators*, *159*, 111667. DOI: 10.1016/j.ecolind.2023.111489

Martelli, P. A., Bassini, E., & Ugues, D. (2024). Effect of hot isostatic pressing pressure and solution annealing cooling rate on CM247 LC nickel-based superalloy processed by laser-based powder bed fusion. *Progress in Additive Manufacturing*, *10*(1), 605–618. DOI: 10.1007/s40964-024-00645-6

Martin, C., Carter, L. N., Withers, P. J., & Attallah, M. M. (2014). Laser scan strategy effects on superalloy microstructure. *Journal of Alloys and Compounds*, *615*, 338–347. DOI: 10.1016/j.jallcom.2014.06.172

Martínez-Gómez, J., & Morán, P. (2022). Integration of social life cycle assessment into sustainability decisions in additive manufacturing. *The International Journal of Life Cycle Assessment*, *27*(8), 1580–1592. DOI: 10.1016/B978-0-323-90386-8.00057-7

Meier, C., Weissbach, R., Weinberg, J., Wall, W. A., & Hart, A. J. (2018). *Critical influences of particle size and adhesion on the powder layer uniformity in metal additive manufacturing*. DOI: 10.1016/j.jmatprotec.2018.10.037

Mertens, R., Kullmann, J., Treitz, M., & Wycisk, E. (2017). Laser additive manufacturing processes : Thermodynamic and ecological evaluation. *Procedia CIRP*, *61*, 143–148. DOI: 10.1016/j.procir.2016.11.202

Motalebi, A., Khondoker, M. A. H., & Kabir, G. (2024). A systematic review of life cycle assessments of 3D concrete printing. *Sustainable Operations and Computers*, *5*, 14–25. DOI: 10.1016/j.susoc.2023.08.003

Murr, L. E., Gaytan, S. M., & Ramirez, D. A. (2012). Preheating process optimization in EBM for nickel superalloys. *Journal of Materials Science and Technology*, *28*(1), 1–14. DOI: 10.1016/S1005-0302(12)60016-4

Murtaza, H. A., Mukhangaliyeva, A., Golman, B., Perveen, A., & Talamona, D. (2023). Quantitative characterization of metal powder morphology, size distribution, and flowability for additive manufacturing. *Journal of Materials Engineering and Performance*, *33*(20), 10853–10867. DOI: 10.1007/s11665-023-08761-0

Ngo, T. D., Kashani, A., Imbalzano, G., Nguyen, K. T., & Hui, D. (2018). Additive manufacturing (3D printing) : A review of materials, methods, applications and challenges. *Composites. Part B, Engineering*, *143*, 172–196. DOI: 10.1016/j.compositesb.2018.02.012

Nidumolu, R., Prahalad, C. K., & Rangaswami, M. R. (2009). Why sustainability is now the key driver of innovation. *Harvard Business Review*, *87*(9), 56–64.

Park, S., & Kim, J. (2022). Digital twins in sustainable manufacturing. *Computers & Industrial Engineering*, *168*, 108030. DOI: 10.1016/j.cie.2022.108030

Patel, D. (2023). Exploring the Frontiers of Microfluidics : Challenges and Future Prospects. In *Advances in MEMS and Microfluidic Systems* (p. 11-31).

Patel, D. (2024). Sustainable Renewable Energy Sources and Emerging Technologies. In *Optimization Techniques for Hybrid Power Systems* (pp. 343–361). Renewable Energy, Electric Vehicles, and Smart Grid. DOI: 10.4018/979-8-3693-0492-1.ch015

Peng, T., Kellens, K., Tang, R., Chen, C., & Chen, G. (2018). Powder efficiency and sustainability. *Additive Manufacturing*, *21*, 694–704. DOI: 10.1016/j.addma.2018.04.022

Peng, T., Tang, R., & Kellens, K. (2018). Sustainability assessment of powder reuse and recycling in additive manufacturing. *Journal of Cleaner Production*, *192*, 801–813. DOI: 10.1016/j.jclepro.2018.05.173

Pollock, T. M., & Tin, S. (2006). Nickel-based superalloys for advanced turbine engines : Chemistry, microstructure, and properties. *Journal of Propulsion and Power*, *22*(2), 361–374. DOI: 10.2514/1.18239

Ramsperger, M., Markl, M., Berroth, K., Körner, C., & Singer, R. F. (2019). Hot cracking mechanism in nickel-based superalloys processed by selective electron beam melting. *Acta Materialia*, *159*, 77–87. DOI: 10.1016/j.actamat.2018.08.002

Reed, R. C. (2006). *The superalloys : Fundamentals and applications*. Cambridge University Press. DOI: 10.1017/CBO9780511541285

Reeves, P., Tuck, C., & Hague, R. (2011). Additive manufacturing for mass customisation. In *Customization 2.0* (pp. 275–289). Springer., DOI: 10.1007/978-1-84996-489-0_13

Reith, T., Bichler, L., Kopnarski, M., & May, S. (2020). Life cycle inventory database for additive manufacturing : Bridging data gaps. *Journal of Cleaner Production*, *247*, 119142. DOI: 10.1016/j.jclepro.2019.119142

Rejeski, D., Zhao, F., & Huang, Y. (2018). Research needs and recommendations on environmental implications of additive manufacturing. *Additive Manufacturing*, *19*, 21–28. DOI: 10.1016/j.addma.2017.10.019

Rickenbacher, L., Spierings, A. B., & Wegener, K. (2013). High temperature application of SLM-processed IN738LC parts. *Rapid Prototyping Journal*, *19*(4), 282–290. DOI: 10.1108/13552541311323281

Sames, W. J., List, F. A., Pannala, S., Dehoff, R. R., & Babu, S. S. (2016). The metallurgy and processing science of metal additive manufacturing. *International Materials Reviews*, *61*(5), 315–360. DOI: 10.1080/09506608.2015.1116649

Schmidt, O., Chadwick, E., & Tress, B. (2021). Life Cycle Costing in Additive Manufacturing : Unlocking economic insights. *Additive Manufacturing*, *45*, 102015. DOI: 10.1016/j.addma.2021.102015

Simchi, A. (2006). Direct laser sintering of metal powders : Mechanism, kinetics, and microstructural features. *Materials Science and Engineering A*, *428*(1-2), 148–158. DOI: 10.1016/j.msea.2006.04.117

Sims, C. T., Stoloff, N. S., & Hagel, W. C. (1987). *Superalloys II: High-temperature materials for aerospace and industrial power*. Wiley.

Singh, M., Fuenmayor, E., Hinchy, E. P., Qiao, Y., Murray, N., & Devine, D. M. (2021). Digital Twin : Origin to Future. *Applied System Innovation*, *4*(2), 36–36. DOI: 10.3390/asi4020036

Slotwinski, J. A., Garboczi, E. J., & Hebenstreit, K. M. (2014). Characterization of metal powders used for additive manufacturing. *Journal of Research of the National Institute of Standards and Technology*, *119*(), 460–493. DOI: 10.6028/jres.119.018

Smith, T., & Williams, R. (2022). Competitive advantage through sustainability in manufacturing. *Sustainable Production and Consumption*, *31*, 101–110. DOI: 10.1016/j.spc.2022.01.004

Spierings, A. B., Voegtlin, M., Bauer, T., Wegener, K., & Kerres, M. (2013). Powder flowability characterization methodology for powder-bed-based metal additive manufacturing. *Additive Manufacturing*, *1–4*, 87–98. DOI: 10.1007/s40964-015-0001-4

Sun, Z., Jiang, X., Dong, J., Wang, L., Xu, Z., Li, K., & Wang, M. (2025). Recovery of creep performance at different temperature and stress levels of a directionally solidified Ni-based superalloy by hot isostatic pressing and rejuvenation heat treatments. *Metallurgical and Materials Transactions A*. DOI: 10.1007/s11661-025-07855-9

Szmelter-Jarosz, A. (2026). Energy-efficient management policies in smart cities. In *Energy-efficient transformative technologies for data-driven smart cities* (pp. 161–178). Elsevier., DOI: 10.1016/B978-0-443-27618-7.00007-8

Tanaka, Y., & Singh, R. (2019). Economic barriers to solar adoption in urban areas. *Energy Economics Journal*, *41*(3), 310–322.

Thompson, M. K., Moroni, G., Vaneker, T., Fadel, G., Campbell, R. I., Gibson, I., & Martina, F. (2016). Design for Additive Manufacturing : Trends, opportunities, considerations, and constraints. *CIRP Annals*, *65*(2), 737–760. DOI: 10.1016/j.cirp.2016.05.004

Tiwari, A., & Turner, T. A. (2019). Additive manufacturing and sustainability : A review. *Materials Today: Proceedings*, *18*, 2121–2128. DOI: 10.1016/j.matpr.2019.07.166

Výtisk, J., Kočí, V., Honus, S., & Vrtek, M. (2019). Current options in the life cycle assessment of additive manufacturing products. *Open Engineering*, *9*(1), 674–682. DOI: 10.1515/eng-2019-0073

Wang, J., Wang, Z., Sui, Q., Xu, S., Yuan, Q., Zhang, D., & Liu, J. (2023). A comparison of the microstructure, mechanical properties, and corrosion resistance of the K213 superalloy after conventional casting and selective laser melting. *Materials (Basel)*, *16*(4), 1331. DOI: 10.3390/ma16041331

Wong, K. V., & Hernandez, A. (2012). A review of additive manufacturing. *ISRN Mechanical Engineering*, *2012*, 1–10. DOI: 10.5402/2012/208760

Xu, X. (2023). *Vat photopolymerisation 3D printing of controlled drug delivery devices.*

Yang, L., Martina, F., Colegrove, P., Williams, S., Wang, F., & Serres, H. (2019). Microstructural evolution and mechanical properties of Inconel 718 alloy processed by directed energy deposition additive manufacturing. *Materials Science and Engineering A*, *755*, 170–180. DOI: 10.1016/j.msea.2019.03.110

Zafar, M. Q., Sajjad, R., Anwar, M. T., & Mustafa, M. S. (2025). A review on metal additive manufacturing: Types, applications and future trends. *Recent Progress in Materials*, *7*(1). Advance online publication. DOI: 10.21926/rpm.2501006

Zhang, Z., Tan, Z. J., Wang, Y. F., Ren, D. X., & Li, J. Y. (2023). The relationship between microstructures and mechanical properties in friction stir lap welding of titanium alloy. *Materials Chemistry and Physics*, *296*, 127251. DOI: 10.1016/j.matchemphys.2022.127251

Zhao, Y., He, S., & Li, L. (2022). Application of hot isostatic pressing in nickel-based single crystal superalloys. *Crystals*, *12*(6), 805. DOI: 10.3390/cryst12060805

Zhu, H. H., Fuh, J. Y. H., & Lu, L. (2007). The influence of powder apparent density on the density in direct laser-sintered metallic parts. *International Journal of Machine Tools & Manufacture*, *47*(2), 294–298. DOI: 10.1016/j.ijmachtools.2006.03.019

Zhu, W., & Dhokia, V. (2021). A review of hybrid manufacturing processes—State-of-the-art and future perspectives. *Journal of Manufacturing Processes*, *68*, 279–297. DOI: 10.1016/j.jmapro.2021.05.042

Zuo, T., & Zhao, H. (2019). Life cycle assessment and external audit in aerospace manufacturing : Enhancing sustainability. *Journal of Cleaner Production*, *237*, 117729. DOI: 10.1016/j.jclepro.2019.117729

Chapter 6
The Fourth Dimension Transforming Manufacturing With 4D Printing

K. Madhukumar
https://orcid.org/0000-0002-7174-9110
Sir M. Visvesvaraya Institute of Technology, India

ABSTRACT

This chapter explores the revolutionary concept of 4D printing, a transformative extension of traditional additive manufacturing that incorporates the dimension of time. Unlike 3D printing, which constructs objects layer by layer, 4D printing leverages smart materials that can alter their shape and properties in response to various external stimuli such as temperature, moisture, and light. This adaptive capability, often referred to as programmable matter (PM), opens new avenues for designing materials that can autonomously adjust to their environment, thereby enhancing functionality and performance. The chapter discusses the principles of 4D printing, its diverse applications across industries and the implications for sustainable development. Furthermore, it addresses the challenges faced in material innovation, design processes, and manufacturing integration. Through a comprehensive review of ongoing research and future prospects, this chapter highlights the potential of 4D printing to reshape manufacturing paradigms and contribute to the creation of dynamic responsive products

DOI: 10.4018/979-8-3373-5473-6.ch006

1. INTRODUCTION

The advent of 4D printing is deeply rooted in the progression of 3D printing technology. To fully appreciate the transformative leap from 3D to 4D printing, it is essential to understand the foundational developments in 3D printing. This historical context not only elucidates the evolution of printing technologies but also highlights the technological advancements that paved the way for 4D printing. 3D printing, also known as additive manufacturing (AM) or rapid prototyping (RP), represents a revolutionary shift from traditional manufacturing methods. This technology involves the creation of three-dimensional objects through a layer-by-layer deposition process based on computer-aided design (CAD) models. The inception of 3D printing can be traced back to the groundbreaking work of Chuck Hull, whose patent for stereolithography (SLA) in 1986 laid the groundwork for the widespread interest and development in the field. Although Hull's stereolithographic process was a significant milestone, the term "3D printing" was later associated with the development of adhesive jetting technology and powder bed techniques at the Massachusetts Institute of Technology (MIT). This era marked a pivotal moment when 3D printing technology began to gain traction in various domains, including industrial and consumer applications (Gross et al., 2014; Miao et al., 2020).

Over the years, 3D printing technology has diversified into several distinct techniques, each with its unique advantages and applications. Key methods include stereolithography (SLA), which utilizes ultraviolet light to cure resin layer by layer and is known for its high precision and smooth surface finish; selective laser sintering (SLS), which employs a laser to fuse powdered materials, allowing for the creation of complex geometries and durable parts; and fused deposition modeling (FDM), which involves extruding thermoplastic filaments through a heated nozzle, widely used for its affordability and ease of use. Additionally, jet 3D printing (3DP) utilizes inkjet technology to deposit layers of material, often used for creating multi-material objects; selective laser melting (SLM) uses a laser to melt metallic powders, ideal for producing strong and intricate metal parts; direct ink writing (DIW) involves extruding viscous materials through a nozzle, suitable for creating functional structures with various materials; and electron beam melting (EBM) uses an electron beam to melt metal powders in a vacuum, offering high precision and efficiency in metal part production. These methods have significantly contributed to the rapid advancements in 3D printing, which have included improvements in speed, accuracy, and cost-effectiveness. The technology's interdisciplinary nature, involving material science, mechanical engineering, and data processing, has enabled the development of complex and innovative designs previously unattainable with traditional manufacturing techniques (Chen et al., 2024; Shirazi et al., 2015; Tofail et al., 2018)

The evolution of manufacturing technologies has reached a pivotal moment with the emergence of 4D printing, a groundbreaking advancement that transcends the capabilities of traditional three-dimensional (3D) printing. While 3D printing revolutionized the way objects are created by enabling the layer-by-layer construction of complex geometries, 4D printing introduces the concept of time as the fourth dimension, allowing printed objects to change shape, function, or properties over time in response to environmental stimuli. This innovative approach is set to transform industries ranging from healthcare and aerospace to automotive and consumer goods by enabling the production of dynamic, self-transforming materials that can adapt to their surroundings(Tibbits et al., 2014).

At the heart of 4D printing is the integration of smart materials, often referred to as stimuli-responsive or shape-memory materials, which possess the ability to undergo predefined changes when exposed to specific triggers such as heat, light, moisture, pH, or magnetic fields 9. These materials can be engineered to respond in a controlled and predictable manner, enabling the creation of structures that can morph, self-assemble, or self-repair over time. For instance, in the field of biomedical engineering, 4D printing has been used to develop patient-specific implants and devices that can adapt to physiological conditions, offering the potential to significantly improve patient outcomes and reduce recovery times (Baig et al., 2024; Hatami et al., 2024a)

The ability of 4D printed parts to autonomously sense changes in the external environment and change its form and functions presents its potential to replace sensors, actuators, motors in electromechanical systems like robotic arms as it enables the user to programme and re-programme the material itself to behave in a desired manner (Campbell et al., 2014; Tibbits, 2012). The property of 4D printed components to change its form according to the application also eliminated the need for intermediate shaping operations as the capability of shape change is embedded into the material itself (Pei & Loh, 2018).

The term "Programmable Matter (PM)" was introduced to describe how the material can be programmed / controlled by a user to change its form or functions which not only allows the user to decide how the material will respond under different conditions but it also enables the user to control and manipulate each step of the transformation. Recent research has been aimed at developing universally acceptable folding, twisting, expansion and shrinking techniques so that the transformation of a component from one arbitrary shape to another can be carried out (Campbell et al., 2014).

One of the major challenges in this domain would be to exercise a good amount of control over the autonomous transformation of the material and also to ensure that this transformation can be carried out without human or robotic intervention. The aim of this is to eliminate devices like motors, sensors, actuators, etc. which are found to be inaccurate and error prone at times. PM can have a wide variety of applications in aerospace, automotive industry, health, construction and robotics (due to their autonomous sensing capability), etc. (Pei & Loh, 2018).

One of the most promising applications of 4D printing lies in tissue engineering, where the ability to create scaffolds that mimic the dynamic nature of biological tissues is of paramount importance. Traditional 3D-printed scaffolds have been limited by their static nature, which does not fully replicate the complex environment of living tissues. However, the introduction of 4D printing allows for the creation of scaffolds that can change shape or properties in response to the body's environment, promoting better integration with surrounding tissues and enhancing the healing process (Grira et al., 2023; Hasanzadeh et al., 2023; Shahbazi et al., 2023). For example, hydrogels, a class of materials known for their high-water content and biocompatibility, have been widely explored in 4D printing for their ability to respond to changes in temperature and pH, making them ideal candidates for dynamic tissue engineering applications (Chopra et al., 2024; Jolaiy et al., 2024).

Beyond the biomedical field, 4D printing is also making significant strides in the development of smart materials for use in aerospace, automotive, and consumer goods. These industries demand materials that can withstand harsh environments, adapt to changing conditions, and offer enhanced functionality. The use of shape-memory polymers and other stimuli-responsive materials in 4D printing enables the creation of components that can adjust their properties, such as stiffness or shape, in response to external stimuli, leading to more efficient and versatile products (Agarwal et al., 2023; Kumar et al., 2024; Mallakpour et al., 2024; Zhou et al., 2023). In the aerospace sector, for example, 4D printing is being explored for the development of morphing structures that can change shape during flight to optimize aerodynamics and fuel efficiency (Ameta et al., 2022).

The potential of 4D printing extends to the food industry as well, where unconventional sourced proteins and other innovative materials are being combined with 4D printing techniques to create food products that can change texture, shape, or flavor in response to cooking or consumption conditions. This approach not only opens up new possibilities for personalized nutrition but also addresses sustainability challenges by reducing food waste and improving the efficiency of food production processes (Pereira et al., 2024). In material science, the bigel 4D printing technology incorporates spontaneous color change features. This innovative approach, employs dual-unit 3D printers, facilitates real-time monitoring of bioactive substances, thereby expanding the functionality and versatility of printed materials in areas such as bioengineering and food monitoring(R. Qiu et al., 2024). Additionally, molecular-level manipulation of viscoelastic silicone inks, allows for precise control over the spatiotemporal behavior of cellular silicones in 4D printing. This development underscores the capability to create complex, responsive structures with customized properties (Liao et al., 2024).

Moreover, 4D printing is contributing to the development of sustainable and environment-friendly manufacturing practices. By enabling the creation of materials that can self-repair or adapt to prolong their lifespan, 4D printing reduces the need for frequent replacements and minimizes waste. This aligns with the growing emphasis on sustainability in manufacturing, where the goal is to create products that are not only high-performing but also environmentally responsible (Agarwal et al., 2023; Khalid et al., 2023).

The integration of biomimetic approaches into 3D/4D printing technologies has opened new avenues for enhancing the functionality of printed products. Incorporating biomimetic principles into 3D/4D printing can significantly advance drug delivery systems. By mimicking natural biological processes, these approaches aim to improve the precision and effectiveness of drug delivery, making treatments more targeted and efficient (Dowaidar, 2024). Furthermore, advancements in materials science are crucial for expanding the capabilities of 3D/4D printing. The use of advanced materials such as ceramic/zeolite reinforced photopolymers in 3D/4D printing can be employed in applications requiring high durability and specific functional properties, such as gas adsorption. This development not only broadens the range of printable materials but also enhances their performance in specialized applications (Y. Zhang et al., 2022a).

Recent research has made notable advances in optimizing 4D printing processes, materials and role of machine learning in enhancing 4D and 3D printing technologies. The transformative impact of integrating advanced data analytics and predictive modeling, improves the efficiency and precision of printing processes. This integration enables the production of more accurate and functional printed objects, marking a significant advancement in the field (Alli et al., 2024).

The development of 4D printing technologies is not limited to one region but is a global phenomenon. (Lu et al., 2024)provides an overview of the advancements in ceramic 3D/4D printing in China, highlighting regional innovations and their implications for the field. This global perspective enriches our understanding of the diverse approaches and applications of 4D printing technologies.

4D printing represents a paradigm shift in manufacturing, offering unprecedented opportunities to create dynamic, responsive, and sustainable materials across a wide range of industries. By integrating smart materials with advanced additive manufacturing techniques, 4D printing is poised to unlock new levels of innovation, enabling the creation of products that can adapt to their environment, improve functionality, and contribute to a more sustainable future. As research continues to advance and the technology matures, the full potential of 4D printing is likely to reshape the manufacturing landscape, ushering in a new era of possibilities.

2. BASIC CONCEPTS RELATED TO 4D PRINTING

The following concepts form the foundation of understanding 4D printing and its emerging role in advanced manufacturing.

2.1. Self-Assembly

One of the primary challenges of Additive Manufacturing (AM) lies in the limitation imposed by the print volume of 3D printers, restricting the size of the producible objects. However, 4D printing offers a significant advantage by enabling the production of simple components that can be printed within the available print volume and subsequently programmed to undergo self-transformation into more complex geometries and structures as per user requirements. This transformative capability, known as Self-Assembly, effectively removes the constraints posed by the initial print volume.

Self-Assembly not only enhances design flexibility but also drastically reduces manufacturing time and facilitates efficient storage. The initial structures are simple and compact, making them easier to package and store. By allowing individual parts to be 3D printed and then autonomously assembling into larger and more intricate structures, Self-Assembly simplifies the challenges associated with complex assemblies. This technique minimizes the need for manual intervention and reduces the complexity and time involved in the manufacturing process (Nkomo, 2018)

Moreover, Self-Assembly leverages programmable materials that respond to specific stimuli, such as temperature, moisture, or light, to trigger the transformation. This adds an additional layer of sophistication, enabling the creation of dynamic, adaptive products that can respond to environmental conditions or user interaction, further extending the potential applications of 4D printing in fields like aerospace, biomedical engineering, and robotics.

2.2. Self-Repair

Conventional manufacturing methods often suffer from significant material wastage and recycling challenges, leading to inefficiencies and environmental concerns. In contrast, the advanced responsive materials used in 4D printing offer a groundbreaking solution through their intrinsic self-healing capabilities. These materials can be utilized, reused, and recycled with minimal or no material wastage, thereby addressing one of the critical disadvantages of traditional manufacturing.

A notable example of this innovation is the use of self-healing hydrogels as extruder materials in 3D printing. Components printed with these hydrogels exhibit self-repairing properties, allowing them to autonomously heal after sustaining damage.

The self-repairing process can address a wide range of damage, from superficial scratches to deeper defects within the bulk of the material, thereby extending the lifecycle of the printed products and reducing the need for replacements.

This self-repair capability is driven by the material's ability to respond to environmental stimuli, such as changes in temperature or humidity, which triggers the healing process. Such materials are particularly advantageous in applications where durability and longevity are critical, such as in biomedical devices, aerospace components, and other high-performance industries. By reducing material waste and enhancing the sustainability of manufacturing processes, self-repairing materials in 4D printing represent a significant leap forward in sustainable manufacturing practices.

2.3. Voxels

In the realm of materials for 4D printing, the concept of the Volumetric Pixel or Voxel is paramount. A voxel serves as the fundamental unit of printed matter, akin to how a pixel is the basic unit of an image. Just as biological life is composed of 22 basic amino acids that combine in various ways to create diverse forms of life, the current focus in 4D printing is on defining and utilizing a set of basic voxel types. These foundational voxels can be combined to produce a wide array of responsive materials tailored for specific applications in 4D printing.

For instance, incorporating soft voxels into a structure can result in soft, flexible materials, while rigid voxels can produce hard, durable materials. Additionally, sensor voxels can be integrated into materials to enable autonomous sensing capabilities, which are particularly valuable in robotics, potentially replacing traditional electronic components such as sensors and actuators.

This voxel-based approach not only enhances the versatility of 4D printing but also opens up new possibilities for creating complex, multifunctional materials. By strategically combining different voxel types, it is possible to engineer materials with properties tailored to specific applications, whether in biomedical devices, aerospace components, or smart textiles. This voxel framework is key to advancing the development of responsive and adaptive materials, paving the way for innovations in numerous fields.

3. RECENT ADVANCEMENTS IN 4D PRINTING

The term 4D printing was introduced by Skylar Tibbits, the Founder and Co-Director of the Self-Assembly Lab at the Massachusetts Institute of Technology (MIT), during his 2013 TED talk (Campbell et al., 2014; Tibbits, 2012). This groundbreaking concept emerged from a series of experiments conducted in collaboration with Stratasys, Ltd., and Autodesk, Inc. Utilizing Stratasys' advanced Connex printer, which is capable of multi-material deposition, these experiments explored the potential of creating dynamic structures that could transform over time.

The material selected for these experiments was an innovative polymer designed to expand by up to 150 percent when submerged in water. This unique characteristic demonstrated the feasibility of creating objects that could autonomously change shape or function in response to environmental stimuli, marking a significant advancement in the field of additive manufacturing. Tibbits' work laid the foundation for 4D printing, a technology that leverages time as an additional dimension in the printing process, opening new possibilities for responsive and adaptive materials in various industries.

One of the early and notable experiments in 4D printing involved a single multi-material strand composed of both rigid and flexible materials. When submerged in water, this strand autonomously transformed into the acronym "MIT" as illustrated in Figure 1. This experiment was among the first to clearly demonstrate the potential of 4D-printed components to change shape in response to external stimuli, with water serving as the activating trigger in this case (Campbell et al., 2014).

Figure 1. Transformation of a single multi-material strand (Campbell et al., 2014)

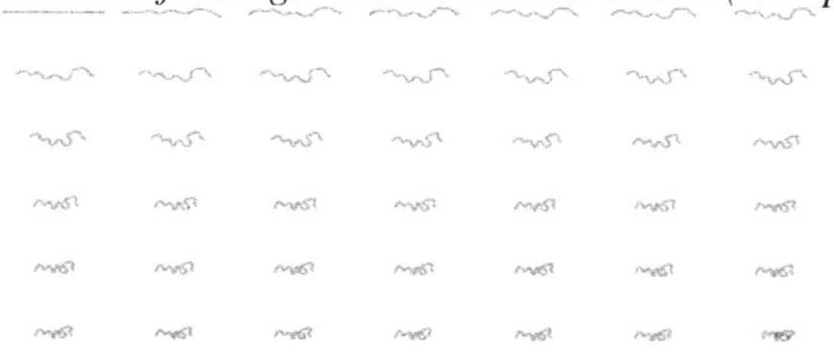

This pioneering demonstration showcased the practical application of 4D printing, illustrating how materials could be programmed to undergo specific transformations, thus laying the groundwork for future innovations in self-assembling and adaptive materials.

The Self-Assembly Lab at MIT has advanced the field of 4D printing by developing a self-folding structure that transforms into a cube when submerged in water (Tibbits et al., 2014). This innovative design was achieved by leveraging the capabilities of multi-material 3D printing, using materials with distinct porosities and water absorption characteristics.

The structure was meticulously engineered with one side composed of a highly porous material that absorbed water rapidly, leading to a significant increase in volume. In contrast, the opposite side was made from a rigid, waterproof material that remained unaffected by water exposure. The differential swelling between the two sides created mechanical stress within the structure, causing it to bend and fold.

As a result, the structure transitioned from a flat, single strand into a three-dimensional cube, a transformation driven entirely by the interaction between the material properties and the surrounding water. This experiment, not only demonstrated the feasibility of creating self-assembling structures through 4D printing but also underscored the potential of this technology for applications where autonomous shape-shifting is required.

This self-folding cube exemplifies the integration of material science with advanced manufacturing techniques, paving the way for future innovations in fields such as robotics, where components may need to self-assemble or change shape in response to environmental conditions. The experiment also highlights the precision and control possible with 4D printing, offering insights into how complex, adaptive structures can be designed and manufactured for various industrial and research applications.

Tetherless thermobiochemically actuated microgrippers represent a significant advancement in the field of micro scale biomedical devices, particularly for applications in targeted drug delivery and minimally invasive surgical procedures. As the name implies, these microgrippers operate autonomously, without the need for physical tethers or external power sources. Their actuation is triggered by specific thermal and biochemical cues present in biological environments, enabling precise and controlled functionality within the human body (Leong et al., 2009). The key innovation of these devices lies in their ability to operate in confined and delicate spaces, responding to physiological conditions without requiring continuous external manipulation. This feature greatly enhances their utility in tasks such as gripping small tissue samples, handling cells, or delivering therapeutic agents in situ with high precision. A notable demonstration of this technology is illustrated in Figure 5A–C, where a microgripper was remotely guided into a capillary tube and thermally actuated to capture a portion of a living cell mass stained with Neutral red, a dye that accumulates in lysosomes of viable cells. After successfully grasping the cell cluster, the gripper was retracted from the capillary tube (Figure 5D and E), still holding the viable cells, as evidenced by their retained red coloration. The recovered cell cluster is shown in Figure 5F. To further validate biocompatibility, multiple trials were conducted using LIVE/DEAD cell staining techniques (Figure 5G), confirming that the thermally actuated capture process did not compromise cell viability. Moreover, when the grippers containing the retrieved cells were incubated in growth media for 72 hours, the cells remained viable (Figure 5H). This result indicates that both the materials used in fabricating the grippers and the mechanical operations involved in capture and release are non-toxic to living cells. These findings underscore the potential of such microgrippers not only as tools for cell manipulation but also as temporary biological storage systems, capable of maintaining cell viability post-retrieval until further analysis is performed. The integration of thermal and chemical responsiveness at the microscale opens promising avenues in the development of intelligent biomedical devices, paving the way for advancements in smart material systems and autonomous therapeutic technologies.

Figure 2. Thermobiochemically actuated tetherless microgripper capturing and retrieving viable cells. (A–C) Remote navigation of the microgripper into a capillary tube and thermal activation to grasp a portion of a living cell mass stained with Neutral red. (D–E) Withdrawal of the microgripper from the capillary tube while securely holding the captured cells. (F) Retrieved viable cell cluster, indicated by red coloration. (G) LIVE/DEAD staining confirming cell viability after thermal capture. (H) Cells remain viable after 72-hour incubation, demonstrating biocompatibility of the gripper materials and procedures (Leong et al., 2024)

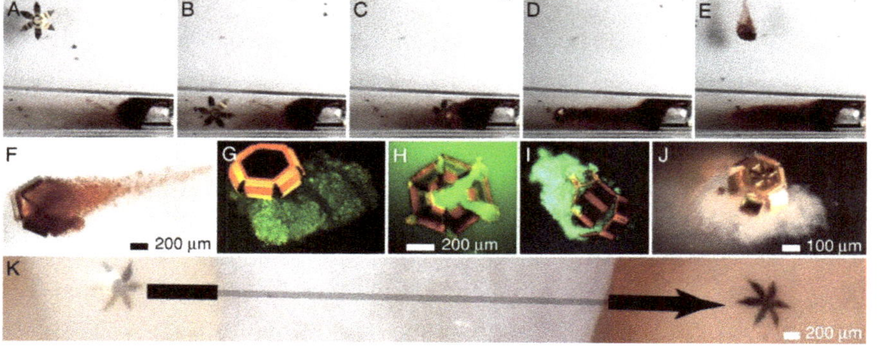

A new generation of 4D-printed multi-material grippers has been developed using multiple Shape Memory Polymers (SMPs), each engineered with distinct rubbery moduli and glass transition temperatures. These SMPs exhibit failure strains up to 300% higher than conventional materials used in additive manufacturing, making them particularly suitable for dynamic, reconfigurable, and responsive biomedical or soft robotic applications (Ge et al., 2016a). One of the major advancements in this approach is the ability of multiple-SMP systems to memorize more than two temporary shapes, significantly extending the functionality and programmability of 4D-printed structures. However, most traditional implementations rely on direct thermal actuation, limiting the selective and sequential responsiveness that multi-SMP systems could potentially offer(Ge et al., 2016b; Wan et al., 2022).

To address these limitations, a novel 4D printing strategy was employed to fabricate multi-material structures composed of poly (D,L-lactide-co-trimethylene carbonate) (PLMC), poly(trimethylene carbonate) (PTMC), and Fe_3O_4 nanoparticles. This design enables multi-stimuli-responsive shape transformations through remotely applied magnetic fields and thermal triggers. Initially, pure SMP compositions were optimized to precisely tune the multiple shape memory effect. Shape recovery behavior was quantitatively analyzed via stepwise thermal programming. Subsequently, the integration of Fe_3O_4 nanoparticles allowed for localized magnetic responsiveness, enabling selective actuation of specific regions within the 4D-printed structure (Wan et al., 2022).

The combination of multi-material additive manufacturing and multiple-SMP programming enabled the fabrication of structures capable of quintuple shape transformations, with high spatial and temporal control. These transformations could be independently actuated by applying targeted magnetic or thermal stimuli, offering exceptional functional diversity and design flexibility. Importantly, in addition to their mechanical performance, the multi-material systems demonstrated excellent biocompatibility, further expanding their potential for biomedical devices, soft robotics, and minimally invasive surgical tools(Meloni et al., 2021).

The gripper, illustrated in the time-lapse sequence shown in Figure 3, is programmed to contract and grip an object upon contact. When subjected to heat, the gripper releases the object, showcasing the material's ability to change shape in response to thermal stimuli. This function is crucial for applications requiring precise, repeatable control over gripping and releasing actions.

Figure 3. Time-lapse of a 4D printed multi-material gripper contracting to grip an object and releasing it when subjected to heat, illustrating the responsive behavior of Shape Memory Polymers (Ge et al., 2016a)

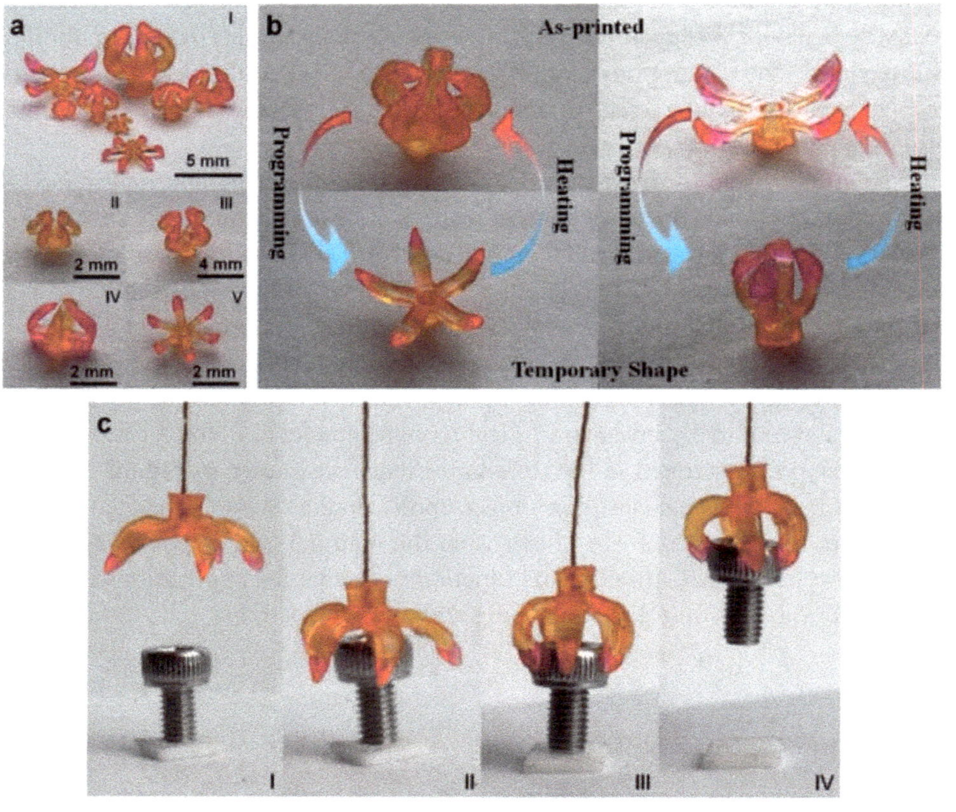

To achieve this level of control, the dynamic properties of the SMPs must be finely tuned, allowing for a time-dependent, step-by-step shape recovery process. This capability opens up new possibilities in fields such as robotics and biomedical devices, where responsive and adaptable materials are essential for complex operations.

Further advancements in 4D printing have led to the development of a robotic finger (Fig. 4) and an integrated sensing and actuating wing flap (Figure 5). These components are fabricated using Additive Manufacturing techniques and are designed to be actuated by signals from external sources.

Figure 4. Autonomous sensing robotic finger developed using Additive Manufacturing techniques (Campbell et al., 2014)

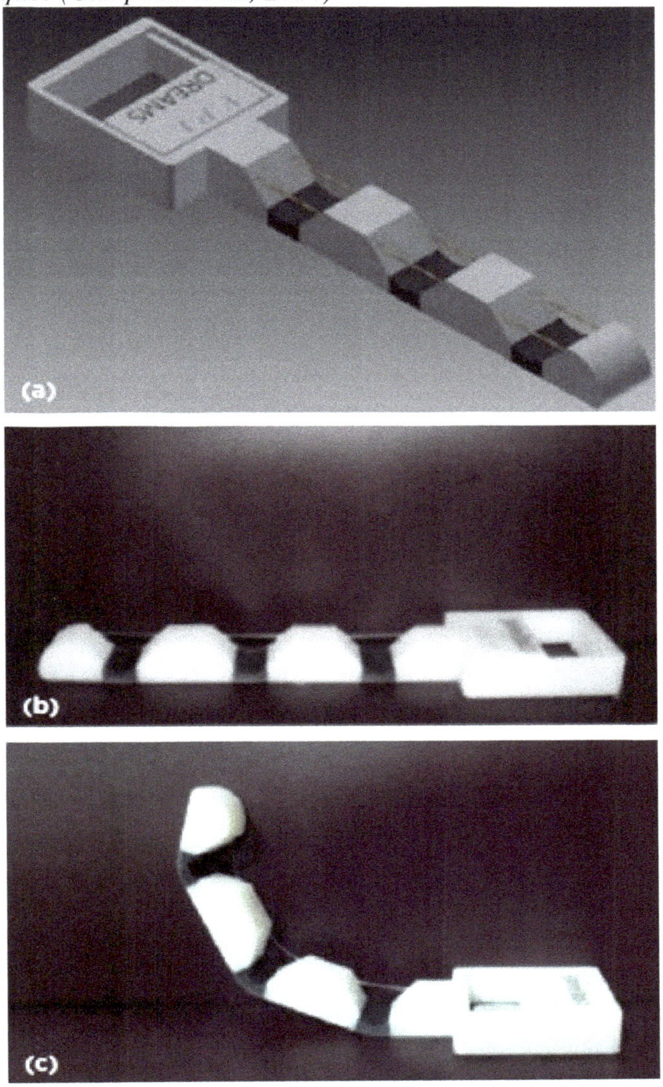

Figure 5. Integrated sensing and actuating wing flap, fabricated with Additive Manufacturing, designed to respond to external signals without traditional actuators or sensors (Campbell et al., 2014)

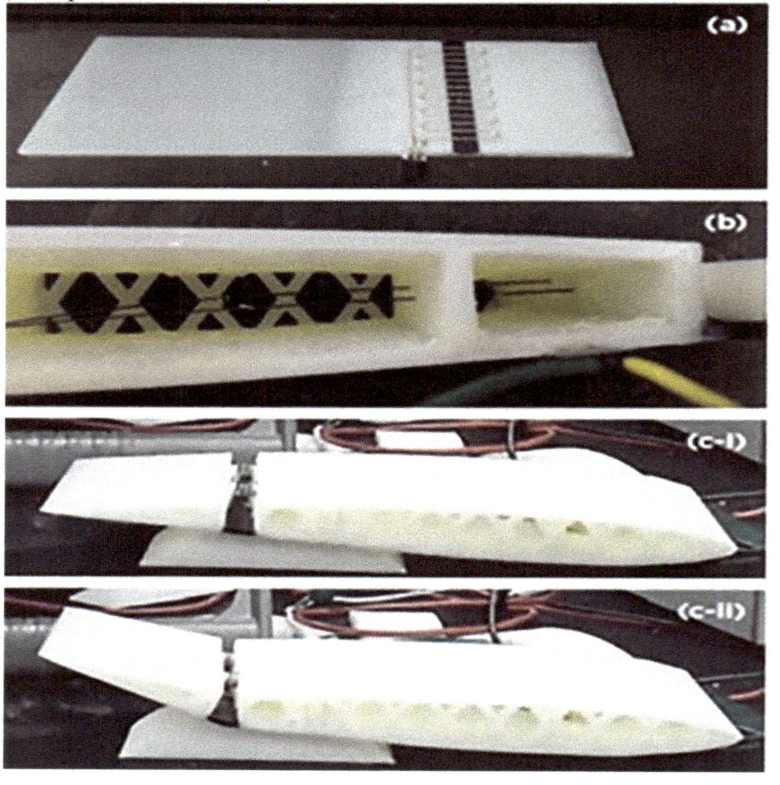

Notably, these innovative parts eliminate the need for traditional, error-prone devices such as actuators, motors, and sensors. Instead, their autonomous sensing and actuation capabilities are intrinsically embedded within the materials themselves, activated upon receiving external signals. This design approach significantly enhances the reliability and functionality of these components, making them particularly useful in fields such as robotics, as well as in architecture and construction industries. The integration of these smart materials into functional components highlights the potential for creating more efficient and less complex systems, paving the way for broader applications where responsive, self-sufficient behavior is crucial.

Given the critical need for shape transformation, material programming, and simulation in 4D printing, the development of specialized software is essential. Project Cyborg, a software developed by Autodesk, Inc. in collaboration with the MIT research group, addresses these requirements. This software is designed to simulate various shape transformations, optimize the processes, and control the sequential folding of components during 4D printing. Project Cyborg facilitates the intricate programming and simulation necessary for creating adaptive and responsive 4D-printed materials. By providing tools to visualize and refine the transformation sequences, it ensures that designs can be accurately implemented and tested, paving the way for more advanced and reliable 4D printing applications.

Software solutions like Origami have been developed to facilitate the creation of intricate origami shapes by generating crease patterns. These patterns enable solid models to fold into complex polyhedral forms, with the number of seams serving as a key parameter in the design process. Another notable tool, the E-Origami System, developed by the Symbolic Computation Research Group (SCORE), allows users to perform sheet folding using mathematical and symbolic methods. This system supports the precise manipulation of folds and creases, enhancing the ability to design and simulate complex origami structures. Both Origami and E-Origami illustrate the application of computational techniques to advance the design and functionality of foldable structures, expanding the possibilities for both artistic and practical applications in fields such as architecture and product design(Kshirsagar et al., 2023; Meloni et al., 2021; Pei & Loh, 2018)

4. MATERIALS USED IN 4D PRINTING

4D printing is an advanced extension of 3D printing technology, where materials not only create static objects but also possess the capability to change their shape, structure, or properties over time in response to environmental stimuli. This dynamic behavior introduces a new dimension to material science and manufacturing, allowing for the creation of adaptive and responsive products. In this section, we explore the various materials utilized in 4D printing, each designed to react to specific stimuli such as temperature, light, or magnetic fields. Understanding these materials is crucial for leveraging their unique properties in innovative applications across industries like robotics, healthcare, and smart textiles (Hazrat Ali et al., 2019; Leist & Zhou, 2016; Mohammad Vaezi Srisit Chianrabutra & Yang, 2013). Detailed Overview is presented in Table 1.

Table 1. Materials used in 4D printing: Detailed overview (Ahmed et al., 2021a)

Material Type	Description	Stimulus	Applications
Shape Memory Polymers (SMPs)	Polymers that return to a pre-defined shape upon heating or cooling.	Temperature	Self-healing materials, adaptive structures, actuators
Hydrogels	Polymers that absorb water and swell or contract, altering their shape.	Water (hydration)	Drug delivery systems, soft robotics, tissue engineering
Thermo-responsive Materials	Materials that change properties or shape in response to temperature variations.	Temperature	Temperature-sensitive coatings, smart textiles
Photo-responsive Polymers	Polymers that change shape or properties when exposed to light.	Light (UV or visible)	Light-driven actuators, responsive surfaces
Electro-active Polymers	Polymers that alter shape or properties in response to an electric field.	Electric field	Artificial muscles, sensors, flexible electronics
Magneto-responsive Materials	Materials that change shape or properties in response to magnetic fields.	Magnetic field	Magnetic actuators, controlled drug release systems

4.1 Shape memory polymers

Shape Memory Polymers (SMPs) are a class of smart materials capable of returning to a predetermined shape upon exposure to specific external stimuli, such as thermal energy or moisture (Behl & Lendlein, 2007). This property is harnessed in 4D printing to facilitate dynamic transformation of objects over time, thereby introducing an additional dimension of adaptability and functionality. In the context of 4D printing, SMPs enable the creation of structures that can autonomously assemble or alter their form in response to environmental changes. This capability is pivotal for applications in advanced fields such as biomedical devices, aerospace engineering, and smart consumer products. The synergy of SMPs with 4D printing technologies allows for the design of materials that not only achieve a desired shape post-printing but also exhibit adaptive behavior based on environmental stimuli (Bonetti et al., 2024; Lalegani Dezaki & Bodaghi, 2024; Rahmatabadi et al., 2024; Zhang et al., 2022b).

SMPs exhibit several advantageous properties:

- SMPs are lightweight, which is beneficial for applications requiring minimal mass.
- They can be recycled efficiently and economically.

- They can undergo significant deformation and revert to their original shape upon stimulus application.
- They exhibit resistance to chemical degradation.
- SMPs are amenable to various processing techniques.
- They are suitable for medical applications due to their compatibility with biological systems and potential for controlled degradation (Bond et al,. 2023).

The fundamental mechanism of SMPs involves a dual-shape effect. Initially, the polymer is deformed into a temporary shape (Shape A) through mechanical processing. Upon application of an appropriate stimulus, the SMP transitions to its permanent shape (Shape B). This transition can be triggered by thermal, moisture, or other stimuli, depending on the specific SMP formulation. Research on SMPs began with a focus on thermally induced shape memory effects, but has since expanded to include a range of stimuli. This includes indirect thermal activation and direct actuation through molecular-level interactions with stimuli-sensitive groups (Behl & Lendlein, 2007).

Applications of SMPs are diverse and impactful. In everyday life, they are utilized in smart textiles, heat-shrinkable tubes for electronics, packaging films, self-deployable solar panels for space missions, intelligent medical devices, and implants for minimally invasive surgical procedures. SMPs, classified as 'actively moving' polymers, are capable of significant shape changes from Shape A to Shape B, demonstrating their dynamic and responsive nature. This integration of SMPs with 4D printing technologies heralds advancements in material science, enabling the development of smart materials with enhanced functionality and adaptability.

4.1.1 Key Attributes and Mechanisms of Shape Memory Polymers (SMPs)

Shape Memory Polymers (SMPs) are distinguished by two primary attributes: their ability to retain a temporary programmed shape and their capacity to revert to an original shape upon activation by an external stimulus. The shape-memory effect in SMPs is fundamentally dependent on the polymer's molecular architecture rather than on specific chemical structures within the repeating units of the polymer chain. This allows for the customization of material properties, including mechanical characteristics, by altering molecular parameters to meet the needs of various applications.

The shape-memory effect is not an inherent property of the polymers themselves but arises from a combination of polymer morphology and processing techniques. The process involves two main stages:

1. **Initial Shaping:** The polymer is first processed into its permanent shape (Shape B) using conventional methods such as injection molding or extrusion. This shape is stable under normal conditions.
2. **Programming and Activation:** In the subsequent programming phase, the polymer is deformed into a temporary shape (Shape A) and fixed into this form through cooling or other means. When an external stimulus is applied—such as heat, chemical changes, or light—the polymer returns to its original permanent shape (Shape B) (Lalegani Dezaki & Bodaghi, 2024).

Common stimuli (Behl & Lendlein, 2007) for activating SMPs include:

- **Temperature Variation (Thermo-responsive):** This is the most common stimulus, where heat triggers the transition between shapes (Kholkhoev et al., 2024).
- **Chemical Changes (Chemo-responsive):** This includes changes in water content or pH levels.
- **Light (Photo-responsive):** Certain SMPs can be activated by light without temperature changes (Nikishina et al., 2024).

Types of Shape Memory Effects:

1. **One-Way Shape Memory Effect:** This is the most commonly observed type of shape memory effect in SMPs. In this phenomenon, the polymer undergoes an irreversible transition when exposed to an external stimulus. The stimulus triggers the polymer to move from its temporary shape (Shape A) to its permanent shape (Shape B). Once the polymer has transitioned to the permanent shape, it cannot return to the temporary shape without undergoing a reprogramming process (Alshebly et al., 2021). This effect is illustrated in Figure 6.

Figure 6. Irreversible transition of one-way shape memory effect in SMPs

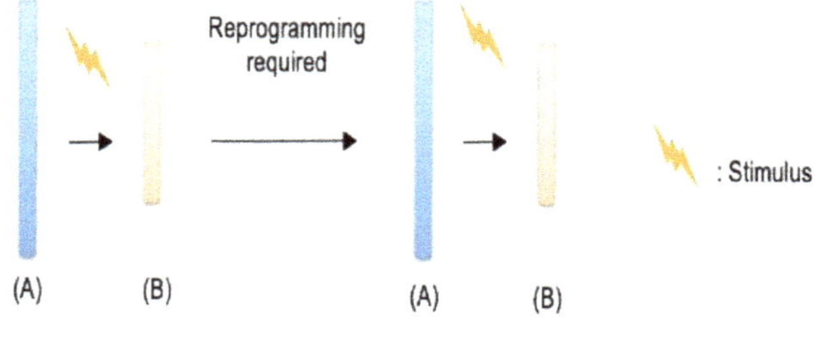

2. **Two-Way Shape Memory Effect:** Shape Memory Polymers (SMPs) exhibiting a two-way shape memory effect (TWSME) possess the capability to retain and revert between two distinct shapes in response to external stimuli. Unlike one-way shape memory polymers, which only return to their original shape from a deformed state (Fig.7), TWSMEs can transition between two programmed shapes. This reversible transformation occurs through thermal or other stimuli-induced changes.

Figure 7. Two-way shape memory effect in SMPs

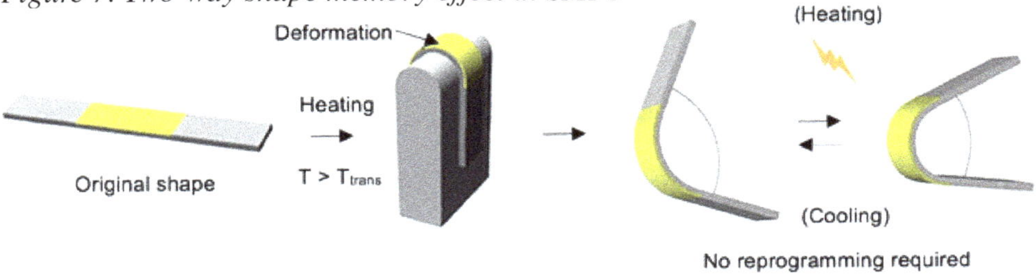

The Two-Way Shape Memory Effect (TWSME) is prominently observed in liquid crystalline elastomers (LCEs) and photo-actuated polymers. A polyurethane-based shape memory polymer laminate was effectively demonstrated to exhibit TWSME. The laminate bent upon heating from 25°C to 60°C and reversed its bend upon cooling from 60°C to 25°C, with no need for reprogramming. This indicates that the polymer retains its two-way shape memory capabilities over multiple thermal cycles (Bonetti et al., 2024; Pilate et al., 2016).

4.1.2 Thermally Induced Shape Memory Polymers (SMPs)

Thermally induced shape memory polymers exploit heat as the activating stimulus, operating through a dual-component system. Typically, these systems consist of an elastic matrix and temperature-responsive fibers. The fibers undergo reversible changes in stiffness based on temperature, functioning as the switch mechanism. In such SMPs, the polymer network comprises molecular switches and netpoints (Fig. 8). Netpoints, which define the permanent shape of the polymer network, can be either chemical (e.g., covalent bonds) or physical (e.g., intermolecular interactions). The domains corresponding to the highest thermal transition temperature (T_perm) serve as netpoints, while chain segments in domains with a lower thermal transition temperature (T_trans) act as molecular switches. When the temperature exceeds T_trans, the switching domains become flexible, and the polymer network exhibits irreversible elastic behavior. Upon deformation under external stress, the polymer can revert to its initial shape once the stress is removed, due to the flexible nature of the domains above T_trans. Heat-induced SMPs primarily rely on thermal activation for shape memory. This involves the cleavage of additional cross-links that stabilize the deformed state. If these cross-links are physical, they exhibit a distinct T_trans, either as a glass transition temperature (T_g) or a melting temperature (T_m). Glass transitions span a broader temperature range, while melting occurs over a narrower interval. SMPs can also use light as a stimulus when cross-links are based on photo-reversible reactions. Additionally, electrical currents or electromagnetic fields can indirectly heat the material, expanding the range of stimuli beyond just thermal activation (Moradi et al., 2024; Uyan & Celiktas, 2024; H. Zhang et al., 2024).

Figure 8. Schematic of polymer network structure in shape memory polymers: Molecular switches and netpoints

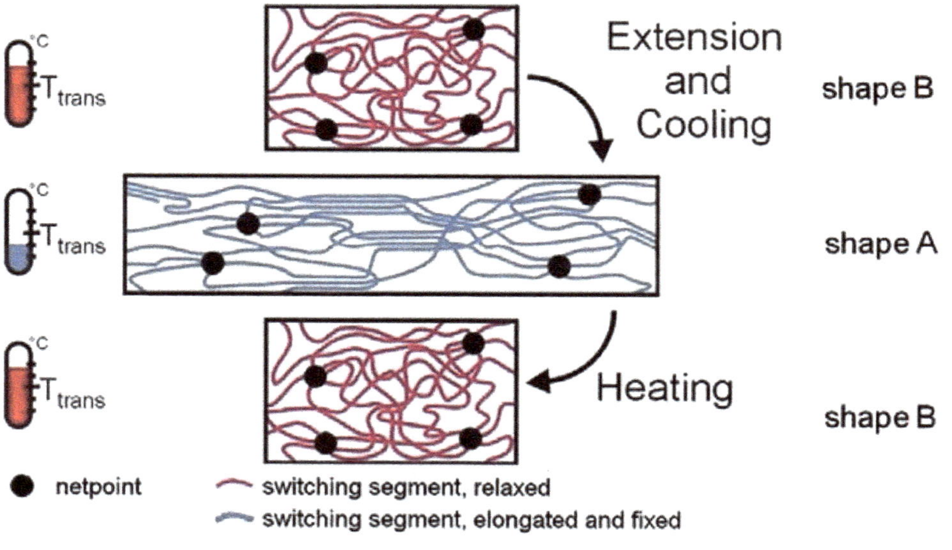

4.1.3 Advantages of Using Shape Memory Polymers (SMPs) Over Shape Memory Alloys (SMAs)

Shape Memory Polymers (SMPs) and Shape Memory Alloys (SMAs) are both materials that exhibit the unique ability to return to a pre-defined shape when exposed to specific stimuli. While SMAs have long been the dominant choice in shape memory materials due to their robust mechanical properties and responsiveness to various stimuli, SMPs have gained attention due to their distinct advantages in flexibility, cost, and potential for customization. This section explores the comparative advantages of SMPs over SMAs and highlights the contexts in which SMPs may offer superior performance.

- **High Recoverable Strain**: SMPs exhibit a remarkably high recoverable strain, reaching up to 400%, whereas SMAs typically achieve only 8%-12%. This substantial strain recovery makes SMPs ideal for applications requiring large, reversible deformations, such as deployable structures in aerospace or morphing surfaces in medical devices (Uyan & Celiktas, 2024).
- **Lightweight Composition**: SMPs, owing to their polymeric structure, have a lower density compared to SMAs, significantly reducing the overall weight of components. This property is advantageous in weight-sensitive industries such as automotive, aerospace, and robotics, where minimizing mass can lead to increased efficiency, lower fuel consumption, and enhanced performance (Pilate et al., 2016).
- **Tunable Properties**: Unlike SMAs, which are often limited to specific alloy compositions, SMPs can have their mechanical and thermal properties precisely tuned by modifying their polymer chemistry, cross-link density, and filler content. This adaptability allows for custom-designed materials that can be optimized for specific applications, including varying temperature actuation ranges (H. Zhang et al., 2023, 2024).
- **Biodegradability and Biocompatibility**: SMPs offer significant advantages in biomedical applications due to their ability to degrade within the body after fulfilling their function. Their biocompatibility makes them suitable for medical devices such as biodegradable stents, drug delivery systems, and sutures. The degradability reduces the need for secondary surgeries to remove the device, improving patient outcomes (Bond et al., 2023).
- **Cost Efficiency**: In comparison to SMAs, SMPs are generally more cost-effective to produce due to the lower cost of raw materials and simpler processing techniques, such as injection molding and extrusion, which are well-established in the polymer industry.

4.1.4 Limitations of Using Shape Memory Polymers (SMPs)

Despite the promising advantages of Shape Memory Polymers (SMPs), they also have limitations that can restrict their application in certain fields. These limitations arise from the inherent properties of polymers compared to metals and alloys, particularly in areas such as mechanical strength, thermal conductivity, and stimuli responsiveness. Understanding these challenges is crucial for determining where SMPs may be unsuitable and where alternative materials, such as SMAs, may provide better solutions. The following points outline the key limitations of SMPs in technical detail.

- **Lower Mechanical Strength and Stiffness**: SMPs have significantly lower tensile strength, modulus, and stiffness when compared to SMAs. This limits their use in high-load-bearing applications, where materials like titanium- or nickel-based SMAs are favored due to their superior mechanical properties and ability to withstand large stresses without permanent deformation.
- **Low Thermal Conductivity**: SMPs exhibit low thermal conductivity, which can impede their actuation speed and efficiency when responding to temperature changes. This can be problematic in applications where rapid temperature-induced shape recovery is critical, such as in aerospace structures or robotics.
- **Limited Stimuli Responsiveness**: SMPs are primarily responsive to thermal stimuli, whereas SMAs can be actuated by a range of stimuli, including electrical currents, magnetic fields, and mechanical loads. Recent advancements in SMP technology have introduced light- and magnetically responsive variants, but these remain less common and not as efficient as SMAs in multi-stimuli environments.
- **Slow Actuation and Recovery Time**: SMPs typically exhibit slower actuation and recovery times compared to SMAs, which can change shape rapidly under the influence of external stimuli. The slower response of SMPs is due to their reliance on the diffusion of thermal energy through the material, which limits their use in fast-moving, dynamic applications like robotics or active damping systems.
- **Narrow Working Temperature Range**: SMPs operate within a relatively narrow temperature range, typically determined by their glass transition temperature (Tg) or melting temperature (Tm). Exceeding this range can result in either incomplete recovery (at low temperatures) or permanent deformation (at high temperatures), thus restricting their use in extreme environments.
- **Fatigue and Durability Issues**: Repeated cycling of SMPs between their programmed shapes and their original form can lead to fatigue and gradual degradation of their shape memory properties over time. This contrasts with SMAs, which generally demonstrate better fatigue resistance, making SMAs more suitable for applications requiring long-term cyclic actuation.

4.2 Hydrogels and hydrogel composites in 4D printing

Hydrogels and hydrogel composites have emerged as critical materials for advanced 4D printing applications due to their distinctive properties. Hydrogels are hydrophilic polymers capable of absorbing and retaining large amounts of water, forming a soft, flexible network that closely mimics biological tissues. This makes them particularly valuable in biomedical fields, such as tissue engineering, where their biocompatibility and ability to support cell growth are essential. The ability to engineer hydrogels with customizable properties, such as varying levels of stiffness, porosity, and degradation rates, allows them to be tailored for specific applications, including drug delivery systems and wound dressings. However, the structural integrity of printed hydrogels remains a challenge, as they can collapse under their own weight due to their soft nature. Hydrogel composites address some of the limitations of pure hydrogels by integrating them with other materials to enhance their mechanical properties and functionality. By incorporating fillers such as nanomaterials, ceramic particles, or fibers, the mechanical strength and durability of hydrogel composites can be significantly improved. Additionally, hydrogel composites can be designed as smart materials, capable of responding to external stimuli like temperature, pH, or light. This property is particularly relevant in 4D printing, where printed objects are designed to change shape or function over time in response to environmental triggers, adding a dynamic aspect to the traditional 3D printed structures (Aufa et al., 2023; Lai et al., 2021; Y. Li et al., 2022; Song et al., 2023).

In 4D printing, hydrogels and hydrogel composites enable the creation of objects that can alter their form or behavior after being printed, often in response to moisture, heat, or other stimuli. For example, hydrogels that swell, shrink, or bend when exposed to water are being explored for applications such as soft robotics, smart textiles, and self-assembling structures. This responsive nature also lends itself to self-healing materials, which can repair themselves when damaged, offering new possibilities for long-lasting, adaptable technologies. In biomedical applications, 4D printed hydrogels are particularly promising for dynamic tissue scaffolds and smart drug delivery systems that adjust to the body's needs over time (S. Liu et al., 2020).

The use of hydrogels in 4D printing has opened up a wide range of applications, particularly in the biomedical sector, where they are used to create patient-specific implants, wound healing devices, and soft tissue scaffolds. Their similarity to biological tissues makes them ideal for regenerative medicine, where the goal is to create structures that can integrate seamlessly with the body. Outside of healthcare, hydrogels are being explored for use in soft robotics, where their flexibility and responsiveness make them suitable for creating robots that can move and adapt to their surroundings (Liu et al., 2020).

Despite the many advantages, several challenges remain. One major limitation is the mechanical strength of hydrogels, which can restrict their use in load-bearing applications. Enhancing their durability while maintaining their unique properties is a key area of ongoing research. Additionally, there is a need for scalable methods of production that can bring hydrogel-based 3D and 4D printing to larger commercial markets. Integrating hydrogels with electronic components to create bioelectronics or smart medical devices is another exciting frontier, combining the responsive nature of hydrogels with modern technology to create innovative solutions for healthcare and beyond.

4.2.1 Thermo-responsive hydrogels

Thermo-responsive hydrogels are advanced materials that exhibit significant changes in their physical properties in response to temperature variations. These hydrogels can undergo dramatic transitions, such as swelling or shrinking, depending on the temperature. They are generally composed of polymers with specific thermal response characteristics, including a lower critical solution temperature (LCST) or an upper critical solution temperature (UCST). LCST hydrogels, like poly(N-isopropylacrylamide) (PNIPAAm), typically swell and remain hydrophilic below a certain temperature, while they contract and become hydrophobic above this threshold. Conversely, UCST hydrogels absorb water and swell when the temperature exceeds a certain point, and they expel water and contract when it falls below this temperature. These thermo-responsive properties make hydrogels particularly valuable for applications such as controlled drug delivery, where the release of drugs can be regulated by body temperature or environmental changes, and tissue engineering, where they can serve as adaptive scaffolds that support cell growth. Additionally, the dynamic nature of these materials enables their use in soft robotics, where they can create responsive devices that alter their shape with temperature fluctuations (Tran et al., 2022).

4.2.2 Magnetic Responsive Hydrogels

Magnetic responsive hydrogels represent a significant advancement in the field of smart materials, combining the unique properties of hydrogels with magnetic responsiveness. These hydrogels offer the ability to undergo controlled transformations or actuation in response to magnetic fields, providing a powerful tool for applications requiring precise, remote-controlled behaviors. This section explores how magnetic fields can be utilized to manipulate the properties of hydrogels, the types of magnetic additives used, and the practical applications and mechanisms of these smart materials.

The use of magnetic fields as an external stimulus to control the properties of hydrogels offers several advantages, including rapid response times, high biocompatibility, and effective remote actuation. This is particularly valuable in biomedical applications, such as drug delivery systems, where non-invasive control is essential. By incorporating magnetic additives into the hydrogel matrix, researchers can tailor the material's behavior to respond to magnetic fields, allowing for sophisticated control mechanisms in various environments.

To achieve magnetic responsiveness, hydrogels are embedded with exogenous magnetic additives. These additives can include paramagnetic or ferromagnetic materials, which influence the hydrogel's properties when subjected to a magnetic field. Common magnetic additives include metal alloys (such as iron and neodymium alloys), metal oxides (like ferrous ferric oxide), and functionalized magnetic nanoparticles. These additives are integrated into polymer networks such as poly(N-isopropylacrylamide) (pNIPAM), poly(acrylamide) (pAAm), and gelatin, either through chemical or physical cross-linking. More advanced hydrogels employ covalent and coordination bonds to enhance the interactions between the magnetic particles and the polymer network, improving the material's responsiveness and functionality.

The synthesis of magnetic responsive hydrogels generally involves mixing magnetic additives with the polymer matrix. Unlike traditional hydrogels that require additional cross-linkers, magnetic responsive hydrogels often gel spontaneously upon mixing. This simplifies the preparation process and makes it easier to produce hydrogels with consistent and reliable properties (Tran et al., 2022). Each method has its own set of advantages and limitations, so the choice of preparation method depends on the desired properties of the final magnetic-responsive hydrogel and the specific application. Table 2 provides a detailed overview of different preparation methods for magnetic-responsive hydrogels. It outlines the key components, advantages, and disadvantages of each method, offering insights into the process and helping guide the selection of the most suitable technique based on specific requirements and applications.

Table 2. Preparation methods for magnetic-responsive hydrogels

Method	Description	Key Components	Advantages	Disadvantages
Direct Mixing	Magnetic nanoparticles are mixed directly with the polymer solution to form a gel.	Polymer (e.g., polyacrylamide), Magnetic nanoparticles (e.g., Fe3O4)	Simple and straightforward; no additional cross-linkers required.	May have uneven distribution of magnetic particles; gelation might be inconsistent.
In-Situ Polymerization	Magnetic nanoparticles are added during the polymerization process.	Polymer precursors (e.g., acrylamide), Magnetic nanoparticles	Ensures uniform distribution of magnetic particles; allows for controlled gelation.	Requires precise control of polymerization conditions; may need additional cross-linkers.
Layer-by-Layer Deposition	Alternating layers of polymer and magnetic nanoparticles are deposited to form the hydrogel.	Polymer solutions, Magnetic nanoparticles	Provides precise control over the distribution and orientation of magnetic particles.	Complex and time-consuming; requires specialized equipment.
Sol-Gel Transition	Magnetic nanoparticles are dispersed in a sol (liquid) form which transitions into a gel upon certain conditions.	Polymer solution, Magnetic nanoparticles	Allows for smooth and uniform gelation; suitable for large-scale production.	Requires specific conditions for sol-gel transition; may need additional steps for stabilization.
Electrospinning	Magnetic nanoparticles are incorporated into fibers during the electrospinning process to create a fibrous gel network.	Polymer solution, Magnetic nanoparticles	Produces nanofibers with high surface area; can be tailored for specific applications.	Requires specialized electrospinning equipment; may involve complex processing steps.
Hydrothermal Synthesis	Magnetic nanoparticles are synthesized and embedded in the polymer matrix under high temperature and pressure.	Polymer, Magnetic precursors	Produces well-dispersed magnetic nanoparticles; can enhance gel properties.	Complex and requires high-temperature equipment; may not be suitable for all polymers.
Freeze-Drying	Hydrogels with magnetic nanoparticles are freeze-dried to form a porous structure that retains magnetic responsiveness.	Polymer solution with magnetic nanoparticles	Creates a porous structure that can enhance responsiveness and functionality.	Requires a freeze-drying step which can be time-consuming; may affect hydrogel properties.

4.2.3 Electrical Responsive Hydrogels

Inspired by the principles of artificial muscle biomimicry, electrically responsive hydrogels are materials that undergo reversible expansion or contraction in response to stimuli such as solvent changes or the application of an external electrical field. These hydrogels hold significant potential in drug delivery systems, primarily due to their ability to offer rapid, precise, and programmable control over the release of therapeutic agents. The functionality of electrically responsive hydrogels in drug delivery is fundamentally linked to the behavior of ions within the hydrogel matrix under an applied electrical field. When an electrical field is introduced, it influences the movement of ions within the hydrogel, leading to a reorganization of the ion concentration at the interface between the hydrogel and the surrounding swelling medium. This reorganization is governed by the balance between the fixed charges present on the polymer chains of the hydrogel and the mobile counterions from the surrounding medium. The non-uniform distribution of ion concentrations inside and outside the hydrogel results in osmotic pressure gradients that drive the swelling or deswelling of the material, thereby modulating drug release kinetics (Brossier et al., 2024; Deng et al., 2024).

Electrically responsive hydrogels are often composed of polyelectrolytes, which are polymers with ionizable groups attached to their side chains or backbone. These groups can dissociate in an aqueous environment, imparting the material with the ability to respond to electrical stimuli. A variety of synthetic polyelectrolytes have been employed in the fabrication of these hydrogels, including:

- **Poly (vinyl alcohol) (PVA)**: Known for its excellent film-forming properties and biocompatibility (Hirai et al., 1993).
- **Poly(sodium maleate-co-sodium acrylate)**: A copolymer that combines the properties of sodium maleate and sodium acrylate, offering enhanced ion exchange capabilities (Gao et al., 2008).
- **PVA/pAAc (Polyacrylic Acid)**: This combination is used to improve the mechanical strength and swelling behavior of the hydrogel (L. Li & Hsieh, 2005).
- **pAAc/poly(N-vinylpyrrolidone) (PVP)**: PVP is a water-soluble polymer that, when combined with pAAc, forms hydrogels with superior swelling properties and responsiveness (Jin et al., 2013).
- **Sulfonated polystyrene (s-PS)**: This material is known for its strong ionic nature, making it highly responsive to electrical fields (Rahimi et al., 2012; Yao & Krause, 2003)

In addition to synthetic polyelectrolytes, natural polyelectrolytes such as proteins, polysaccharides, and polypeptides have been utilized due to their inherent biocompatibility and biodegradability. Examples include:

- **Proteins**: Fibrin, a natural protein, has been blended with pAAc to create hydrogels with enhanced mechanical and responsive properties(Rahimi et al., 2012)
- **Polysaccharides**: Chitosan, a natural polysaccharide, has been coupled with poly(N,N-dimethylacrylamide) to produce hydrogels that are both biocompatible and electrically responsive(Sarmad et al., 2013).
- **Polypeptides**: Alginate, a naturally occurring polypeptide, is combined with poly(methacrylic acid) to form hydrogels suitable for 4D printing applications in drug delivery systems (Kim et al., 2004).

These hybrid hydrogels, combining both natural and synthetic components, represent a promising frontier in the development of smart materials for controlled drug delivery, offering the advantages of both classes of materials. The integration of electrically responsive hydrogels into drug delivery platforms enables the design of systems with tunable release profiles, responsive to real-time physiological cues, and adaptable to the specific needs of various therapeutic applications.

4.2.4 Photo-Responsive Hydrogels

Light as a stimulus for controlling the behavior of 4D printed hydrogels represents a cutting-edge approach in the field of smart materials, particularly for the controlled delivery of therapeutic agents. The ability to remotely induce the expansion and contraction of these hydrogels through light exposure enables precise control over drug release, which is essential for advanced therapeutic applications.

4.2.4.1 Mechanisms of Light-Responsive Hydrogels

The two primary mechanisms by which light-responsive hydrogels function are **reversible crosslinking** and **photothermal excitation**. Both mechanisms rely on the inclusion of photoactive moieties within the hydrogel structure, which respond to specific wavelengths of light.

1. **Reversible Crosslinking:** In reversible crosslinking, the hydrogel matrix is functionalized with photoactive groups such as azobenzene or o-nitrobenzyl. Azobenzene, for example, undergoes a reversible photoisomerization between its trans and cis isomers when exposed to UV or visible light. This photoisomerization alters the spatial configuration of the polymer chains, leading to a

reversible change in the hydrogel's crosslinking density. Similarly, o-nitrobenzyl groups can undergo photocleavage upon UV light exposure, resulting in the breaking and reformation of crosslinks within the hydrogel matrix.

The reversible changes in crosslinking density directly affect the hydrogel's volume, swelling behavior, and mechanical properties. For instance, the trans-to-cis isomerization of azobenzene typically causes the polymer chains to contract, reducing the overall volume of the hydrogel. Conversely, reversing the isomerization (cis-to-trans) under visible light leads to the expansion of the hydrogel as the crosslinking density decreases. These reversible changes are crucial for applications where precise and dynamic control over drug release is needed.

Research has shown that hydrogels containing azobenzene derivatives can achieve significant volume changes of up to 200% in response to light, enabling the controlled release of drugs at rates that can be finely tuned by adjusting the light intensity and exposure time (Kloxin et al., 2009; Wu et al., 2011).

2. **Photothermal Excitation:** Photothermal excitation involves the incorporation of photothermal nanomaterials, such as gold nanoparticles, carbon nanotubes, or black phosphorus, into the hydrogel matrix. These materials absorb light, particularly in the near-infrared (NIR) spectrum, and efficiently convert it into heat. The localized heat generated by these nanomaterials induces a reversible dehydration and rehydration cycle within the hydrogel, leading to its contraction or expansion. The temperature increase caused by photothermal excitation affects the polymer network in two ways: it can trigger the collapse of temperature-sensitive polymers (e.g., poly(N-isopropylacrylamide) (pNIPAM)) or cause a phase transition that leads to a reduction in the hydrogel's water content. For instance, pNIPAM hydrogels, which have a lower critical solution temperature (LCST) around 32°C, shrink when the temperature exceeds this threshold due to the collapse of the polymer network. By precisely controlling the light exposure, researchers can modulate the extent of this thermal response, thereby controlling the drug release rate. In studies, hydrogels containing gold nanorods have demonstrated rapid and reversible volume changes in response to NIR light, with response times as short as a few seconds. These hydrogels have shown potential for applications where immediate drug release is critical, such as in emergency medicine or pain management (Deng et al., 2024; Fujigaya et al., 2008).

4.2.4.2 Integration of Nanomaterials in Photo-Responsive Hydrogels

Numerous types of nanomaterials have been incorporated into photo-responsive hydrogels to enhance their light sensitivity and mechanical properties:

- **Inorganic Nanomaterials:** Gold nanoparticles, known for their excellent biocompatibility and strong photothermal conversion efficiency, are frequently used in these hydrogels. Neodymium oxide, another inorganic material, has been used for its ability to absorb NIR light and induce a controlled thermal response within the hydrogel. Research has shown that gold nanoparticle-loaded hydrogels can achieve temperature increases of up to 50°C upon NIR irradiation, leading to significant and rapid hydrogel contraction (Sershen et al., 2005).
- **Carbon-Based Materials:** Single-walled carbon nanotubes (SWNTs) and single-walled nanohorns (SWNHs) are widely used for their ability to absorb NIR light and convert it into heat with high efficiency. These materials are integrated into hydrogels like agarose and pNIPAM to create smart hydrogels that undergo phase transitions upon NIR exposure. Studies have reported that SWNT-containing hydrogels can achieve a 60% reduction in volume within seconds of NIR irradiation, demonstrating their potential for rapid drug release applications (Yang et al., 2019).
- **Black Phosphorus:** As an emerging photothermal nanomaterial, black phosphorus exhibits strong absorption across a broad range of the light spectrum, including NIR. Its incorporation into hydrogels has enabled the development of materials that not only respond to light but also degrade in a controlled manner, which is particularly useful for applications requiring the timed release of drugs or the self-destruction of the hydrogel matrix after fulfilling its function. Experimental data indicate that black phosphorus-containing hydrogels can achieve a controlled degradation rate over days to weeks, depending on the light exposure parameters.

The additive manufacturing of these photo-responsive hydrogels for 4D printing is still a nascent field but holds tremendous potential. The ability to print hydrogels with spatially controlled distributions of photoactive moieties or photothermal nanomaterials opens new avenues for designing complex drug delivery systems with programmable and localized release profiles. For example, 4D printed hydrogels can be designed to respond to specific light patterns, allowing for the sequential release of different drugs from a single hydrogel matrix.

Recent advancements in 3D and 4D printing technologies have enabled the fabrication of hydrogels with intricate microarchitectures that enhance their light responsiveness and mechanical properties. These printed hydrogels can be tailored to release drugs in response to patient-specific needs, making them highly suitable for personalized medicine.

4.2.5 pH Responsive Hydrogels

pH-responsive hydrogels are a sophisticated class of smart materials that undergo significant physical changes—such as swelling, degradation, or controlled drug release—in response to fluctuations in environmental pH (Jeong & Gutowska, 2002; F. Liu & Urban, 2010; Roy & Gupta, 2003). These hydrogels are particularly valuable in biomedical fields, including drug delivery, tissue engineering, and diagnostics, due to their ability to precisely respond to the pH variations found in different body tissues and cellular environments (Gupta et al., 2002; K. S. Soppimath T. M. Aminabhavi & Rudzinski, 2002; Y. Qiu & Park, 2001; Roy & Gupta, 2003; Schmaljohann, 2006).

The human body exhibits a wide range of pH values across different tissues and organs:

- **Stomach:** Highly acidic with a pH ranging from 1.5 to 3.5.
- **Small Intestine:** Neutral to slightly alkaline, with a pH between 6 and 7.4.
- **Colon:** Slightly acidic to neutral, with a pH range of 5.5 to 7.
- **Tumor Microenvironment:** Tumors typically exhibit a slightly acidic extracellular pH of 6.5 to 6.9, unlike normal tissues.
- **Intracellular Compartments:** Lysosomes and endosomes have an acidic pH (around 4.5 to 5.5), whereas the cytoplasm maintains a more neutral pH of approximately 7.2.

These variations in pH provide strategic opportunities for designing hydrogels that can specifically respond to the local environment, ensuring that drug release occurs only at the intended site.

The responsiveness of these hydrogels is largely due to the presence of ionizable groups within their polymer network. These groups can accept or donate protons (H^+ ions) depending on the pH of the surrounding environment, leading to changes in the hydrogel's charge density, hydrophilicity, and crosslinking. The primary types of ionizable groups include:

1. **Acidic Groups (e.g., Carboxylic Acid, Sulfonic Acid):** At low pH (acidic conditions), these groups remain protonated and uncharged, resulting in reduced electrostatic repulsion between polymer chains and minimal swelling. As the pH increases, these groups ionize, becoming negatively charged, which leads to increased electrostatic repulsion and the expansion of the hydrogel. Poly (acrylic acid) (PAAc) is a common polymer in pH-responsive hydrogels. At acidic pH, the carboxyl groups (-COOH) are protonated, causing the hydrogel to remain

in a collapsed state. As the pH shifts to neutral or alkaline, the carboxyl groups ionize ($-COO^-$), significantly increasing the hydrogel's swelling.
2. **Basic Groups (e.g., Amino Groups):** Basic groups are protonated and positively charged in acidic environments, which increases swelling due to electrostatic repulsion. As the pH rises, these groups deprotonate, lose their charge, and cause the hydrogel to contract. Poly (N,N-dimethylaminoethyl methacrylate) (PDMAEMA) exhibits pH-responsive behavior where its amino groups are protonated at low pH, leading to swelling. At higher pH levels, deprotonation occurs, causing the hydrogel to shrink.

With the diverse pH environments found throughout various tissues and cellular compartments in the human body, 4D printed pH-responsive hydrogels offer a significant advantage for precisely controlled drug release in specific areas under particular physiological conditions. These advanced materials leverage the dynamic pH variations in different regions, making them highly effective for targeted therapeutic interventions offering several key advantages:

1. **Precision in Drug Release:** By designing hydrogels that respond to the specific pH of a target tissue, drugs can be released precisely where they are needed. For example, a pH-responsive hydrogel can be programmed to remain stable in the acidic stomach environment and then release its payload in the more neutral pH of the small intestine. 4D printing allows for the fabrication of hydrogels with time-dependent responses. For instance, a hydrogel could be designed to gradually swell and release its drug load over time, responding to pH changes that occur as a disease progresses or in response to metabolic activities.
2. **Customization and Personalization:** 4D printing enables the customization of hydrogels based on individual patient profiles. This means that the drug delivery system can be tailored to the specific pH variations in a patient's body, optimizing treatment outcomes. For instance, in personalized cancer therapy, a hydrogel could be designed to release chemotherapeutics in response to the unique pH profile of a patient's tumor. The dynamic nature of 4D printed hydrogels allows them to adapt to changing conditions within the body. If a tissue's pH changes due to disease progression or treatment, the hydrogel can respond accordingly, adjusting the rate or timing of drug release.
3. **Enhanced Stability and Protection:** The structure of pH-responsive hydrogels can protect drugs from premature degradation in unfavorable pH environments, such as the acidic conditions of the stomach. This ensures that the drug remains intact until it reaches the target site. The degradation of the hydrogel itself can be controlled by the pH of its environment. In tissue engineering or wound healing,

for example, a hydrogel scaffold can gradually degrade as the surrounding tissue heals, releasing bioactive molecules that promote regeneration.

4. **Reduced Side Effects:** By releasing drugs specifically at the target site, pH-responsive hydrogels minimize systemic exposure, reducing the risk of side effects. For example, a hydrogel designed to release anti-inflammatory drugs in the colon can reduce the potential for gastrointestinal side effects that occur with oral administration.

4.2.6 Water Responsive Hydrogels

Water-responsive hydrogels, often referred to as superabsorbent polymers, are three-dimensional, crosslinked macromolecular networks known for their remarkable ability to absorb and retain large amounts of liquid. This high swelling capacity makes them ideal for encapsulating and releasing therapeutic agents in biological environments. Typically composed of ionic monomers and featuring a weak cross-linking structure, these hydrogels can absorb significant quantities of water. As a result, they have become highly popular in commercial applications, especially in the healthcare sector, where they are used in products such as pharmaceuticals, personal hygiene items, wound dressings, and drug delivery systems (Cipriano et al., 2014; Sharma et al., 2016).

Historically, most water-responsive hydrogels have been synthesized from petrochemical sources, with a heavy reliance on acrylic monomers like acrylamide (AAm), acrylic acid (AAc), and their copolymers (Karadağ et al., 2001; Z. S. Liu & Rempel, 1997; Zhu et al., 2012). However, there has been a growing shift towards using more environmentally friendly and biocompatible materials, driven by concerns over the limited degradability and potential biocompatibility issues of traditional synthetic hydrogels. In response to these concerns, bio-based water-responsive hydrogels have been developed, utilizing renewable resources such as cellulose, soy protein, starch, natural gums, chitin, and various hybrids and composites. These advancements offer a sustainable and customizable approach to creating 4D printed hydrogels for drug delivery applications, combining improved environmental impact with enhanced performance in biological settings (Athukoralalage et al., 2019; Ma et al., 2015; Yoshimura et al., 2005).

5 COMPARISONS BETWEEN TRADITIONAL METHODS, 3D PRINTING AND 4D PRINTING

The table below provides a detailed comparison between traditional manufacturing methods, 3D printing, and 4D printing, highlighting their unique attributes in terms of materials used, manufacturing processes, customization potential, costs, scalability, and more. This comparison is essential for understanding which method best suits different manufacturing needs, whether for large-scale production, specialized designs, or adaptive and responsive applications (Campbell et al., 2014; Hatami et al., 2024b; Tibbits et al., 2014).

Table 3.

Criteria	Traditional Methods	3D Printing	4D Printing
Materials Used	Typically uses a broad range of raw materials like metals, plastics, wood, and composites. Requires large amounts of material, often leading to significant waste.	Utilizes a wide variety of materials, including plastics, resins, metals, ceramics, and even bio-materials. The additive process uses only the material needed, significantly reducing waste.	Employs smart materials that can change shape, structure, or function over time in response to stimuli such as heat, moisture, or light. This allows for dynamic objects that can adapt to their environment.
Manufacturing Process	Involves subtractive processes such as machining, casting, forging, and molding. These methods are often time-consuming, require skilled labor, and involve multiple steps, including significant post-processing.	An additive manufacturing process that builds objects layer by layer directly from a digital model. Allows for rapid prototyping and the creation of complex geometries with less manual labor.	Builds on 3D printing by introducing time as a factor, where the printed objects can self-assemble or change configuration after printing. The process is still in the experimental phase but offers innovative possibilities.
Customization	Customization is typically difficult and expensive, especially for complex shapes. Modifications often require new tooling or significant rework, increasing costs and time.	Highly customizable, allowing for the production of intricate and unique designs that would be challenging or impossible with traditional methods. Modifications can be made directly in the digital model without the need for new tools.	Adds a new dimension of customization by creating objects that can adapt or change over time. Ideal for applications requiring objects that must respond to environmental conditions or user interactions.
Cost & Speed	Generally cost-effective for large-scale mass production but can be slow and expensive for small batches or custom designs. Setup costs are high, but per-unit costs decrease significantly with volume.	More cost-effective for low-volume production, custom designs, and prototypes. Speed varies depending on the object's size and complexity but is generally faster for creating prototypes and small-scale production.	Currently more expensive and slower compared to traditional and 3D printing due to the complexity of materials and processes involved. Offers unique capabilities that justify the cost in specific, high-value applications.

continued on following page

Table 3. Continued

Criteria	Traditional Methods	3D Printing	4D Printing
Waste & Sustainability	Often produces significant waste, particularly in subtractive processes like machining where excess material is removed. Sustainability depends on the material used and the efficiency of the process.	The additive process minimizes waste by using only the necessary amount of material to build the object. It is considered more sustainable, especially when combined with recyclable or biodegradable materials.	Shares the waste-reduction benefits of 3D printing and further enhances sustainability through the creation of objects that can adapt, self-repair, or have longer lifespans. Potential to reduce the need for replacement or repairs, thereby reducing long-term waste.
Applications	Widely used across various industries, including automotive, aerospace, consumer goods, and construction. Suitable for high-volume production where consistency and durability are critical.	Commonly used in rapid prototyping, medical devices, custom parts, aerospace, automotive industries, and even emerging fields like food and fashion. Ideal for applications requiring complex, customized designs.	Emerging in high-tech fields such as aerospace, robotics, and medical devices where objects that change shape or properties over time provide significant advantages. Potential applications include self-assembling structures, adaptive devices, and responsive systems.
Scalability	Highly scalable for mass production, with significant economies of scale. Costs per unit decrease substantially as production volume increases.	Less scalable for mass production, making it more suitable for low-volume, custom, or prototype production. Scalability challenges exist, particularly in maintaining quality and speed in large-scale production.	Still in the experimental and developmental stages, with significant scalability challenges. However, it shows promise for specialized applications where adaptability and functionality are prioritized over production volume.

continued on following page

Table 3. Continued

Criteria	Traditional Methods	3D Printing	4D Printing
Post-Processing	Often requires significant post-processing, including finishing, assembling, and treating parts to meet final specifications. Can involve multiple steps, adding time and cost to the production process.	May require post-processing such as cleaning, sanding, or applying coatings, depending on the material and desired finish. Generally simpler and faster than traditional methods but still necessary for achieving the final product quality.	Post-processing can include programming the smart materials to respond to specific stimuli, adding an additional layer of complexity. The process can be intricate, requiring precise control over material behavior after printing.
Flexibility	Less flexible; changes in design or production often require retooling, which can be time-consuming and costly. Not well-suited for dynamic environments where frequent changes are needed.	Extremely flexible, allowing for rapid changes in design without the need for new tools or significant rework. Ideal for iterative design processes and environments that demand quick adaptability.	Introduces adaptability, allowing printed objects to change shape, structure, or function over time in response to environmental triggers. This flexibility is particularly valuable in dynamic or unpredictable environments where traditional static designs would fail.

6 NEXT-GENERATION APPLICATIONS OF 4D PRINTING TECHNOLOGY

4D printing technology is set to revolutionize various industries by enabling the creation of adaptive, self-healing, and self-assembling materials and products. In the defense and protection sector, 4D printing is being explored for the development of materials that can change their properties in response to environmental conditions. For instance, military vehicles could be coated with 4D-printed materials that adapt to different environments, while uniforms made from these materials could alter their shape or provide enhanced protection upon contact with sharp objects. Additionally, self-assembling materials could streamline logistics and deployment in warfare, allowing for rapid assembly of critical components in the field.

In consumer products, 4D printing offers the potential to create everyday items that adapt to changing environmental conditions, such as temperature and humidity, based on user needs. This adaptability could lead to products that are more durable and reduce waste, as they can be reconfigured or repaired after initial use. For example, shoes made with 4D-printed materials could become waterproof when exposed to rain or change their shape to provide better support under different conditions.

One of the most promising applications of 4D printing lies in the development of pipes that can alter their shape and orientation in response to fluid flow. These pipes can dynamically adjust to varying flow conditions, optimizing fluid transport efficiency. Additionally, they possess self-healing properties, allowing them to automatically repair any damage, significantly reducing maintenance costs and preventing potential failures in critical infrastructure.

The construction industry stands to benefit significantly from 4D printing through the development of self-adapting building materials. These materials could sense and respond to environmental changes such as temperature, humidity, and wind conditions. For instance, building facades could adjust their thermal properties based on the weather, enhancing energy efficiency and occupant comfort. This adaptability could lead to smarter, more resilient infrastructure.

The medical field is another area where 4D printing holds immense promise. 4D bioprinting can be used to create adaptive stents, implants, and medical devices that respond to the body's environment, improving patient outcomes. This technology also enables the development of nanoscale devices that can self-assemble within the body to perform specific medical functions, such as targeted drug delivery or minimally invasive surgery. Moreover, 4D printing can advance tissue engineering, potentially enabling the fabrication of organs for transplants that are more compatible with the patient's body (Castro et al., 2017; Javaid & Haleem, 2019; Y.-C. Li et al., 2016).

In the realm of robotics, 4D printing facilitates the creation of flexible, adaptive robots that can adjust to environmental changes. Soft robotics, which requires materials that can deform and respond to stimuli, benefits greatly from 4D-printed components, such as electroactive polymers (EAPs). These materials can be used to fabricate soft actuators and other robotic components that improve the versatility and efficiency of robotic systems, allowing them to perform tasks that traditional rigid robots cannot ((Raviv et al., 2014).

Aerospace is another industry where 4D printing is making significant strides. The technology allows for the manufacturing of lightweight, self-assembling components that are ideal for space missions. These materials can adjust to extreme environmental conditions in space, reducing the need for heavy and rigid structures. For example, shape memory alloys like Nitinol can be 4D-printed to create components that are both durable and adaptable, potentially revolutionizing spacecraft design and operation.

Self-evolving structures, which change shape in response to external stimuli such as water, represent another promising application of 4D printing. These materials could be used in deep-sea exploration, where traditional materials might fail, by enabling the self-assembly of tools and repair systems in underwater environments. Such technology could also enhance the capabilities of deep-sea exploration vehicles and diving suits (Raviv et al., 2014).

Active origami, an innovative application of 4D printing, involves creating foldable structures that can autonomously change shape. This technology is particularly useful in industries requiring compact, deployable solutions, such as aerospace and packaging. By integrating smart materials like shape memory polymers (SMPs), 4D-printed origami structures can transition from flat forms to complex 3D shapes in response to environmental triggers, providing a versatile approach to space-saving design and deployment (Ge et al., 2014; Nadgorny et al., 2016).

Lastly, 4D printing is advancing the development of sensors and flexible electronics. The technology enables the creation of sensors that are highly responsive to environmental changes, such as temperature, moisture, and pH levels. These sensors are not only cost-effective but also lightweight, accurate, and capable of self-healing, which extends their operational lifespan. Additionally, 4D printing is being used to fabricate flexible electronics, including smart textiles that can adapt to user needs and environmental conditions, further enhancing the functionality and sustainability of consumer products and industrial applications (Ge et al., 2014; White et al., 2001).

These next-generation applications of 4D printing technology highlight its transformative potential across various fields, emphasizing the versatility, adaptability, and innovative capabilities that this emerging technology brings to modern manufacturing and design.

7 CHALLENGES AND FUTURE PROSPECTS OF 4D PRINTING

4D printing, a cutting-edge research area that emerged from 3D printing technology, is still in its infancy, with 3D printing itself being a developing field. Given its nascent stage, 4D printing faces several limitations and challenges that need to be addressed for it to reach its full potential. The three primary challenges in 4D printing pertain to technology, materials, and design ((Ahmed et al., 2021b).

One of the major hurdles in 4D printing is the limited availability of suitable 3D printing technologies. Currently, only a few research institutes globally have access to 3D printers capable of producing 4D structures. Presently, 4D printing is mainly conducted using PolyJet and Selective Laser Melting (SLM) technologies. PolyJet printers require the smart material to be in a liquid state, where it solidifies upon exposure to UV light. SLM technology, on the other hand, uses a laser to melt powder particles and create the desired structure. Both methods impose specific requirements on the smart materials—liquid and UV-responsive for PolyJet printing, and powder form for SLM. Consequently, only a limited range of smart materials can currently be used in 4D printing. There is a pressing need for technological innovation to develop new printing technologies that can accommodate a broader range of smart materials.

The materials used in 4D printing also present significant challenges. For example, shape memory alloys (SMAs) like NiTi require precise control during printing to avoid defects, as these materials are sensitive to composition. Some smart materials may degrade with continuous deformation, while others, such as hydrogels, may deteriorate after repeated cycles of wetting and drying. In certain cases, materials must be mixed with other components to make them printable, which can complicate the printing process. Furthermore, the interaction between smart materials and 3D printing technology can be problematic, necessitating further research to discover new smart materials that can be directly printed, are durable under continuous use, and interact seamlessly with existing printing technologies.

Designing smart structures for 4D printing is another critical challenge. These structures must be capable of adjusting to their environment and performing the required changes. Proper design is essential to achieve the desired results, especially when producing actuators that function optimally in a pre-strain mode. However, only a few studies have successfully achieved this pre-strain mode, and no 3D printing technique currently exists that can directly print pre-strain membranes. Therefore, more research and development are needed in designing smart structures tailored for 4D printing. Additionally, the environmental impact of 4D printing has not been thoroughly explored. However, a 2020 study suggested that energy consumption in 4D printing of polymers could be reduced by 37.33% by optimizing shape programming and recovery temperatures, increasing the concentration of multifunctional crosslinkers, and adjusting printing parameters.

Once these limitations are overcome, 4D printing has the potential to revolutionize manufacturing and product design. It could lower capital requirements, reduce time-to-market, facilitate easier transportation of products, and enable the production of compact items, leading to a new and efficient business model. In aerospace, 4D printed structures could be highly beneficial for space missions, as they can withstand harsh environmental conditions. Components for satellites and aircraft printed using 4D printing could be designed to self-assemble in space, simplifying assembly and deformation processes. Additionally, 4D printing could enhance the design of 3D structures by enabling user-responsive products that adapt to environmental conditions and user requirements.

Furthermore, many smart materials remain unexplored, offering significant opportunities for advancing 4D printing and expanding its applications. New materials, such as custom textile composites, carbon fibers, and printed wood grain, could open up new possibilities in 4D printing. While current applications focus primarily on the shape-changing capabilities of 4D printed structures, future research could reveal additional functionalities, making these structures truly multi-functional.

The diverse responses of 4D printed materials to different stimuli could be further explored, adding to their potential uses.

It is estimated that by 2025, the global market for 4D printing will reach a value of approximately $537.8 million. As 4D printing is inherently multidisciplinary, future collaborations among various fields could enhance control over the shape evolution of these structures. Ultimately, 4D printing is poised to have a profound impact on everyday life, driving innovation across multiple industries.

CONCLUSION

The chapter has provided an in-depth exploration of 4D printing, showcasing its potential to revolutionize manufacturing by integrating time as an active element in the production process. Unlike traditional 3D printing, which creates static objects, 4D printing introduces smart materials that can alter their shape, properties, and functionality in response to environmental stimuli. This dynamic capability positions 4D printing as a transformative technology across multiple industries, including healthcare, aerospace, construction, and textiles.

One of the key takeaways from this discussion is the role of smart materials—such as shape memory polymers, hydrogels, and other stimuli-responsive substances—in enabling the adaptive behaviors characteristic of 4D-printed objects. These materials have opened new avenues for innovation, particularly in creating products that can self-assemble, self-repair, and respond autonomously to environmental changes. Such advancements promise to enhance the functionality, efficiency, and sustainability of products across various sectors.

Moreover, the chapter highlighted the broad range of applications that 4D printing could impact. In healthcare, for instance, 4D-printed scaffolds that mimic the dynamic nature of biological tissues represent a significant leap forward in tissue engineering. In aerospace, the development of materials that can adapt their properties during flight to optimize performance under varying conditions exemplifies the practical benefits of 4D printing. Additionally, the potential to create sustainable, self-repairing materials aligns with global efforts to reduce waste and extend the lifecycle of products.

However, the chapter also identified several challenges that must be addressed to fully realize the potential of 4D printing. These include the need for further advancements in material science, the development of more sophisticated printing technologies, and the refinement of design methodologies that can accommodate the complexities of programmable matter. Addressing these challenges will be crucial for scaling up the technology from experimental phases to widespread industrial adoption.

4D printing stands at the frontier of manufacturing innovation, with the potential to fundamentally alter how we design, produce, and interact with materials. As research progresses and technology matures, 4D printing could usher in a new era of smart, responsive, and sustainable products that meet the evolving needs of various industries, thereby shaping the future of manufacturing.

REFERENCES

Agarwal, T., Chiesa, I., Costantini, M., Lopamarda, A., Tirelli, M. C., Borra, O. P., Varshapally, S. V. S., Kumar, Y. A. V., Koteswara Reddy, G., De Maria, C., Zhang, L. G., & Maiti, T. K. (2023). Chitosan and its derivatives in 3D/4D (bio) printing for tissue engineering and drug delivery applications. *International Journal of Biological Macromolecules*, *246*, 125669. DOI: 10.1016/j.ijbiomac.2023.125669

Ahmed, A., Arya, S., Gupta, V., Furukawa, H., & Khosla, A. (2021a). 4D printing: Fundamentals, materials, applications and challenges. *Polymer*, *228*, 123926. DOI: 10.1016/j.polymer.2021.123926

Ahmed, A., Arya, S., Gupta, V., Furukawa, H., & Khosla, A. (2021b). 4D printing: Fundamentals, materials, applications and challenges. *Polymer*, *228*, 123926. https://doi.org/https://doi.org/10.1016/j.polymer.2021.123926. DOI: 10.1016/j.polymer.2021.123926

Alli, Y. A., Anuar, H., Manshor, M. R., Okafor, C. E., Kamarulzaman, A. F., Akçakale, N., Mohd Nazeri, F. N., Bodaghi, M., Suhr, J., & Mohd Nasir, N. A. (2024). Optimization of 4D/3D printing via machine learning: A systematic review. *Hybrid Advances*, *6*, 100242. DOI: 10.1016/j.hybadv.2024.100242

Alshebly, Y. S., Nafea, M., Mohamed Ali, M. S., & Almurib, H. A. F. (2021). Review on recent advances in 4D printing of shape memory polymers. *European Polymer Journal*, *159*, 110708. https://doi.org/https://doi.org/10.1016/j.eurpolymj.2021.110708. DOI: 10.1016/j.eurpolymj.2021.110708

Ameta, K. L., Solanki, V. S., Singh, V., Devi, A. P., Chundawat, R. S., & Haque, S. (2022). Critical appraisal and systematic review of 3D & 4D printing in sustainable and environment-friendly smart manufacturing technologies. *Sustainable Materials and Technologies*, *34*, e00481. DOI: 10.1016/j.susmat.2022.e00481

Athukoralalage, S. S., Balu, R., Dutta, N. K., & Choudhury, N. R. (2019). 3D bioprinted nanocellulose-based hydrogels for tissue engineering applications: A brief review. In *Polymers* (Vol. 11, Issue 5). MDPI AG. DOI: 10.3390/polym11050898

Aufa, A. N., Ismail, Z., & Zaki Hassan, M. (2023). Emerging trends in 4d printing of hydrogels in the biomedical field: A review. *Materials Today: Proceedings*. https://doi.org/https://doi.org/10.1016/j.matpr.2023.01.101

Baig, A., Barse, R., Paryekar, A., & Jagtap, V. (2024). Evolution of pharmaceuticals using 3D and 4D printing. *Intelligent Pharmacy*. https://doi.org/https://doi.org/10.1016/j.ipha.2024.05.008

Behl, M., & Lendlein, A. (2007). Shape-memory polymers. *Materials Today*, *10*(4), 20–28. https://doi.org/https://doi.org/10.1016/S1369-7021(07)70047-0. DOI: 10.1016/S1369-7021(07)70047-0

Bond, G., Mahjoubnia, A., Zhao, W., King, S. D., Chen, S., & Lin, J. (2023). 4D printing of biocompatible, hierarchically porous shape memory polymeric structures. *Biomaterials Advances*, *153*, 213575. https://doi.org/https://doi.org/10.1016/j.bioadv.2023.213575. DOI: 10.1016/j.bioadv.2023.213575

Bonetti, L., Natali, D., Pandini, S., Messori, M., Toselli, M., & Scalet, G. (2024). 4D printing of semi-crystalline crosslinked polymer networks with two-way shape-memory effect. *Materials & Design*, *238*, 112725. https://doi.org/https://doi.org/10.1016/j.matdes.2024.112725. DOI: 10.1016/j.matdes.2024.112725

Brossier, T., Habib, M., Benkhaled, B. T., Volpi, G., Lapinte, V., & Blanquer, S. (2024). 4D printing of hydrogels based on poly(oxazoline) and poly(acrylamide) copolymers by stereolithography. *Materials Advances*, *5*(7), 2750–2758. DOI: 10.1039/D3MA00665D

Campbell, T. A., Tibbits, S., & Garrett, B. (2014). *THE NEXT WAVE: 4D PRINTING PROGRAMMING THE MATERIAL WORLD*.

Castro, N. J., Meinert, C., Levett, P., & Hutmacher, D. W. (2017). Current developments in multifunctional smart materials for 3D/4D bioprinting. *Current Opinion in Biomedical Engineering*, *2*, 67–75. https://doi.org/https://doi.org/10.1016/j.cobme.2017.04.002. DOI: 10.1016/j.cobme.2017.04.002

Chen, J., Virrueta, C., Zhang, S., Mao, C., & Wang, J. (2024). 4D printing: The spotlight for 3D printed smart materials. *Materials Today*, *77*, 66–91. https://doi.org/https://doi.org/10.1016/j.mattod.2024.06.004. DOI: 10.1016/j.mattod.2024.06.004

Chopra, V., Fuentes-Velasco, V., Nacif-Lopez, S. R., Melendez-Malpicca, J., Mendez-Hernandez, A. S., Ramos-Mendez-Iris, L. F., Arroyo-Jimenez, D. A., Reyes-Segura, D. G., Gonzalez-Y-Mendoza, P., Sanchez-Hernandez, K. A., Spinola-Corona, E., Vazquez-del-Mercado-Pardiño, J. A., & Chauhan, G. (2024, October). Gonzalez-Y-Mendoza, P., Sanchez-Hernandez, K. A., Spinola-Corona, E., Vazquez-del-Mercado-Pardiño, J. A., & Chauhan, G. (2024). Advancements in 3D-4D printing of hydroxyapatite composites for bone tissue engineering. *Ceramics International*, *50*(20), 38819–38840. Advance online publication. DOI: 10.1016/j.ceramint.2024.07.266

Cipriano, B. H., Banik, S. J., Sharma, R., Rumore, D., Hwang, W., Briber, R. M., & Raghavan, S. R. (2014). Superabsorbent Hydrogels That Are Robust and Highly Stretchable. *Macromolecules*, *47*(13), 4445–4452. DOI: 10.1021/ma500882n

Deng, C., Qu, J., Dong, J., Guo, Y., Wu, X., Fang, Y., Sun, X., Wei, Y., & Li, Z. (2024). 4D printing of magnetic smart structures based on light-cured magnetic hydrogel. *Chemical Engineering Journal*, *494*, 152992. https://doi.org/https://doi.org/10.1016/j.cej.2024.152992. DOI: 10.1016/j.cej.2024.152992

Dowaidar, M. (2024). Guidelines for biomimetic 3D/4D printing in drug delivery. *Materials Chemistry and Physics*, *325*, 129692. DOI: 10.1016/j.matchemphys.2024.129692

Fujigaya, T., Morimoto, T., Niidome, Y., & Nakashima, N. (2008). NIR Laser-Driven Reversible Volume Phase Transition of Single-Walled Carbon Nanotube/Poly(N-isopropylacrylamide) Composite Gels. *Advanced Materials*, *20*(19), 3610–3614. https://doi.org/https://doi.org/10.1002/adma.200800494. DOI: 10.1002/adma.200800494

Gao, Y., Xu, S., Wu, R., Wang, J., & Wei, J. (2008). Preparation and characteristic of electric stimuli responsive hydrogel composed of polyvinyl alcohol/poly (sodium maleate-co-sodium acrylate). *Journal of Applied Polymer Science*, *107*(1), 391–395. https://doi.org/https://doi.org/10.1002/app.26375. DOI: 10.1002/app.26375

Ge, Q., Dunn, C. K., Qi, H. J., & Dunn, M. L. (2014). Active origami by 4D printing. *Smart Materials and Structures*, *23*(9), 94007. DOI: 10.1088/0964-1726/23/9/094007

Ge, Q., Sakhaei, A. H., Lee, H., Dunn, C. K., Fang, N. X., & Dunn, M. L. (2016a). Multimaterial 4D Printing with Tailorable Shape Memory Polymers. *Scientific Reports*, *6*(1), 31110. Advance online publication. DOI: 10.1038/srep31110

Ge, Q., Sakhaei, A. H., Lee, H., Dunn, C. K., Fang, N. X., & Dunn, M. L. (2016b). Multimaterial 4D Printing with Tailorable Shape Memory Polymers. *Scientific Reports*, *6*(1), 31110. DOI: 10.1038/srep31110

Grira, S., Khalifeh, H. A., Alkhedher, M., & Ramadan, M. (2023). 3D printing algae-based materials: Pathway towards 4D bioprinting. *Bioprinting (Amsterdam, Netherlands)*, *33*, e00291. DOI: 10.1016/j.bprint.2023.e00291

Gross, B. C., Erkal, J. L., Lockwood, S. Y., Chen, C., & Spence, D. M. (2014). Evaluation of 3D printing and its potential impact on biotechnology and the chemical sciences. *Analytical Chemistry*, *86*(7), 3240–3253. DOI: 10.1021/ac403397r

Gupta, P., Vermani, K., & Garg, S. (2002). Hydrogels: From controlled release to pH-responsive drug delivery. *Drug Discovery Today*, *7*(10), 569–579. https://doi.org/https://doi.org/10.1016/S1359-6446(02)02255-9. DOI: 10.1016/S1359-6446(02)02255-9

Hasanzadeh, R., Mihankhah, P., Azdast, T., Rasouli, A., Shamkhali, M., & Park, C. B. (2023). Biocompatible tissue-engineered scaffold polymers for 3D printing and its application for 4D printing. *Chemical Engineering Journal*, *476*, 146616. DOI: 10.1016/j.cej.2023.146616

Hatami, H., Almahmeed, W., Kesharwani, P., & Sahebkar, A. (2024a). Exploring the potential of 3D and 4D printing in advancing stent manufacturing for cardiovascular diseases. *European Polymer Journal*, *212*, 113035. https://doi.org/https://doi.org/10.1016/j.eurpolymj.2024.113035. DOI: 10.1016/j.eurpolymj.2024.113035

Hatami, H., Almahmeed, W., Kesharwani, P., & Sahebkar, A. (2024b). Exploring the potential of 3D and 4D printing in advancing stent manufacturing for cardiovascular diseases. *European Polymer Journal*, *212*, 113035. DOI: 10.1016/j.eurpolymj.2024.113035

Hazrat Ali, M., Abilgaziyev, A., & Adair, D. (2019). 4D printing: A critical review of current developments, and future prospects. *International Journal of Advanced Manufacturing Technology*, *105*(1-4), 701–717. DOI: 10.1007/s00170-019-04258-0

Hirai, T., Nemoto, H., Suzuki, T., Hayashi, S., & Hirai, M. (1993). Actuation of Poly(vinyl alcohol) Gel by Electric Field. *Journal of Intelligent Material Systems and Structures*, *4*(2), 277–279. DOI: 10.1177/1045389X9300400219

Javaid, M., & Haleem, A. (2019). 4D printing applications in medical field: A brief review. *Clinical Epidemiology and Global Health*, *7*(3), 317–321. https://doi.org/https://doi.org/10.1016/j.cegh.2018.09.007. DOI: 10.1016/j.cegh.2018.09.007

Jeong, B., & Gutowska, A. (2002). Lessons from nature: Stimuli-responsive polymers and their biomedical applications. *Trends in Biotechnology*, *20*(7), 305–311. https://doi.org/https://doi.org/10.1016/S0167-7799(02)01962-5. DOI: 10.1016/S0167-7799(02)01962-5

Jin, S., Gu, J., Shi, Y., Shao, K., Yu, X., & Yue, G. (2013). Preparation and electrical sensitive behavior of poly (N-vinylpyrrolidone-co-acrylic acid) hydrogel with flexible chain nature. *European Polymer Journal*, *49*(7), 1871–1880. https://doi.org/https://doi.org/10.1016/j.eurpolymj.2013.04.022. DOI: 10.1016/j.eurpolymj.2013.04.022

Jolaiy, S., Yousefi, A., Hosseini, M., Zolfagharian, A., Demoly, F., & Bodaghi, M. (2024). Limpet-inspired design and 3D/4D printing of sustainable sandwich panels: Pioneering supreme resiliency, recoverability and repairability. *Applied Materials Today*, *38*, 102243. DOI: 10.1016/j.apmt.2024.102243

Karadağ, E., Saraydın, D., & Güven, O. (2001). Radiation Induced Superabsorbent Hydrogels. Acrylamide/Itaconic Acid Copolymers. *Macromolecular Materials and Engineering*, *286*(1), 34–42. https://doi.org/https://doi.org/10.1002/1439-2054(20010101)286:1<34:AID-MAME34>3.0.CO;2-J. DOI: 10.1002/1439-2054(20010101)286:1<34::AID-MAME34>3.0.CO;2-J

Khalid, M. Y., Arif, Z. U., Noroozi, R., Hossain, M., Ramakrishna, S., & Umer, R. (2023). 3D/4D printing of cellulose nanocrystals-based biomaterials: Additives for sustainable applications. *International Journal of Biological Macromolecules*, *251*, 126287. DOI: 10.1016/j.ijbiomac.2023.126287

Kholkhoev, B. Ch., Nikishina, A. N., Bardakova, K. N., Matveev, Z. A., Sapozhnikov, D. A., Efremov, Y. M., Timashev, P. S., & Burdukovskii, V. F. (2024). 4D-printing of high-temperature shape-memory polymers based on polyimide, N,N-dimethylacrylamide and photoactive cross-linkers. *Polymer*, *299*, 126978. https://doi.org/https://doi.org/10.1016/j.polymer.2024.126978. DOI: 10.1016/j.polymer.2024.126978

Kim, S. J., Yoon, S. G., Lee, Y. H., & Kim, S. I. (2004). Bending behavior of hydrogels composed of poly(methacrylic acid) and alginate by electrical stimulus. *Polymer International*, *53*(10), 1456–1460. https://doi.org/https://doi.org/10.1002/pi.1560. DOI: 10.1002/pi.1560

Kloxin, A. M., Kasko, A. M., Salinas, C. N., & Anseth, K. S. (2009). Photodegradable Hydrogels for Dynamic Tuning of Physical and Chemical Properties. *Science*, *324*(5923), 59–63. DOI: 10.1126/science.1169494

Kshirsagar, M., Ambike, S. D., Jaya Prakash, N., Kandasubramanian, B., & Deshpande, P. (2023). Origami engineering: Creating dynamic functional materials through folded structures. *Hybrid Advances*, *4*, 100092. DOI: 10.1016/j.hybadv.2023.100092

Kumar, P., Suryavanshi, P., Kumar Dwivedy, S., & Banerjee, S. (2024). Stimuli-responsive materials for 4D Printing: Mechanical, Manufacturing, and Biomedical Applications. *Journal of Molecular Liquids*, *410*, 125553. https://doi.org/https://doi.org/10.1016/j.molliq.2024.125553. DOI: 10.1016/j.molliq.2024.125553

Lai, J., Ye, X., Liu, J., Wang, C., Li, J., Wang, X., Ma, M., & Wang, M. (2021). 4D printing of highly printable and shape morphing hydrogels composed of alginate and methylcellulose. *Materials & Design*, *205*, 109699. https://doi.org/https://doi.org/10.1016/j.matdes.2021.109699. DOI: 10.1016/j.matdes.2021.109699

Lalegani Dezaki, M., & Bodaghi, M. (2024). 4D printing and programming of continuous fibre-reinforced shape memory polymer composites. *European Polymer Journal*, *210*, 112988. https://doi.org/https://doi.org/10.1016/j.eurpolymj.2024.112988. DOI: 10.1016/j.eurpolymj.2024.112988

Leist, S. K., & Zhou, J. (2016). Current status of 4D printing technology and the potential of light-reactive smart materials as 4D printable materials. *Virtual and Physical Prototyping*, *11*(4), 249–262. DOI: 10.1080/17452759.2016.1198630

Leong, T. G., Randall, C. L., Benson, B. R., Bassik, N., Stern, G. M., & Gracias, D. H. (2009). Tetherless thermobiochemically actuated microgrippers. *Proceedings of the National Academy of Sciences of the United States of America*, *106*(3), 703–708. DOI: 10.1073/pnas.0807698106

Leong, T. G., Randall, C. L., Benson, B. R., Bassik, N., Stern, G. M., & Gracias, D. H. (2024). *Tetherless thermobiochemically actuated microgrippers.*

Li, L., & Hsieh, Y.-L. (2005). Ultra-fine polyelectrolyte hydrogel fibres from poly(acrylic acid)/poly(vinyl alcohol). *Nanotechnology*, *16*(12), 2852–2860. DOI: 10.1088/0957-4484/16/12/020

Li, Y., Zheng, W., Li, B., Dong, J., Gao, G., & Jiang, Z. (2022). Double-layer temperature-sensitive hydrogel fabricated by 4D printing with fast shape deformation. *Colloids and Surfaces. A, Physicochemical and Engineering Aspects*, *648*, 129307. https://doi.org/https://doi.org/10.1016/j.colsurfa.2022.129307. DOI: 10.1016/j.colsurfa.2022.129307

Li, Y.-C., Zhang, Y. S., Akpek, A., Shin, S. R., & Khademhosseini, A. (2016). 4D bioprinting: The next-generation technology for biofabrication enabled by stimuli-responsive materials. *Biofabrication*, *9*(1), 12001. DOI: 10.1088/1758-5090/9/1/012001

Liao, E., Zhang, Y., Su, Y., Zhang, C., Geng, C., Li, C., Liu, X., Liu, Y., & Lu, A. (2024). Active mixing of two-component viscoelastic silicone ink at molecular level for spatiotemporally controlled 3D/4D printing of cellular silicones. *Polymer*, *302*, 127053. DOI: 10.1016/j.polymer.2024.127053

Liu, F., & Urban, M. W. (2010). Recent advances and challenges in designing stimuli-responsive polymers. *Progress in Polymer Science*, *35*(1), 3–23. https://doi.org/https://doi.org/10.1016/j.progpolymsci.2009.10.002. DOI: 10.1016/j.progpolymsci.2009.10.002

Liu, S., Chen, X., & Zhang, Y. (2020). Chapter 14 - Hydrogels and hydrogel composites for 3D and 4D printing applications. In K. K. Sadasivuni, K. Deshmukh, & M. A. Almaadeed (Eds.), *3D and 4D Printing of Polymer Nanocomposite Materials* (pp. 427–465). Elsevier. https://doi.org/https://doi.org/10.1016/B978-0-12-816805-9.00014-4

Liu, Z. S., & Rempel, G. L. (1997). Preparation of superabsorbent polymers by crosslinking acrylic acid and acrylamide copolymers. *Journal of Applied Polymer Science*, *64*(7), 1345–1353. https://doi.org/https://doi.org/10.1002/(SICI)1097-4628(19970516)64:7<1345:AID-APP14>3.0.CO;2-W. DOI: 10.1002/(SICI)1097-4628(19970516)64:7<1345::AID-APP14>3.0.CO;2-W

Lu, X., Liu, G., & Lu, J. (2024). Development of Ceramic 3D/4D Printing in China. *Additive Manufacturing Frontiers*, 200158. https://doi.org/https://doi.org/10.1016/j.amf.2024.200158

Ma, J., Li, X., & Bao, Y. (2015). Advances in cellulose-based superabsorbent hydrogels. *RSC Advances*, *5*(73), 59745–59757. DOI: 10.1039/C5RA08522E

Mallakpour, S., Tabesh, F., & Hussain, C. M. (2024). Entering a new era in pharmaceutics through 3D/4D printing technologies. In *Medical Additive Manufacturing* (pp. 267–280). Elsevier., DOI: 10.1016/B978-0-323-95383-2.00006-8

Meloni, M., Cai, J., Zhang, Q., Sang-Hoon Lee, D., Li, M., Ma, R., Parashkevov, T. E., & Feng, J. (2021). Engineering Origami: A Comprehensive Review of Recent Applications, Design Methods, and Tools. *Advanced Science (Weinheim, Baden-Wurttemberg, Germany)*, *8*(13), 2000636. Advance online publication. DOI: 10.1002/advs.202000636

Miao, Z., Du, J., Dong, F., Liu, Y., & Wang, X. (2020). Identifying technology evolution pathways using topic variation detection based on patent data: A case study of 3D printing. *Futures*, *118*, 102530. https://doi.org/https://doi.org/10.1016/j.futures.2020.102530. DOI: 10.1016/j.futures.2020.102530

Mohammad Vaezi Srisit Chianrabutra, B. M., & Yang, S. (2013). Multiple material additive manufacturing – Part 1: A review. *Virtual and Physical Prototyping*, *8*(1), 19–50. DOI: 10.1080/17452759.2013.778175

Moradi, A., Ansari, R., Hassanzadeh-Aghdam, M. K., & Jang, S.-H. (2024). Thermomechanical behavior of carbon nanotube/graphene nanoplatelet-reinforced shape memory polymer nanocomposites: A micromechanics-based finite element approach. *European Journal of Mechanics. A, Solids*, *107*, 105360. https://doi.org/https://doi.org/10.1016/j.euromechsol.2024.105360. DOI: 10.1016/j.euromechsol.2024.105360

Nadgorny, M., Xiao, Z., Chen, C., & Connal, L. A. (2016). Three-Dimensional Printing of pH-Responsive and Functional Polymers on an Affordable Desktop Printer. *ACS Applied Materials & Interfaces*, *8*(42), 28946–28954. DOI: 10.1021/acsami.6b07388

Nikishina, A. N., Kholkhoev, B. Ch., Bardakova, K. N., Matveev, Z. A., Kurdanova, Z. I., Shakhmurzova, K. T., Zhansitov, A. A., Khashirova, S. Yu., Timashev, P. S., & Burdukovskii, V. F. (2024). Photopolymerization-based 4D-printing of shape-memory materials containing high-performance polymers. *Reactive & Functional Polymers*, *202*, 106001. https://doi.org/https://doi.org/10.1016/j.reactfunctpolym.2024.106001. DOI: 10.1016/j.reactfunctpolym.2024.106001

Pei, E., & Loh, G. H. (2018). Technological considerations for 4D printing: An overview. *Progress in Additive Manufacturing*, *3*(1–2), 95–107. DOI: 10.1007/s40964-018-0047-1

Pereira, N. I. A., Oliveira, M. dos S., Reis, B. C. C., Nascimento, B. L., Carneiro, C. R., Arruda, T. R., Vieira, E. N. R., & Leite Junior, B. R. C. (2024). Unconventional sourced proteins in 3D and 4D food printing: Is it the future of food processing? *Food Research International*, *192*, 114849. DOI: 10.1016/j.foodres.2024.114849

Pilate, F., Toncheva, A., Dubois, P., & Raquez, J.-M. (2016). Shape-memory polymers for multiple applications in the materials world. *European Polymer Journal*, *80*, 268–294. https://doi.org/https://doi.org/10.1016/j.eurpolymj.2016.05.004. DOI: 10.1016/j.eurpolymj.2016.05.004

Qiu, R., Liu, X., Tian, H., Hu, Z., Wang, K., & Zhao, L. (2024). Exploration of bigel 4D printing with spontaneous colour change for monitoring bio-actives kinetic behaviour based on the dual-units 3D printer. *Journal of Food Engineering*, *367*, 111861. DOI: 10.1016/j.jfoodeng.2023.111861

Qiu, Y., & Park, K. (2001). Environment-sensitive hydrogels for drug delivery. *Advanced Drug Delivery Reviews*, *53*(3), 321–339. https://doi.org/https://doi.org/10.1016/S0169-409X(01)00203-4. DOI: 10.1016/S0169-409X(01)00203-4

Rahimi, N., Molin, D. G., Cleij, T. J., van Zandvoort, M. A., & Post, M. J. (2012). Electrosensitive Polyacrylic Acid/Fibrin Hydrogel Facilitates Cell Seeding and Alignment. *Biomacromolecules*, *13*(5), 1448–1457. DOI: 10.1021/bm300161r

Rahmatabadi, D., Khajepour, M., Bayati, A., Mirasadi, K., Amin Yousefi, M., Shegeft, A., Ghasemi, I., Baniassadi, M., Abrinia, K., Bodaghi, M., & Baghani, M. (2024). Advancing sustainable shape memory polymers through 4D printing of polylactic acid-polybutylene adipate terephthalate blends. *European Polymer Journal*, *216*, 113289. https://doi.org/https://doi.org/10.1016/j.eurpolymj.2024.113289. DOI: 10.1016/j.eurpolymj.2024.113289

Raviv, D., Zhao, W., McKnelly, C., Papadopoulou, A., Kadambi, A., Shi, B., Hirsch, S., Dikovsky, D., Zyracki, M., Olguin, C., Raskar, R., & Tibbits, S. (2014). Active Printed Materials for Complex Self-Evolving Deformations. *Scientific Reports*, *4*(1), 7422. DOI: 10.1038/srep07422

Roy, I., & Gupta, M. N. (2003). Smart Polymeric Materials: Emerging Biochemical Applications. *Chemistry & Biology*, *10*(12), 1161–1171. https://doi.org/https://doi.org/10.1016/j.chembiol.2003.12.004. DOI: 10.1016/j.chembiol.2003.12.004

Sarmad, S., Yenici, G., Gürkan, K., Keçeli, G., & Gürdağ, G. (2013). Electric field responsive chitosan–poly(N,N-dimethyl acrylamide) semi-IPN gel films and their dielectric, thermal and swelling characterization. *Smart Materials and Structures*, *22*(5), 55010. DOI: 10.1088/0964-1726/22/5/055010

Schmaljohann, D. (2006). Thermo- and pH-responsive polymers in drug delivery. *Advanced Drug Delivery Reviews*, *58*(15), 1655–1670. https://doi.org/https://doi.org/10.1016/j.addr.2006.09.020. DOI: 10.1016/j.addr.2006.09.020

Sershen, S. R., Mensing, G. A., Ng, M., Halas, N. J., Beebe, D. J., & West, J. L. (2005). Independent Optical Control of Microfluidic Valves Formed from Optomechanically Responsive Nanocomposite Hydrogels. *Advanced Materials*, *17*(11), 1366–1368. https://doi.org/https://doi.org/10.1002/adma.200401239. DOI: 10.1002/adma.200401239

Shahbazi, M., Jäger, H., Ettelaie, R., Mohammadi, A., & Asghartabar Kashi, P. (2023). Multimaterial 3D printing of self-assembling smart thermo-responsive polymers into 4D printed objects: A review. *Additive Manufacturing*, *71*, 103598. DOI: 10.1016/j.addma.2023.103598

Sharma, K., Kumar, V., Chaudhary, B., Kaith, B. S., Kalia, S., & Swart, H. C. (2016). Application of biodegradable superabsorbent hydrogel composite based on Gum ghatti-co-poly(acrylic acid-aniline) for controlled drug delivery. *Polymer Degradation & Stability*, *124*, 101–111. https://doi.org/https://doi.org/10.1016/j.polymdegradstab.2015.12.021. DOI: 10.1016/j.polymdegradstab.2015.12.021

Shirazi, S. F. S., Gharehkhani, S., Mehrali, M., Yarmand, H., Metselaar, H. S. C., Adib Kadri, N., & Osman, N. A. A. (2015). A review on powder-based additive manufacturing for tissue engineering: Selective laser sintering and inkjet 3D printing. In *Science and Technology of Advanced Materials* (Vol. 16, Issue 3). Institute of Physics Publishing. DOI: 10.1088/1468-6996/16/3/033502

Song, M., Zhu, G., & Guo, J. (2023). 4D printing of biodegradable shape memory double-network hydrogel for highly bionic devices. *Journal of Materials Research and Technology*, *24*, 2935–2945. https://doi.org/https://doi.org/10.1016/j.jmrt.2023.03.180. DOI: 10.1016/j.jmrt.2023.03.180

Soppimath, K. S., Aminabhavi, T. M., Dave, A. M., Kumbar, S. G., & Rudzinski, W. E. (2002). Stimulus-Responsive "Smart" Hydrogels as Novel Drug Delivery Systems. *Drug Development and Industrial Pharmacy*, *28*(8), 957–974. DOI: 10.1081/DDC-120006428

Tibbits, S. (2012). Design to Self-Assembly. *Architectural Design*, *82*(2), 68–73. DOI: 10.1002/ad.1381

Tibbits, S., McKnelly, C., Olguin, C., Dikovsky, D., & Hirsch, S. (2014). *4D Printing and Universal Transformation*. 539–548. DOI: 10.52842/conf.acadia.2014.539

Tofail, S. A. M., Koumoulos, E. P., Bandyopadhyay, A., Bose, S., O'Donoghue, L., & Charitidis, C. (2018). Additive manufacturing: Scientific and technological challenges, market uptake and opportunities. *Materials Today*, *21*(1), 22–37. https://doi.org/https://doi.org/10.1016/j.mattod.2017.07.001. DOI: 10.1016/j.mattod.2017.07.001

Tran, T. S., Balu, R., Mettu, S., Roy Choudhury, N., & Dutta, N. K. (2022). 4D Printing of Hydrogels: Innovation in Material Design and Emerging Smart Systems for Drug Delivery. In *Pharmaceuticals* (Vol. 15, Issue 10). MDPI. DOI: 10.3390/ph15101282

Uyan, M., & Celiktas, M. S. (2024). Novel constitutive models, challenges and opportunities of shape memory polymer composites. *Emerging Materials Research*, *13*(3), 1–20. DOI: 10.1680/jemmr.23.00116

Wan, X., He, Y., Liu, Y., & Leng, J. (2022). 4D printing of multiple shape memory polymer and nanocomposites with biocompatible, programmable and selectively actuated properties. *Additive Manufacturing*, *53*, 102689. DOI: 10.1016/j.addma.2022.102689

White, S. R., Sottos, N. R., Geubelle, P. H., Moore, J. S., Kessler, M. R., Sriram, S. R., Brown, E. N., & Viswanathan, S. (2001). Autonomic healing of polymer composites. *Nature*, *409*(6822), 794–797. DOI: 10.1038/35057232

Wu, Y., Wu, S., Tian, X., Wang, X., Wu, W., Zou, G., & Zhang, Q. (2011). Photoinduced reversible gel–sol transitions of dicholesterol-linked azobenzene derivatives through breaking and reforming of van der Waals interactions. *Soft Matter*, *7*(2), 716–721. DOI: 10.1039/C0SM00330A

Yang, M., Yuan, Z., Liu, J., Fang, Z., Fang, L., Yu, D., & Li, Q. (2019). Photoresponsive Actuators Built from Carbon-Based Soft Materials. *Advanced Optical Materials*, *7*(16), 1900069. https://doi.org/https://doi.org/10.1002/adom.201900069. DOI: 10.1002/adom.201900069

Yao, L., & Krause, S. (2003). Electromechanical Responses of Strong Acid Polymer Gels in DC Electric Fields. *Macromolecules*, *36*(6), 2055–2065. DOI: 10.1021/ma021326q

Yoshimura, T., Uchikoshi, I., Yoshiura, Y., & Fujioka, R. (2005). Synthesis and characterization of novel biodegradable superabsorbent hydrogels based on chitin and succinic anhydride. *Carbohydrate Polymers*, *61*(3), 322–326. https://doi.org/https://doi.org/10.1016/j.carbpol.2005.06.014. DOI: 10.1016/j.carbpol.2005.06.014

Zhang, H., Huang, S., Qian, Z., Zhang, Y., Sheng, J., Zhang, J., Agyenim-Boateng, E., Gao, Y., & Lu, J. (2024). Tunable deformation behavior of 4D printed shape memory polymers with embedded pre-stretched elastomers. *Chemical Engineering Journal*, *492*, 152187. https://doi.org/https://doi.org/10.1016/j.cej.2024.152187. DOI: 10.1016/j.cej.2024.152187

Zhang, H., Huang, S., Sheng, J., Wang, C., Zhang, J., & Zhu, M. Agyenim-Boateng, Emmanuel., Liang, C., Xue, B., Yang, H., Zhou, J., & Lu, J. (2023). 4D printing of shape memory polymer nanocomposites for enhanced performances and tunable response behavior. *European Polymer Journal, 201*, 112568. https://doi.org/https://doi.org/10.1016/j.eurpolymj.2023.112568

Zhang, Y., Gao, Y., Michelin, L., Josien, L., Vidal, L., Schrodj, G., Simon-Masseron, A., & Lalevée, J. (2022a). Photopolymerization of ceramic/zeolite reinforced photopolymers: Towards 3D/4D printing and gas adsorption applications. *European Polymer Journal*, *179*, 111552. https://doi.org/https://doi.org/10.1016/j.eurpolymj.2022.111552. DOI: 10.1016/j.eurpolymj.2022.111552

Zhang, Y., Gao, Y., Michelin, L., Josien, L., Vidal, L., Schrodj, G., Simon-Masseron, A., & Lalevée, J. (2022b). Photopolymerization of ceramic/zeolite reinforced photopolymers: Towards 3D/4D printing and gas adsorption applications. *European Polymer Journal*, *179*, 111552. DOI: 10.1016/j.eurpolymj.2022.111552

Zhou, X., Ren, L., Song, Z., Li, G., Zhang, J., Li, B., Wu, Q., Li, W., Ren, L., & Liu, Q. (2023). Advances in 3D/4D printing of mechanical metamaterials: From manufacturing to applications. *Composites. Part B, Engineering*, *254*, 110585. DOI: 10.1016/j.compositesb.2023.110585

Zhu, Z.-Q., Sun, H.-X., Qin, X.-J., Jiang, L., Pei, C.-J., Wang, L., Zeng, Y.-Q., Wen, S.-H., La, P.-Q., Li, A., & Deng, W.-Q. (2012). Preparation of poly(acrylic acid)–graphite oxide superabsorbent nanocomposites. *Journal of Materials Chemistry*, *22*(11), 4811–4817. DOI: 10.1039/c2jm14210d

Chapter 7
Energy Efficiency and Resource Optimization for Renewable and Sustainable Energy Systems

Dhirendra Patel
https://orcid.org/0000-0003-3308-168X
Amity University, India

ABSTRACT

The transition towards renewable and sustainable energy systems requires not only the deployment of clean energy technologies but also the maximization of energy efficiency and resource optimization throughout generation, conversion, distribution, and end-use. This chapter explores strategies, technologies, and methodologies to enhance energy efficiency and optimize resource utilization in renewable energy systems, including solar, wind, biomass, geothermal, and hydropower. Key focus areas include design optimization, energy storage integration, digitalization through smart grids, advanced control systems, and the circular economy in material usage. The chapter also discusses case studies, future opportunities, and the role of artificial intelligence (AI) in improving performance and sustainability outcomes

INTRODUCTION

Global efforts to combat climate change and reduce dependency on fossil fuels have intensified the adoption of renewable energy sources (RES) such as solar,

DOI: 10.4018/979-8-3373-5473-6.ch007

wind, hydro, biomass, and geothermal energy. However, to ensure these systems are truly sustainable, it is imperative to maximize energy efficiency and optimize resource utilization throughout their lifecycles. Energy efficiency reduces overall energy demand, while resource optimization minimizes environmental impacts and improves economic viability (Smith & Chen, 2020).

The growing global demand for energy, coupled with the urgent need to mitigate climate change, has driven an unprecedented shift toward renewable and sustainable energy systems. Renewable energy sources—such as solar, wind, hydro, geothermal, and biomass—are essential to decarbonizing the world's energy supply. However, simply deploying renewable technologies is not enough. Maximizing energy efficiency and optimizing resource use throughout the entire energy system are critical to achieving true sustainability (R. Patel & Kumar, 2022).

Energy efficiency often described as the "first fuel" is the most cost-effective and immediate way to reduce greenhouse gas emissions, lower energy costs, and improve energy security. It minimizes the energy input required for the same level of output, effectively stretching limited renewable resources further. Resource optimization complements energy efficiency by ensuring that materials, land, water, and other inputs are used in the most sustainable and economically viable way across the lifecycle of renewable energy systems, from production and installation to operation, maintenance, and decommissioning (Ahmed & Wang, 2019).

Moreover, integrating digitalization, advanced control systems, and smart grids enables real-time monitoring, predictive analytics, and demand-side management, all of which enhance the performance and reliability of renewable energy systems. Hybrid systems combining multiple renewable sources with energy storage further optimize resource use by taking advantage of complementary generation profiles, reducing intermittency, and minimizing the need for fossil fuel backup.

Together, energy efficiency and resource optimization form the backbone of a resilient, low-carbon energy transition. They reduce the environmental footprint of energy systems, improve economic feasibility, and enable renewable energy to meet the growing global demand sustainably. This chapter explores the principles, strategies, technologies, and future opportunities for enhancing energy efficiency and resource optimization within renewable and sustainable energy systems (Singh & Patel, 2020).

The world is undergoing a profound transformation in how energy is produced, distributed, and consumed. Driven by the dual imperatives of reducing greenhouse gas emissions and ensuring reliable access to energy, renewable and sustainable energy systems have emerged as central pillars of global energy strategies. Solar photovoltaics, wind turbines, biomass plants, geothermal systems, and hydropower installations are being deployed at an unprecedented scale. Yet despite this rapid growth, significant challenges remain in ensuring these systems operate efficiently and use resources sustainably.

Energy efficiency represents one of the most powerful and accessible solutions to address these challenges. By minimizing the amount of energy required to provide goods and services, energy efficiency directly reduces energy consumption, operating costs, and emissions. For renewable systems, higher efficiency translates into more usable energy extracted from each unit of natural resource whether it be sunlight, wind, or biomass thereby amplifying the positive environmental impact of renewables.

Resource optimization, meanwhile, focuses on minimizing waste and maximizing the value of every material, component, and system involved in renewable energy infrastructure. This encompasses the entire lifecycle of renewable energy technologies, from the extraction of raw materials to manufacturing, installation, operation, maintenance, and eventual decommissioning or recycling. Incorporating resource optimization into renewable systems ensures not only environmental sustainability but also economic competitiveness by lowering resource costs and dependency on critical raw materials.

The increasing complexity of modern energy systems driven by the variability of renewable energy sources, evolving demand patterns, and decentralized generation—requires advanced strategies to optimize both energy use and resource allocation. Smart grids, Internet of Things (IoT) devices, and artificial intelligence (AI) enable dynamic control of energy flows, real-time optimization of generation and consumption, and predictive maintenance of energy assets, all of which contribute to improved efficiency and resource use (Brown, 2021; Chandra, 2023).

Moreover, the adoption of hybrid renewable energy systems combining solar, wind, biomass, hydro, and energy storage provides opportunities to exploit complementary generation profiles, balance supply and demand more effectively, and reduce reliance on fossil fuel-based backup systems. When paired with demand-side management strategies such as dynamic pricing and intelligent load shifting, these hybrid systems can significantly enhance overall system efficiency and resilience.

Another key element in the sustainable transition is the circular economic approach, which seeks to design renewable energy systems for durability, reparability, and recyclability. By recovering valuable materials and components at the end of their service lives, the circular economy reduces waste, conserves resources, and mitigates environmental impacts associated with mining and manufacturing. Collectively, energy efficiency and resource optimization are essential enablers of sustainable energy systems capable of supporting global climate and development goals. As countries work to achieve net-zero emissions targets, these strategies will play an increasingly central role in ensuring renewable energy systems deliver reliable, affordable, and environmentally responsible power for all (Kumar & Rao, 2022; M. Zhang & Li, 2018).

This chapter explores the key concepts, methodologies, technological solutions, and real-world applications related to energy efficiency and resource optimization in renewable energy systems. It highlights the opportunities and challenges involved and discusses how future advancements, including AI and digitalization, can further enhance the sustainability and performance of clean energy technologies.

IMPORTANCE OF ENERGY EFFICIENCY IN RENEWABLE SYSTEMS

Energy efficiency is often described as the "hidden fuel" or "first fuel" in the transition to sustainable energy systems. It delivers the most immediate and cost-effective means of reducing energy consumption and associated greenhouse gas emissions while enhancing the performance, reliability, and economic viability of renewable energy systems. Unlike fossil-fuel-based generation, renewable energy sources are subject to natural variability, and maximizing efficiency ensures that the limited energy captured from intermittent resources like sunlight and wind is used as effectively as possible (Balani et al., 2021; D. Patel et al., 2021; A. Sharma & Patel, 2023; J. F. Silva & Moradi, 2020; Smith & Chen, 2020; Wang & Gupta, 2021). Indeed, the importance of types of Energy Efficiency in Renewable Systems as shown by Figure 1.

Figure 1. Energy efficiency in renewable systems

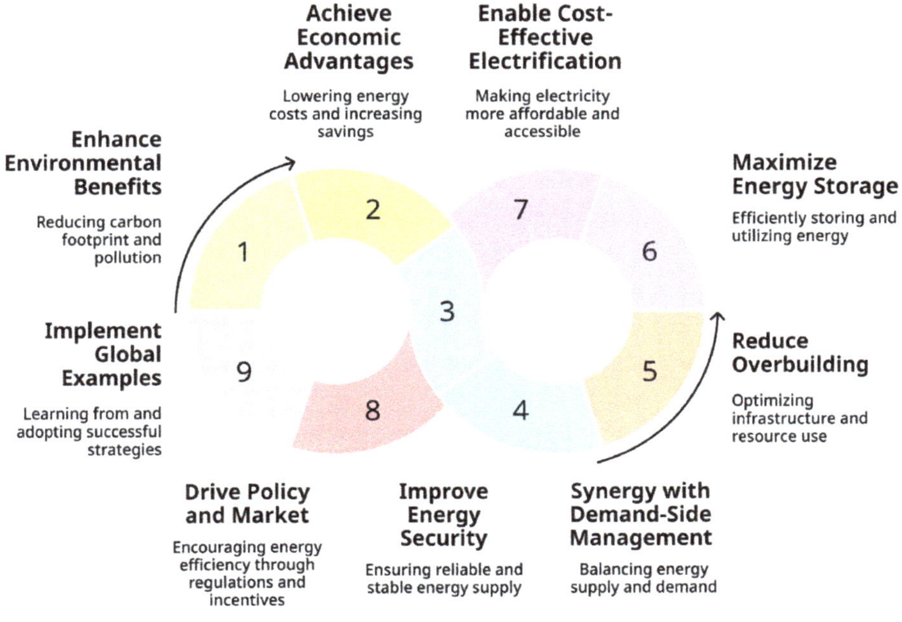

Enhancing Environmental Benefits of Renewables

Renewable energy systems are inherently cleaner than fossil fuel systems, but their true environmental benefit depends heavily on how efficiently they operate. Low-efficiency systems waste valuable energy harvested from natural resources, requiring over-dimensioning of capacity, larger land areas, and greater material usage to meet the same energy demand. This can lead to unnecessary environmental impacts, including increased land use for solar farms or wind parks, and more mining for critical materials like rare earth elements. improving energy efficiency helps renewable systems deliver greater environmental benefits per unit of resource invested, minimizing their ecological footprint (Osei & Zhang, 2019; Das & Reddy, 2022; Gonzalez & Kim, 2021; D. Patel, Mishra, et al., 2022; Mehta, 2020)0. Following step.

- A 10% improvement in the efficiency of a solar photovoltaic system can significantly reduce the number of panels needed, lowering the demand for silicon, aluminum frames, and glass.

- More efficient wind turbine gearboxes reduce energy losses in conversion, which not only improves output but also extends equipment life, reducing the frequency of replacements and associated resource consumption.

Economic Advantages of Energy Efficiency

Energy efficiency improves the economic case for renewable systems in several ways: according to the International Energy Agency (IEA), every dollar invested in energy efficiency can save two dollars in energy supply investments.

- Reduced Levelized Cost of Energy (LCOE): Efficiency improvements lower the LCOE by increasing energy output without proportional increases in capital or operational expenditures.
- Decreased Maintenance Costs: Efficient systems typically operate at lower stress levels, experience less wear and tear, and require fewer repairs or replacements.
- Lowered Grid Infrastructure Investments: Efficient renewable systems reduce peak demand and energy losses, which can defer or eliminate the need for expensive grid reinforcements.

Improved Energy Security and System Reliability

Energy efficiency enhances energy security by reducing overall energy consumption and, therefore, dependence on imported fuels or critical raw materials. In decentralized renewable systems, such as microgrids in rural communities, improving energy efficiency can

- Lower required generation capacity, enabling the use of smaller, less costly renewable installations.
- Increase the resilience of energy supply by extending the duration batteries can meet demand during cloudy, windless, or low-resource periods.
- Reduce the frequency of blackouts by smoothing energy demand and generation fluctuations.

Synergy with Demand-Side Management

Energy efficiency measures complement demand-side management (DSM) strategies by lowering overall demand, shifting consumption to match renewable availability, and reducing grid stress during peak periods. This synergy enhances grid flexibility, lowers the need for fossil-based peaking plants, and supports the integration of higher shares of renewable energy for instance (Rana & Patel, 2021).

- Smart thermostats improve heating and cooling efficiency while shifting energy use to periods of high renewable generation (e.g., midday solar peaks).
- LED lighting reduces energy use in residential, commercial, and industrial sectors, amplifying the positive impact of DSM programs.

Reduced Need for Overbuilding Renewable Capacity

Intermittent renewable resources like solar and wind often require overbuilding capacity to guarantee supply during low-resource periods. By improving the efficiency of systems and end-use applications, energy demand can be reduced to levels that better match renewable supply, minimizing the need for excessive capacity or large-scale energy storage.

- Passive solar building designs and high-performance insulation can reduce heating and cooling loads by up to 50–80%, lowering the need for oversized renewable generation to meet building energy demands.

Maximizing Benefits of Energy Storage

Energy storage is a critical component of renewable energy systems to address the variability of generation. Efficient buildings, appliances, and industrial processes, therefore, maximize the economic and environmental benefits of energy storage investments. Efficiency improvements:

- Reduce the amount of energy that needs to be stored by lowering overall consumption.
- Extending the duration batteries can provide energy, increasing system autonomy during periods without sunlight or wind.
- Decrease the number of charging and discharging cycles, extending battery lifespan and reducing material replacement needs (Choi & Lee, 2023).

Enabling Cost-Effective Electrification

The shift towards electrification of heating, cooling, and transport relies on efficient end-use technologies. Without energy efficiency, electrification could dramatically increase electricity demand, requiring massive investments in generation and grid infrastructure. This efficiency-driven reduction in electricity demand is crucial for a cost-effective, renewable-powered future.

- Electric vehicles (EVs) with high-efficiency drivetrains consume significantly less electricity per kilometer traveled, reducing the strain on renewable generation capacity.
- High-efficiency heat pumps can deliver three to four times more heat energy than the electricity they consume, allowing buildings to heat or cool effectively with minimal energy input.

Policy and Market Drivers for Energy Efficiency in Renewables

Governments, international agencies, and financial institutions increasingly recognize the importance of energy efficiency alongside renewable deployment. These policy and market drivers create a supportive environment for maximizing the benefits of energy efficiency in renewable systems.

- Efficiency performance standards for renewable technologies (e.g., minimum efficiency ratings for solar panels).
- Incentives for energy-efficient building designs that integrate renewables.
- Market mechanisms like white certificates, where efficiency savings can be monetized.
- International targets, such as the United Nations Sustainable Development Goal 7, which promotes both renewable energy and efficiency improvements.

Global Examples of Energy Efficiency in Renewable Systems

- Denmark: A pioneer in wind energy, Denmark couples' high turbine efficiency with district heating systems that recover waste heat from combined heat and power plants, maximizing renewable energy utilization.
- Germany: Energy-efficient building codes like Passivhaus standards reduce heating demand by up to 90%, allowing solar and geothermal systems to meet nearly all energy needs.
- India: In off-grid rural electrification, solar home systems paired with efficient LED lighting and DC appliances extend limited solar generation to power entire households.

RESOURCE OPTIMIZATION CONCEPTS

Resource optimization in renewable and sustainable energy systems refers to the strategic planning, design, and operation of systems to minimize the use of materials, land, water, and other inputs while maximizing energy output and system performance. Unlike traditional resource-intensive energy systems, renewable energy must aim for circular, efficient, and minimal-impact use of finite resources to achieve sustainability (Hernandez & Patel, 2022). This section explores the core concepts, methodologies, and practices of resource optimization in renewable systems.as shown by the figure 2resource optimization in renewable energy.

Figure 2. Resource optimization in renewable energy

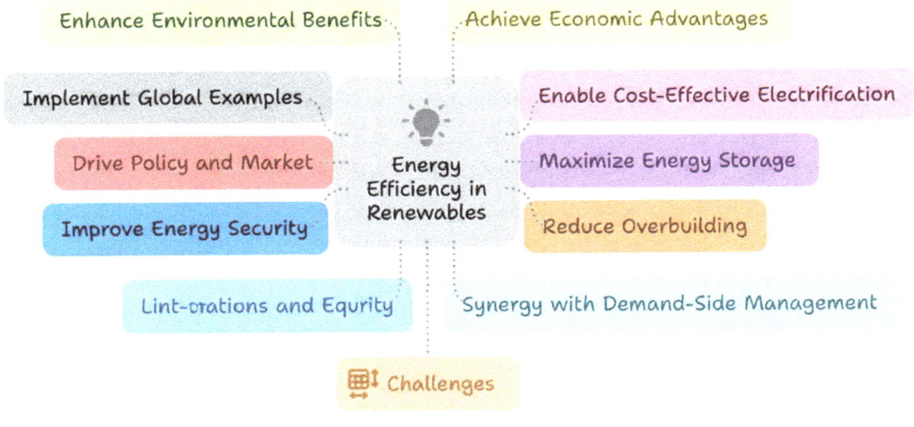

Defining Resource Optimization

Resource optimization seeks to the Reduce the quantity of resources needed to achieve a given energy output. Reuse components or materials wherever possible to extend their lifecycle. Recycle valuable materials at the end of system life to minimize waste and environmental impact. These principles, often summarized by the 3Rs (Reduce, Reuse, Recycle), are critical for making renewable energy systems both environmentally sustainable and economically viable.

Material Efficiency in Renewable Energy Technologies

Renewable systems rely on a variety of critical materials, including metals like silver, copper, aluminum, rare earth elements, and polymers. Key strategies for material efficiency include. Thin-film photovoltaics: Reduces the amount of semiconductor material required compared to conventional silicon panels. Advanced blade materials: Lighter and stronger composites in wind turbine blades improve energy capture while using fewer raw materials. Recycled metals: Reusing aluminum frames, copper wiring, and steel support reduces environmental impact compared to primary material extraction. Material efficiency helps conserve scarce resources, lowers costs, and minimizes the environmental footprint of renewable energy infrastructure (Kapoor & Jain, 2023).

Land and Spatial Resource Optimization

Renewable energy systems, especially solar and wind, have significant land-use requirements. Optimizing land resources involves. Dual land use: Installing solar panels on rooftops, parking lots, or above canals reduces land competition with agriculture and natural habitats. Agrophotovoltaics(APV) Combining agriculture with solar installations allows simultaneous crop production and energy generation on the same land. Wind turbine sitting. Positioning turbines in low-impact areas, such as offshore or existing industrial zones, reduces conflicts with human activities and ecosystems. Such strategies increase land-use efficiency, improve social acceptance, and reduce ecological impacts (D. Patel, Varma, et al., 2022).

Water Resource Optimization

Water is a critical resource in renewable systems such as geothermal, hydropower, and even solar thermal plants. Optimization strategies include. Dry cooling technologies in concentrating solar power (CSP) plants, which use air instead of water to condense steam, reducing water consumption. Run-of-the-river hydropower, which generates electricity without large reservoirs, minimizing water storage impacts and ecosystem disruptions. Efficient geothermal systems that reinject used geothermal fluids back into the reservoir to maintain pressure and reduce freshwater withdrawals. Minimizing water use is essential in arid regions and contributes to the overall sustainability of renewable systems (Jacobson, 2022).

Operational Optimization

Resource optimization extends to the operational phase of renewable systems, where intelligent control can significantly reduce energy and resource waste: Advanced forecasting algorithms predict solar irradiance or wind speeds, enabling operators to schedule energy dispatch, maintenance, and grid balancing efficiently. Predictive maintenance systems identify and fix issues before failures occur, reducing downtime, replacement needs, and material use. Optimal load management ensures that energy generation and consumption are balanced in real time, avoiding unnecessary curtailment or overproduction. These strategies optimize the use of both renewable resources and supporting infrastructure (D. Patel, 2025).

Lifecycle Resource Optimization

Considering the entire lifecycle of renewable energy systems—from raw material extraction to end-of-life disposal—is critical for resource efficiency. Lifecycle optimization involves. Design for disassembly: Products designed for easy dismantling allow components and materials to be recovered and reused or recycled. Extended producer responsibility (EPR): Manufacturers are held responsible for taking back systems at the end of their life, incentivizing sustainable design and recycling. Modular system design: Components can be upgraded individually instead of replacing entire systems, reducing resource use and waste. Lifecycle thinking is essential to building a circular economy in renewable energy (Raj & Mehra, 2022).

Hybrid Renewable Systems for Resource Synergy

Combining multiple renewable sources in hybrid systems optimizes the use of natural resources by leveraging complementary generation profiles. Solar-wind hybrids: Solar produces during sunny days, while wind often peaks at night or during cloudy weather, enabling continuous energy generation with minimal storage. Biomass solar hybrids: Biomass provides dispatchable backup power to balance the variability of solar generation, reducing the need for expensive battery storage. Geothermal-solar hybrids: Shared infrastructure, such as heat exchangers or steam turbines, improves resource efficiency and reduces capital investment. Hybrid systems reduce the redundancy of resources and lower the total material and land requirements.

Digital Tools for Resource Optimization

The rise of digitalization provides powerful tools for optimizing resource use in renewable systems. Digital twins create virtual models of physical assets, allowing operators to simulate scenarios, optimize performance, and extend asset life. Big data analytics enables detailed tracking of resource consumption and identifies inefficiencies across the energy system. Internet of Things (IoT) sensors provide real-time data on energy generation, storage, and consumption, enabling responsive optimization strategies. Digital tools not only improve operational efficiency but also provide insights for long-term planning and design.

Circular Economy Approaches

Integrating circular economy principles in renewable systems ensures continuous resource optimization by designing systems with end-of-life reuse in mind. Recycling rare-earth magnets in wind turbines reduces the environmental impact of mining and processing. Reusing solar modules in secondary markets extends their useful life. Recovering valuable metals from retired batteries reduces the need for virgin raw material extraction. Circular economic strategies help renewable energy systems align with broader sustainability goals beyond clean energy production.

Challenges in Resource Optimization

Despite its benefits, resource optimization faces several barriers. Economic constraints: High upfront costs for efficient materials or systems can deter investment, particularly in developing regions. Technological limitations: Some materials, like certain rare earths, are difficult or costly to recycle with current technology. Policy gaps: Lack of clear regulations on end-of-life management for renewable systems can lead to increased waste and lost resources. Addressing these challenges will require continued innovation, supportive policies, and industry collaboration. Resource optimization is essential for maximizing the environmental and economic benefits of renewable energy systems. By focusing on efficient use of materials, land, water, and operational strategies alongside integrating circular economy principles renewable energy can achieve true sustainability. Effective resource optimization ensures that clean energy systems do not simply shift environmental burdens elsewhere but instead contribute to a regenerative and resilient global energy future (Reddy & Sharma, 2022).

STRATEGIES FOR ENHANCING ENERGY EFFICIENCY

Maximizing energy efficiency is central to sustainable renewable energy systems. Strategies to enhance efficiency span the entire lifecycle from technology design and system integration to operation and user engagement. This section details key strategies across different technologies, their implementation, and their impact on system performance (S. Omar & Ali, 2020).

4.1 Design Optimization of Renewable Energy Technologies

Design improvements are often the first and most impactful strategy for enhancing energy efficiency.

- Solar Photovoltaics (PV)
  - Use of high-efficiency monocrystalline and bifacial panels increases energy conversion rates from ~15% to over 22%.
  - Anti-reflective and self-cleaning coatings minimize energy losses due to dirt or reflection.
  - Maximum Power Point Tracking (MPPT) optimizes inverter operation to extract maximum energy under varying light conditions.
- Wind Turbines
  - Advanced blade aerodynamics reduce drag and increase lift, boosting energy output.
  - Variable-speed operation allows turbines to adapt to fluctuating wind speeds, maintaining efficiency across a wider range.
  - Direct-drive generators eliminate gearbox losses, improving overall system efficiency by 2–4%.
- Hydropower Systems
  - Adjustable-pitch blades in Kaplan turbines maintain efficiency even with varying water flow.
  - Modern micro-hydro designs reduce frictional and turbulence losses in small streams.

enhanced conceptual graph including battery storage alongside solar, wind, and hydropower generation profiles, showing how batteries can help meet evening demand peaks shown figure 3 by the demand graph.

Figure 3. Conceptual generation profiles including battery storage in renewable energy systems (Thomas & Singh, 2021)

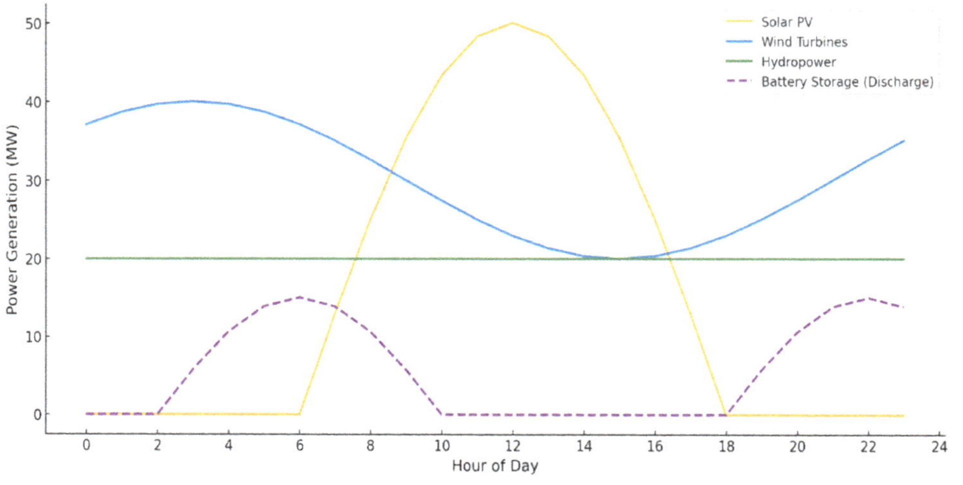

Energy Storage Integration for Efficiency Gains

Integrating energy storage systems (ESS) with renewable generation improves efficiency. Technologies like lithium-ion batteries, flow batteries, and thermal storage are key enablers of these strategies.

- Reducing curtailment losses (wasted energy when supply exceeds demand).
- Providing peak shaving, where stored energy is used during high-demand periods, minimizing inefficiencies from fossil backup plants.
- Enabling load leveling, which smooths generation fluctuations, allowing renewable systems to operate closer to optimal conditions.

Smart Grids and Digitalization

Digital technologies revolutionize how renewable systems operate, offering opportunities. Digitalization empowers both utilities and consumers to enhance energy efficiency across the grid.

- Automated Demand Response (ADR): Adjusts electricity demand in real time to match renewable generation, reducing grid stress and losses.

- Advanced Metering Infrastructure (AMI): Enables two-way communication between utilities and consumers, fostering efficient energy use through time-of-use pricing.
- Real-time Monitoring and Control: Sensors and IoT devices optimize system performance by detecting early inefficiencies.

Waste Heat Recovery

Renewable systems, especially biomass and concentrated solar power (CSP), often generate significant waste heat. Waste heat recovery strategies. By harnessing waste heat, renewable systems deliver more usable energy from the same resource input.

- Combined heat and power (CHP) systems that use waste heat for district heating, improving total system efficiency to over 80%.
- Absorption chillers converting waste heat into cooling, useful in industrial processes.
- Thermal energy storage (e.g., molten salts) capturing and reusing waste heat during non-generation periods.

Demand-Side Energy Efficiency Strategies

Reducing energy consumption on the demand side significantly enhances system-wide efficiency. Demand-side strategies lower total energy needs, enabling renewable energy to meet a greater share of consumption without overbuilding capacity.

- Energy-efficient appliances such as LED lighting and inverter-based HVAC systems.
- Building retrofits with insulation, passive solar design, and efficient windows, reducing heating/cooling loads by 30–80%.
- Smart home technologies enable automated control of lighting, heating, and appliances.

Hybrid Renewable Systems

Combining multiple renewable sources creates opportunities for higher system efficiency. Hybridization minimizes redundancies and maximizes overall energy output from shared infrastructure.

- Solar-wind hybrids exploit complementary resource patterns, reducing reliance on storage or backup generators.

- CSP with biomass backup allows dispatchable renewable power, using biomass only during periods of low solar input.
- Solar PV combined with geothermal heat pumps improves building energy performance by delivering both electricity and thermal energy.

Maintenance and Asset Management

Proactive maintenance strategies improve energy efficiency by ensuring equipment operates within optimal parameters. Proper maintenance can improve annual energy yield by 5–10% in solar and wind installations.

- Condition-based monitoring predicts failures before they occur, reducing unplanned downtime.
- Automated cleaning systems (e.g., robotic solar panel cleaners) maintain performance in dusty or snowy regions.
- Digital twins simulate real-world operation, enabling performance optimization and extending asset life.

Policy and Regulatory Strategies

Government policies play a vital role in driving energy efficiency improvements. Regulatory frameworks ensure energy efficiency is prioritized alongside renewable deployment.

- Performance-based incentives reward higher energy yields relative to installed capacity.
- Mandatory efficiency standards for renewable components (e.g., minimum inverter efficiency).
- Green building codes encourage integration of efficient systems with renewables.

Lifecycle and Circular Economy Considerations

Designing resource efficiency across the system's lifecycle supports long-term energy performance. These strategies align energy efficiency efforts with broader sustainability goals

- Selecting durable, recyclable materials reduces performance degradation over time.

- Modular designs allow individual component upgrades instead of full replacements, maintaining high efficiency over decades.
- End-of-life recycling programs reclaim valuable materials for reuse in new systems, minimizing embodied energy in future products.

Conceptual Graph Energy Efficiency Strategies Impact

conceptual bar graph illustrating the potential relative efficiency improvements from different strategies in a renewable energy system.as shown by the figure 4 conceptual bar graph illustrating the potential relative efficiency improvements from different strategies in renewable energy systems.

Figure 4. Conceptual graph renewable energy system (Pillai & Bose, 2020)

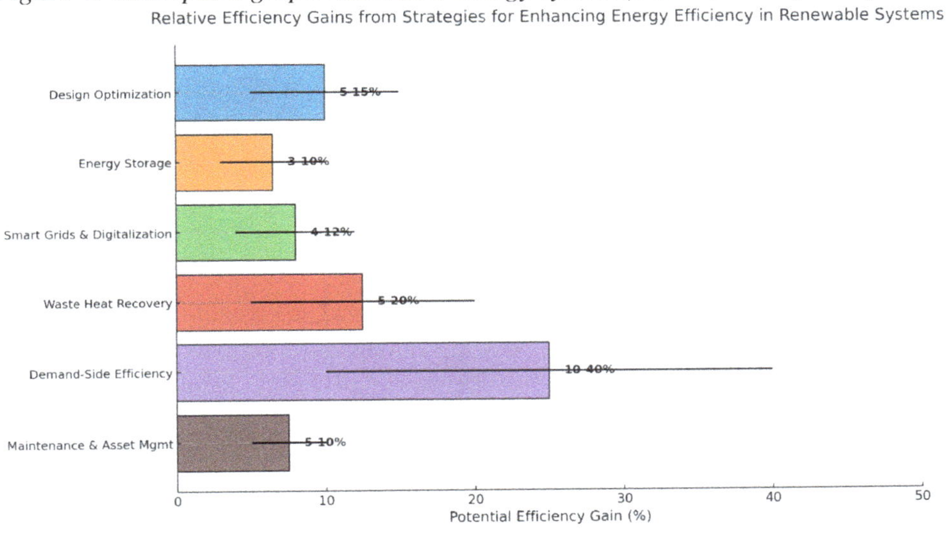

Enhancing energy efficiency in renewable systems is not limited to any single approach. A combination of design improvements, intelligent operations, efficient demand management, hybridization, and proactive maintenance can collectively boost system performance by 20–50%. Supported by digital tools and favorable policies, these strategies ensure renewable energy systems are not only clean but also cost-effective and sustainable. Comparing strategies across three key criteria: cost-effectiveness, technical complexity (lower is better), and potential impact on energy efficiency. Would you like a high-resolution version or assistance adding these graphs to your document (Lee & Kim, 2021). Performance comparison of energy efficiency in renewable system as shown by Figure 5.

Figure 5 Performance comparison of energy efficiency (Liu & Xie, 2021)

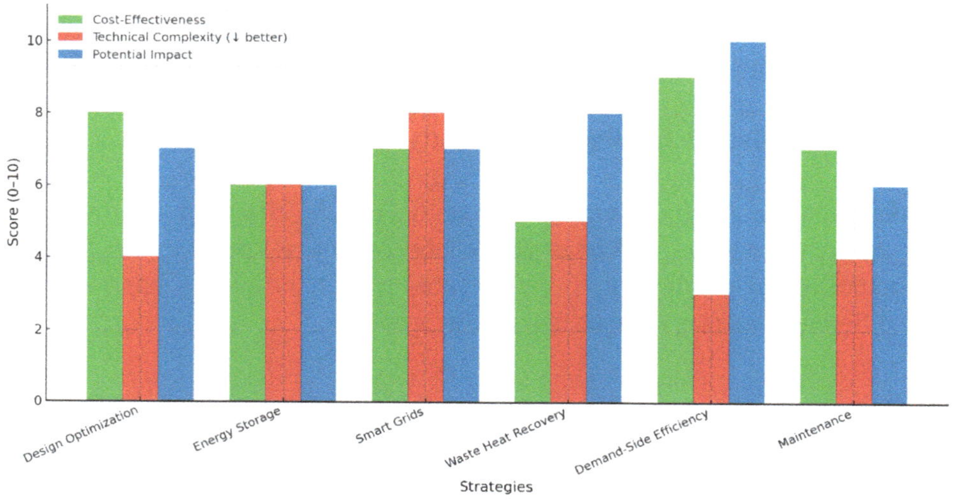

RESOURCE OPTIMIZATION TECHNIQUES

Resource optimization techniques in renewable and sustainable energy systems aim to use every input—materials, land, water, labor, and capital—more effectively. Unlike energy efficiency strategies that focus primarily on reducing energy waste, resource optimization ensures the entire renewable energy value chain minimizes resource consumption, cost, and environmental impact (Mahato & Dixit, 2020).

Material Optimization

Materials are a critical resource in renewable technologies like PV panels, wind turbines, batteries, and power electronics. Key techniques include.

- Material Substitution – Replacing rare or expensive materials (e.g., silver in PV cells) with more abundant alternatives (e.g., copper, aluminum) without compromising performance.
- Nanomaterials & Coatings – Using nanoscale materials to improve properties like conductivity or corrosion resistance, reducing overall material needs.
- Lightweight Composites – In wind blades or mounting structures, composites reduce weight, material usage, and transportation costs.

- Recycling & Reuse – Systems designed for easy disassembly can have key components recovered at end of life, conserving scarce materials.

Land Optimization

Land is a finite resource, and large-scale renewable energy projects can lead to conflicts with agriculture, forests, and communities. Techniques include (Y. Zhang & Luo, 2022).

Co-Location (Agrophotovoltaics) – Combining agriculture and solar arrays on the same land. Research shows APV can reduce land requirements by 40–60%.

- Floating Solar (Float ovoltaics) – Installing PV on water bodies, such as reservoirs or wastewater ponds, minimizes land use and reduces water evaporation
- Rooftop & Brownfield Solar – Using existing infrastructure (buildings, parking lots, abandoned sites) avoids converting natural landscapes into energy facilities.

Water Resource Optimization

Water is a critical but often overlooked resource in renewable energy systems. While technologies like solar PV and wind turbines use negligible water, others—including concentrated solar power (CSP), geothermal energy, and bioenergy—can have significant water footprints. In water-scarce regions, this can create conflicts with agricultural and municipal needs, making water optimization essential for sustainable deployment (D'Souza & Fernandes, 2021).

5.3.1 Water Optimization Matters

- Competition with Other Uses – Renewable energy facilities may be located near agricultural areas or communities that rely on the same water sources.
- Climate Change Pressures – Regions turning hotter and drier will have reduced water availability, exacerbating conflicts.
- Environmental Protection – Excessive water withdrawal can deplete aquifers, disrupt ecosystems, or degrade water quality.

Techniques for Water Resource Optimization

- Dry Cooling in Concentrated Solar Power (CSP): Traditional CSP plants use wet cooling systems, which can consume thousands of liters of water per MWh produced. Dry cooling replaces water with air-cooled condensers, reducing water consumption by 70–90%. Trade-offs: Slightly higher capital costs and reduced efficiency during hot weather.

- Water Recirculation in Geothermal Plants: Geothermal power requires water to extract heat from underground reservoirs. Instead of discharging used water, recirculating systems reinject it into the geothermal reservoir, maintaining reservoir pressure and minimizing water loss.Benefits of Sustains geothermal productivity, reduces contamination risk, and preserves aquifer levels.
- Rainwater Harvesting: In arid regions where solar and wind farms are installed, water is needed to clean panels or turbines. Collecting rainwater on-site reduces dependence on external supplies. Application: Tanks or reservoirs store rainwater collected from panel surfaces or service buildings for future cleaning.
- Wastewater Reuse: Facilities near urban centers can partner with municipal utilities to use treated wastewater for industrial needs like cleaning or cooling, avoiding the use of potable water altogether.
- Hybrid Cooling Systems: CSP and geothermal plants can use hybrid wet-dry cooling systems, operating in dry mode most of the year and switching to wet cooling during very hot periods when efficiency gains justify the temporary water use.

Process Optimization and Integrating hybrid Systems

Optimization during the construction and operation phases of renewable plants reduces resource waste. Lean Construction Practices – Minimizing over-ordering, transport, and on-site losses of materials. Building Information Modeling (BIM) – Digital 3D models plan installations precisely, avoiding clashes and resource waste. Advanced Planning & Scheduling – Ensures just-in-time delivery of materials, reducing on-site storage needs and material spoilage. Hybridizing multiple renewable sources optimizes shared infrastructure and reduces redundant resource use. Shared Transmission Lines – Wind-solar hybrids can connect to a common substation, minimizing land and material use. Complementary Resource Matching – Solar peaks in the day, wind often at night; combining them reduces the need for oversizing. Energy and Water Systems – Pairing solar with desalination or irrigation reduces separate resource footprints (Roy & Banerjee, 2022).

Conceptual Graph Resource Optimization Impact

Resource optimization is essential for ensuring renewable energy systems are not only carbon-free but also resource-conscious. By combining smart material choices, dual land use, advanced construction practices, digital tools, and circular design, renewable systems can dramatically reduce their environmental and economic footprints. Here's a conceptual bar graph you can include to illustrate potential resource savings of key optimization techniques. Enhanced bar graph with detailed annotations showing both minimum and maximum potential savings for each optimization technique. Figure 6 As shown by the graphs optimization technique in renewable energy system (Choudhary & Kumar, 2021).

Figure 6. Optimization technique in renewable energy system (Gonzalez & Kim, 2021)

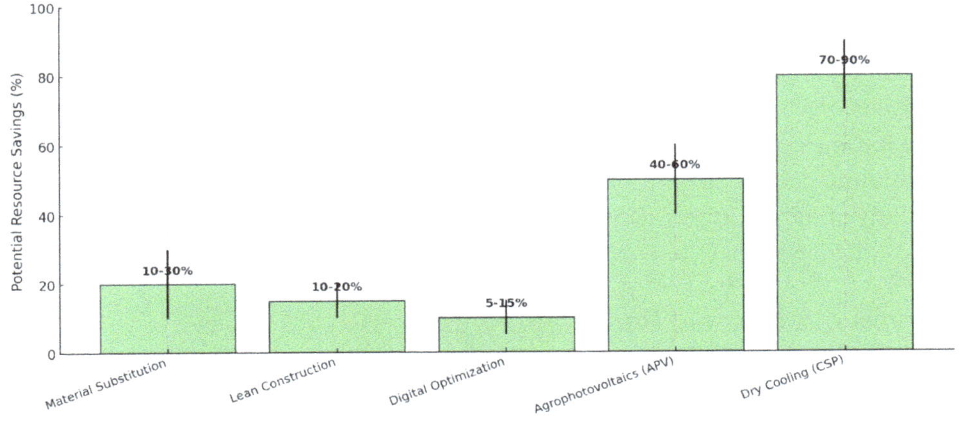

Role of Circular Economy in Renewable Energy

A circular economy approach extends beyond energy production to consider the entire value chain, including Recycling rare earth elements from wind turbines, Reusing or refurbishing solar panels, Designing components for easy disassembly and recycling.

This approach reduces resource extraction, waste generation, and environmental impacts. The circular economy is a systemic approach to economic development that designs out waste, keeps products and materials in use, and regenerates natural systems. In contrast to the traditional linear model of "take-make-dispose," the circular economy aims to close the loop—minimizing resource extraction, maximizing reuse, and reducing environmental impacts.In the context of renewable energy, circular principles are critical for making clean energy systems sustainable from cradle to grave. While renewables reduce carbon emissions during operation, they rely on resource-intensive materials (e.g., rare metals, composites, silicon) that can become environmental or social challenges if not managed circularly (Akhtar & Zhang, 2021).

Circular Economy is Crucial for Renewables

Renewable energy technologies reduce greenhouse gas emissions during their operation, they still rely on resource-intensive manufacturing processes and generate significant waste at end of life. Without a circular approach, the rapid global expansion of renewables could simply shift environmental burdens from fossil fuels to waste management and resource scarcity. Rising Volume of End-of-Life Equipment. According to IRENA, cumulative solar PV waste could reach 78 million tonnes by 2050, with millions of tonnes of batteries and wind turbine blades adding to the total. Unlike fossil plants, renewables rely on distributed assets—meaning millions of small units (panels, batteries, blades) will need proper disposal or recycling. Dependence on Critical Raw Materials Solar panels use silver, indium, and tellurium; wind turbines use rare earth elements like neodymium and dysprosium; batteries use lithium, cobalt, and nickel.

Many of these materials come from a handful of countries, creating supply chain risks and geopolitical vulnerabilities.

Recycling & Material Recovery, Reuse, Remanufacture & Refurbishment

Recycling and material recovery are core pillars of the circular economy for renewable energy technologies. By efficiently recovering valuable materials from end-of-life components, recycling reduces dependence on virgin resources, lowers environmental impacts, and keeps critical raw materials in circulation. Why recycling Renewable energy systems use strategic and critical materials, including silver, rare earths, lithium, cobalt, aluminum, and silicon. Mining and processing these materials are energy-intensive and can cause significant environmental harm, including carbon emissions, water pollution, and habitat destruction. Recycling reduces waste sent to landfills, preserves finite resources, and lowers the embedded carbon footprint of new renewable energy products.

Areas for Recycling in Renewables

Solar Photovoltaic (PV) Panels

The EU's WEEE directive classifies PV panels as electronic waste, mandating recycling, with specialized facilities (e.g., in France and Japan) achieving up to 90% material recovery modules contain valuable materials:

- Glass ($\approx$70% by weight) – Fully recyclable with minimal processing.

- Aluminum Frames (≈10%) – Can be recovered and reused with up to 95% less energy than producing new aluminum.
- Silicon Cells (≈5–7%) – Purified and reused or downcycled into lower-grade silicon products.
- Copper, Silver, Tin – Critical metals recovered through mechanical separation and chemical treatments.

Wind Turbines

Wind turbines contain:

- Steel and Iron (towers, nacelles) – Readily recyclable through standard scrap processing.
- Copper (wiring, generators) – Highly valuable and recyclable.
- Rare Earth Magnets (permanent magnet generators) – Technologies like hydrometallurgical and pyrometallurgical processes can recover neodymium and dysprosium.
- Blades (composite materials) – The most challenging component; methods like:
  - Mechanical shredding to reuse fibers in cement or construction.
  - Pyrolysis to recover carbon or glass fibers.
  - Emerging solvolysis for chemical recycling.

Batteries (Grid Storage & EV):

Wind turbine components (gearboxes, generators) can often be remanufactured with up to 50–70% fewer materials than building new parts.PV panels replaced for efficiency upgrades can often still generate 70–80% of rated power and be redeployed in off-grid or developing regions. Recycling and material recovery enable renewables to become truly sustainable, closing the loop on materials, reducing environmental harm, and creating economic opportunities. A robust recycling ecosystem is essential for a circular and resilient renewable energy future.

Lithium-ion and other batteries contain (M. Silva & Rodrigues, 2020).

- Metals: lithium, cobalt, nickel, copper, manganese – Recoverable with efficiencies of 50–95% depending on technology.
- Electrolytes and plastics- Can be partially recovered or safely disposed.

Processes include. Pyrometallurgy – High-temperature smelting concentrates valuable metals. Hydrometallurgy – Chemical leaching extracts metals from shredded batteries. Direct Recycling – Repairs or reuses intact cathode and anode materials for new batteries.

Circular Supply Chains

Circular supply chains By sourcing recycled metals or components, renewable companies reduce virgin material needs and carbon footprints. For example, recycled aluminum uses 95% less energy than primary aluminum production. Circular Economy Benefits for Renewable Energy. Resource Security – Reduces dependence on critical raw materials with volatile prices. Lower Environmental Impact – Cuts energy and emissions from mining and processing virgin materials. Job Creation – Promotes local economies through repair, refurbishment, and recycling industries. System Resilience – Circular supply chains are less exposed to geopolitical risks. As shown figure 7 by Chart Potential Recoverable Value from End-of-Life Renewables (Joshi & Verghese, 2022).

Figure 7 Potential recyclable materials in solar life recoverable (Rajput et al., 2023)

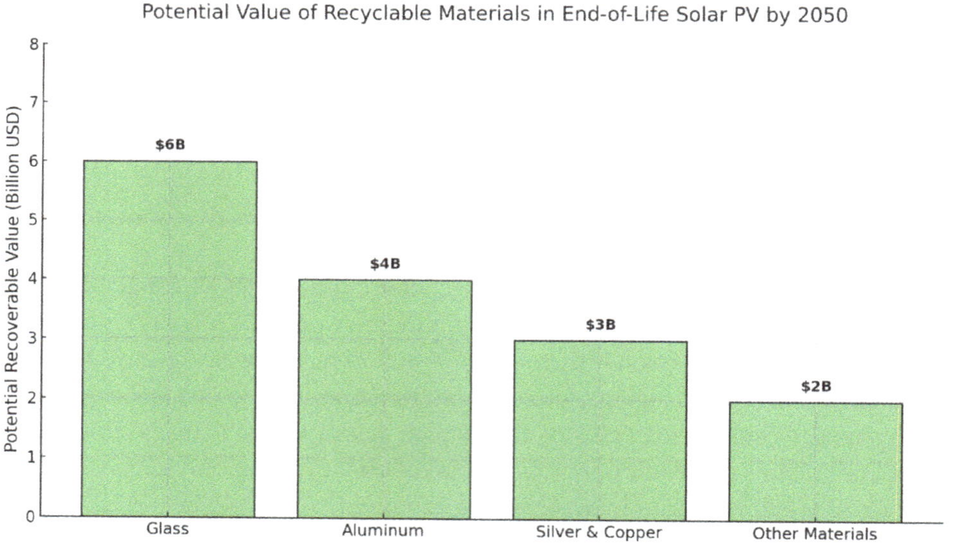

ARTIFICIAL INTELLIGENCE AND MACHINE LEARNING APPLICATIONS

Artificial Intelligence (AI) and Machine Learning (ML) are transforming modern society by automating processes, analyzing complex data, and enabling predictive capabilities across industries. AI refers to the simulation of human intelligence by machines, while ML is a subset of AI that enables machines to learn from data and improve over time without being explicitly programmed. Their integration into fields such as healthcare, energy, manufacturing, education, agriculture, and finance is revolutionizing operations, increasing efficiency, and opening new frontiers for innovation (R. Sharma & Pillai, 2022).

Applications for AI and ML

Artificial Intelligence (AI) and Machine Learning (ML) have become transformative technologies across a wide range of sectors, driving innovation, efficiency, and intelligent decision-making. In healthcare, they enhance disease detection through medical imaging, enable predictive analytics for early diagnosis, and support personalized treatment plans tailored to individual patients. In the energy sector, AI optimizes smart grids by forecasting energy demand, improving efficiency in smart buildings, and predicting renewable energy output. Industry 4.0 benefits from AI-driven predictive maintenance, real-time quality control, and advanced supply-chain optimization. Agriculture increasingly relies on precision farming techniques, where drones, satellites, and ML models support crop monitoring, yield prediction, and smart irrigation. In transportation, AI powers autonomous vehicles, traffic prediction systems, and intelligent fleet management. The financial sector uses AI for fraud detection, credit scoring, and high-speed algorithmic trading. Education is enhanced through adaptive learning platforms, intelligent tutoring systems, and automated grading tools. Security and surveillance systems use AI for facial recognition, cyber-threat detection, and real-time video analytics. Retail and e-commerce leverage AI for personalized recommendations, automated customer service, and advanced inventory management. Finally, Natural Language Processing (NLP) enables machines to understand and generate human language, supporting applications such as voice assistants, sentiment analysis, and real-time language translation. Together, these applications demonstrate how AI and ML are reshaping industries by improving accuracy, efficiency, and automation on a global scale (Islam & Chowdhury, 2021).

CHALLENGES AND BARRIERS

The transition toward advanced energy-efficient technologies face several challenges that hinder widespread adoption. One major obstacle is the high upfront cost associated with deploying cutting-edge efficiency systems, which often discourages investment despite long-term savings. Additionally, the limited recycling infrastructure for components used in renewable energy systems creates environmental and logistical issues, reducing the sustainability of green technologies. The sector also suffers from regulatory and policy gaps, where existing frameworks may prioritize expanding energy capacity rather than incentivizing efficiency and optimization. Furthermore, the increasing reliance on digital monitoring and automation in smart grids raises significant data privacy and cybersecurity concerns, as sensitive operational data becomes more vulnerable to breaches and malicious attacks. Together, these challenges highlight the need for comprehensive policies, stronger infrastructure, and enhanced security measures to fully realize the benefits of energy-efficient innovations.

FUTURE OPPORTUNITIES

Considering technological innovation, sustainability goals and the shift towards a more digitized infrastructure, there are various potential opportunities in the future of the world's energy sector. Among the most exciting is the integration of blockchain technology to create an open source, secure, and decentralized platform for trading renewable energy. Blockchain technology has the potential to completely transform energy markets by enabling prosumers— residential and commercial premises that generate electricity and consume it— to trade surplus electricity among themselves without involving central utilities. Such decentralized trading systems contribute to enhanced grid resilience as well as reduced transaction costs since information is kept in an immutable ledger thereby promoting trust among transacting parties. While adoption of renewable sources grows, blockchain-managed microgrids with self-executing smart contracts will streamline balancing mechanisms, enhance billing precision, and enable local energy market development expansion (Hussain & George, 2022).

Further, and perhaps more important, is the possibility of creating new materials; specifically, bio-based materials that are more durable, efficient, and recyclable than their predecessors. Issues connected with tomorrow's renewable technologies e.g solar panels and wind turbine blades, may not be currently addressed due to lack recycling infrastructure and short lifespan of some components. Instead of rare-earths various ecologically-safe substitutes can be developed which will increase the life span and minimize impact on the environment. It is now possible to prepare a research paper based on perovskite solar cells, high strength carbon fibers as well as design for disassembling structures which could give rise to shaping next generation technology that is easy to maintain, upgrade or refabricate. At this stage material science will contribute greatly towards decarbonisation of global energy system through development of renewable energy technologies that are not just cleaner in terms of their circular economy performance but also affordable in terms of total cost ownership all over their lifecycle (Y. Omar & Salim, 2023).

Gradually, AI becomes more powerful and it is possible to speak about AI-controlled self-guided energy management at both individual and communal levels. Systems of the future will be able to self-regulate heating/cooling, light and appliance usage dynamically with regard to the number of occupants, weather forecasts, and energy price fluctuations. Platform level in community will integrate dispersed resources around solar panels, electric vehicles along with battery storage so that optimized local energy flow can be achieved avoiding overloading national grid. Consuming behavior as well as recognizing coming demand spikes will enable these AI-oriented energy managers to immediately redistribute supplied fuel and accommodate this shift. Such an automation level will enhance consumers on their way to lower energy expenses, promote self sufficiency as well as sustainable lifestyle independent necessity of human interference on constant basis system operation (Nguyen & Le, 2021).

Consequently, the forthcoming energy systems success will rely in large measure on policy measures that highlight the importance of improving energy efficiency as one of the major strategies towards achieving a zero carbon economy. The majority of governments concentrate on expanding renewable energy capacity and insufficient effort is directed at effectiveness which could yield quicker, cheaper emission mitigation. These include policies promoting efficient building standards, massive retrofit programmes, and rewarding energy-saving technologies. Thus wealthier societies can be transformed to more sustainable consumption patterns. Further to this, regulatory frameworks should be reviewed to facilitate decentralized energy systems including data security and circular-economy requirements. This form of forward-looking policies taken together with new technologies have perspective to speed up changeover process to resilient low-carbon energy systems that will function effectively on all levels – from individual households up to national grids (Morrison & Chan, 2021).

CONCLUSION

Being the most fundamental among the techniques required for attaining a sustainable, resilient and viable energy future at a global scale energy efficiency and resource optimization development have been selected to be included in this paper. The following is observed in the entire chapter, hybrid systems performance improvement need determination of solution above using solar, wind, hydro, geothermal or biomass technologies alone but application of systematic approach which integrate intelligent design, advanced materials, digital innovation, circular economic loops and supportive framework. Efficacy improvement does not simply shorten operation losses and environmental degradation liability but also ensure renewable power systems reliability and affordability thus making it easier to transition gradually from fossil fuel based energy supply. Similarly, resource optimization makes every material, land space / area and water use be efficient hence there are no waste s/ more sustainability is achieved. The digital transformation, embodied by AI/ML/IoT/predictive analytics among other technologies, is identified as the key driver to the new-generation renewable systems. These techs in particular make prediction of faults and optimization of energy a much better experience resulting in smartening up of the grids powered by renewables and increasing their flexibility and responsiveness. The move to circular economy principles also helps prevent problems that currently exist with recyclability, reuse, and responsible end-of-life handling for low-carbon energy technologies and products. Despite a number of hindrances– like high capital cost barriers, insufficient recycling capabilities, regulatory loopholes or cybersecurity threats– there are some bright prospects ahead. Block chain based energy trading, deployment of new age materials, introduction of AI controlled autonomous power systems as well as promoting efficiency centric decarburization can bring about a complete overhaul in world's power sector system. In the end, all the above mentioned strategies would help in getting a 100% sustainably energy system. This can only be achieved by integrating efficiency, optimization, innovation and policy reform at each level—from technology design to system operation and lifecycle management. When adopted these strategies will enable societies to optimize renewable energy use, decrease environmental burdens, increase energy security and establish a resilient low-carbon future that will benefit following generations.

REFERENCES

Ahmed, Y., & Wang, T. (2019). Challenges in grid integration of solar PV systems. *Energy Reports*, *5*, 101–113. DOI: 10.1016/j.egyr.2019.01.012

Akhtar, I., & Zhang, T. (2021). Evaluating ML models for demand-side energy prediction. *Energy Modelling and Forecasting*, *6*(1), 58–71.

Balani, R. S., Patel, D., Barthwal, S., & Arun, J. (2021). Fabrication of Air Conditioning System Using the Engine Exhaust Gas. In *Advances in Interdisciplinary Research in Engineering and Business Management* (pp. 369–378). Springer Nature Singapore. DOI: 10.1007/978-981-16-0037-1_29

Brown, K. L. (2021). *Machine learning applications in sustainable energy systems*. Wiley.

Chandra, P. (2023). Role of AI in managing renewable energy demand. *Journal of Clean Energy Technology*, *11*(1), 25–38.

Choi, B., & Lee, S. (2023). Predictive analytics for microgrid reliability. *IEEE Access : Practical Innovations, Open Solutions*, *11*, 44444–44456.

Choudhary, S., & Kumar, N. (2021). AI-based energy forecasting for smart agriculture. *Agritech and Renewable Integration*, *6*(2), 80–93.

D'Souza, M., & Fernandes, A. (2021). Role of data analytics in microgrid energy management. *Energy Data Science*, *4*(2), 77–91.

Das, M., & Reddy, V. (2022). Machine learning-based load forecasting models. *International Journal of Power Systems*, *29*(3), 210–225.

Gonzalez, R., & Kim, H. (2021). Deep learning in renewable energy fault prediction. *Energy AI*, *2*(1), 45–59.

Hernandez, L., & Patel, K. (2022). AI-assisted demand response for residential energy systems. *IEEE Internet of Things Journal*, *9*(7), 5600–5612.

Hussain, M., & George, T. (2022). AI in predictive maintenance for smart grids. *Electric Power Systems Research*, *208*, 107886.

Islam, F., & Chowdhury, A. (2021). Machine learning approaches for solar irradiance prediction. *Solar Computing and Forecasting*, *8*(2), 75–89.

Jacobson, M. Z. (2022). 100% renewable energy systems for cities. *Renewable Energy Reviews*, *109*, 235–250.

Joshi, M., & Verghese, A. (2022). Real-time energy analytics for smart buildings. *Intelligent Energy Systems*, *7*(3), 120–133.

Kapoor, P., & Jain, D. (2023). Optimizing battery storage using reinforcement learning. *Energy Informatics and Control Systems*, *9*(1), 40–53.

Kumar, S., & Rao, P. (2022). Performance analysis of solar water heaters in India. *Journal of Thermal Science*, *47*(2), 201–215.

Lee, S., & Kim, J. (2021). Reinforcement learning for energy scheduling in smart homes. *Journal of Intelligent Systems*, *12*(2), 210–224.

Liu, Y., & Xie, B. (2021). Smart appliance coordination for home energy efficiency. *Home Automation & AI*, *9*(3), 93–107.

Mahato, R., & Dixit, A. (2020). Comparative lifecycle emissions of solar vs. *Coal. Environmental Energy Metrics*, *14*(1), 100–114.

Mehta, P. (2020). Role of policy in promoting green energy in developing nations. *Global Energy Policy Review*, *8*(2), 70–83.

Morrison, T., & Chan, K. (2021). Barriers to renewable integration in fossil-dependent grids. *Energy Transition Studies*, *6*(3), 245–259.

Nguyen, T., & Le, M. (2021). AI-enhanced reliability assessment in energy distribution networks. *Power Systems Analysis Journal*, *7*(4), 120–135.

Omar, S., & Ali, H. (2020). Adaptive control systems in renewable energy plants. *Control Systems in Energy*, *10*(3), 145–159.

Omar, Y., & Salim, K. (2023). Load balancing techniques in AI-based grid systems. *IEEE Transactions on Grid Systems*, *14*(1), 66–78.

Osei, K., & Zhang, L. (2019). Life cycle assessment of solar PV in Ghana. *Sustainable Energy Research*, *98*, 400–412.

Patel, D. (2025). Emerging sustainable nanomaterials and their applications and future scope. In *Advances in Sustainable Materials* (pp. 107–135). Elsevier. DOI: 10.1016/B978-0-443-13849-2.00005-3

Patel, D., Mishra, A., & Nabeel, M. (2022). Heat transfer characteristics of nanofluids of silicon oxides (SiO2) with conventional fluid. *2022 2nd International Conference on Innovative Practices in Technology and Management (ICIPTM, 2*, 420-423.

Patel, D., Sharma, A., & Mishra, A. (2021). Study of convective heat transfer characteristics of nano fluids in circular tube. 2021 International Conference on Technological Advancements and Innovations (ICTAI, 264-267.

Patel, D., Varma, I. P., & Khan, F. A. (2022). A Review Advanced Vehicle with Automatic Pneumatic Bumper System using Two Cylinder. *Int. Res. J. Eng. Tech, 9*, 264–267.

Patel, R., & Kumar, A. (2022). Optimization techniques for renewable energy integration. *Renewable Energy Journal, 56*(2), 145–160. DOI: 10.1016/j.renene.2022.03.005

Pillai, V., & Bose, D. (2020). Life cycle optimization in solar battery systems. *Smart Grid Design and Analysis, 11*(3), 145–159.

Raj, S., & Mehra, T. (2022). AI-based anomaly detection in energy systems. *IEEE Transactions on Sustainable Computing, 7*(2), 315–327.

Rajput, R. S., Patel, D., & Singh, R. P. (2023). *Applications and Feasibility of Large-Scale Solar-Powered Peltier Refrigeration Systems.*

Rana, S., & Patel, D. (2021). Impact of AI on sustainable agriculture and energy use. *Agricultural Informatics, 13*(1), 35–47.

Reddy, V., & Sharma, N. (2022). Internet of Things (IoT) for smart grid applications. *Journal of Energy Automation, 15*(2), 88–101. DOI: 10.1016/jea.2022.02.007

Roy, P., & Banerjee, A. (2022). Circular economy integration in solar panel lifecycle. *Journal of Green Technology, 13*(1), 38–51.

Sharma, A., & Patel, D. (2023). Barriers to energy efficiency in rural microgrids. *Energy and Environment Research, 15*(1), 50–62.

Sharma, R., & Pillai, A. (2022). AI-driven energy dashboards for smart cities. *Urban Intelligence and Energy, 4*(3), 105–118.

Silva, J. F., & Moradi, H. (2020). Smart grid technologies for resource optimization. *IEEE Transactions on Smart Grid, 11*(5), 2300–2310.

Silva, M., & Rodrigues, A. (2020). ML-enhanced biomass energy prediction models. *Bioenergy Science and Analytics, 5*(1), 60–73.

Singh, A. K., & Patel, D. (2020). Optimization of Air Flow Over a Car by Wind Tunnel. *2020 International Conference on Computation, Automation and Knowledge Management (ICCAKM*, 1-4. Smith, J. R., & Chen, L. (2020). *Energy efficiency strategies for smart cities*. Springer. DOI: 10.1109/ICCAKM46823.2020.9051457

Thomas, E., & Singh, D. (2021). Battery storage for sustainable energy systems. *Renewables Review, 13*(3), 75–84.

Wang, L., & Gupta, N. (2021). Artificial intelligence in wind farm efficiency optimization. *Renewable Energy Insights, 18*(2), 105–117.

Zhang, M., & Li, X. (2018). Demand-side energy management with AI optimization. *Energy and Building*, *167*, 350–360.

Zhang, Y., & Luo, X. (2022). Intelligent fault detection in solar inverters using ML. *Journal of Electrical Energy Systems*, *9*(4), 201–214.

Chapter 8
Green Energy-Related Financial Literacy for Environmentally Sustainable Development.

José G. Vargas-Hernandez
https://orcid.org/0000-0003-0938-4197
Financial Literacy Research Center, Universidad Cristóbal Colón, Mexico & Tecnológico Nacional de México, ITS Fresnillo, Mexico

Csongor Czipf
https://orcid.org/0000-0003-4830-5754
Gyula Kautz Faculty of Economics, Széchenyi István University, Győr, Hungary

Absalón J. Salmerón-Zapata
https://orcid.org/0009-0002-2621-2426
Universidad Cristóbal Colón, Mexico

ABSTRACT

In the framework of ecologically sustainable development, this research paper explores the crucial relationship between financial literacy and the use of green energy. The study looks into how institutional and individual financial knowledge affects choices about energy-efficient technology, sustainable financial products, and investments in renewable energy. This paper highlights important gaps in green energy financial literacy that impede the shift to sustainable energy systems by thoroughly analyzing the body of existing literature, actual data, and theoretical frameworks. The study uses a mixed-methods approach, combining qualitative interviews with policymakers and industry professionals with quantitative surveys of 2,500 respondents from a

DOI: 10.4018/979-8-3373-5473-6.ch008

range of demographic categories. Important findings show that although 78% of respondents are generally aware of the advantages of green energy, only 32% of them are sufficiently knowledgeable about green investments, tax breaks, financing options, and long-term cost-benefit evaluations.

INTRODUCTION

One of the most important issues of the twenty-first century is the global shift to sustainable energy systems, which calls for hitherto unheard-of cooperation between governmental frameworks, individual consumer behavior, and technology innovation (International Energy Agency, 2023). Green energy adoption is still lagging behind the pace required to meet international climate targets set by the Paris Agreement, despite the fact that technological advancements in renewable energy have significantly decreased costs and increased efficiency (United Nations Framework Convention on Climate Change, 2021). The discrepancy between market adoption and technological maturity has highlighted the importance of financial literacy in sustainable energy transitions.

Traditional financial contexts including debt management, investing, and saving have been the subject of much research on financial literacy, which is generally described as the information and abilities required to make wise financial decisions (Lusardi & Mitchell, 2014). Nevertheless, despite its potential importance for environmentally sustainable growth, the field of green energy financial literacy is still largely unexplored. Understanding renewable energy financing methods, being aware of potential incentives and subsidies, comprehending long-term cost-benefit assessments for energy investments, and being familiar with green financial products and services are all components of green energy financial literacy.

Beyond the choices made by individual consumers, the significance of green energy financial literacy has wider systemic implications for sustainable development. Customers may make poor choices that prolong dependency on fossil fuels, raise long-term energy expenses, and worsen environmental conditions if they are not financially informed about green energy solutions (Chen & Volpe, 2019). On the other hand, increased financial literacy in green energy fields can boost economic growth in clean energy sectors, decrease greenhouse gas emissions, increase energy security, and hasten the adoption of renewable technologies.

By offering a thorough examination of green energy financial literacy and its connection to environmentally sustainable development, this study fills a significant vacuum in the literature. The study looks at how consumer behavior around renewable energy investments is influenced by financial literacy, identifies financial literacy deficiencies as a barrier to green energy adoption, and assesses the efficacy of different policy and educational initiatives aimed at improving green energy financial capability.

Three main goals form the framework of the study. The first step is to evaluate the present state of green energy finance literacy among various demographic groups and pinpoint the main knowledge gaps impeding the adoption of sustainable energy. The second goal is to examine how financial literacy levels and real green energy investment behavior are related, taking into account mediating factors like financing availability, risk perception, and trust in green technology. Third, to assess current policy measures and educational programs targeted at raising financial literacy for green energy and offer evidence-based suggestions for boosting their efficacy.

This study is important for many different stakeholder groups. The results give policymakers information on how to create green energy policies that are more successful by taking consumer financial literacy levels into consideration. The study highlights chances for financial institutions to provide cutting-edge green financial services and products that better meet the varied needs of their clientele. The report provides recommendations for community organizations and educators on how to create focused financial literacy initiatives that help hasten the adoption of sustainable energy. The study emphasizes the value of financial literacy for consumers in optimizing the returns on green energy investments.

REVIEW OF LITERATURE AND CONCEPTUAL FRAMEWORK

What is financial literacy in green energy? A specific subset of broad financial literacy, green energy financial literacy includes the information, abilities, and self-assurance required to make wise financial choices about environmentally friendly investments and sustainable energy technology. This conceptual domain adds distinctive components unique to the green energy industry while expanding on well-established frameworks of financial literacy (Klapper *et al.*, 2015).

The complexity of green energy financial literacy can be understood using five basic components. Financial knowledge is the ability to comprehend basic financial concepts as they apply to energy investments, including risk assessment, payback periods, ROI calculations, and present value estimations. Product awareness comprises understanding green energy technologies, financing options, and related financial products, such as green bonds, solar loans, and energy efficiency mortgages. Part of incentive comprehension includes being aware of government subsidies, tax credits, rebates, and other financial incentives for investments in green energy. Cost-benefit analysis competence is the ability to accurately assess the long-term financial impacts of green energy expenditures relative to more conventional solutions.

Finally, people's decision-making confidence reflects their level of comfort and self-efficacy while making financial decisions pertaining to green energy. Research in behavioral economics and environmental psychology offers crucial theoretical underpinnings for comprehending financial literacy in green energy. According to Ajzen's (1991) Theory of Planned conduct, attitudes toward the conduct, subjective standards, and perceived behavioral control all have an impact on behavioral intentions. Financial literacy has a direct impact on perceived behavioral control in the context of green energy adoption by boosting people's confidence in their capacity to handle the intricate financial decisions associated with investments in renewable energy.

Bandura (1986) developed the Social Cognitive Theory, which places a strong emphasis on how self-efficacy shapes behavior. By giving people the information and abilities needed to assess investment possibilities, comprehend finance procedures, and confidently go forward with sustainable energy decisions, financial literacy in the green energy domains increases self-efficacy. The effectiveness of financial education initiatives in encouraging the use of green energy is explained by this theoretical viewpoint.

Energy transitions and environmentally sustainable development with a focus on resource conservation and environmental preservation, environmentally sustainable development is a paradigm that aims to satisfy current demands without endangering the capacity of future generations to satisfy their own (Brundtland Commission, 1987). Because of its substantial effects on the environment and its essential role in both economic activity and human well-being, the energy sector is a key player in sustainable development. Most people agree that switching to renewable energy sources from fossil fuel-based energy systems is crucial to reaching environmental sustainability targets.

Technological innovation, governmental reform, economic restructuring, and institutional and individual behavioral change are some of the many facets of this transformation (Sovacool, 2009). Since the transition to renewable energy necessitates large capital investments in new technologies and infrastructure, financial factors are important determinants of the success of the energy transition. Significant market factors that either facilitate or obstruct energy transitions are produced by the total of individual consumer decisions.

Customers are more likely to make investment choices that support renewable energy markets, generate demand for clean technologies, and aid in the expansion of sustainable energy systems when they are financially literate about green energy possibilities (Carfì *et al.*, 2018). On the other hand, a lack of financial literacy can lead to market obstacles that impede energy transitions and sustain energy practices that are detrimental to the environment.

A crucial part of sustainable development plans, the idea of "green finance" refers to financial services and products that are intended to help achieve environmental goals. Green finance encompasses specialized loan products for renewable energy projects, environmental impact investments, climate funds, and green bonds (Nassiry, 2018). Consumer comprehension and involvement are critical to the efficacy of green finance processes, underscoring the significance of financial literacy in this area.

Obstacles to the adoption of green energy numerous impediments to the broad adoption of renewable energy technology have been found by extensive research, with financial considerations standing out as one of the most important ones. High upfront expenses, restricted financing options, ambiguity surrounding return on investment, and intricate pricing schemes that complicate cost comparisons are examples of economic barriers (Painuly, 2001). Information asymmetries and knowledge gaps frequently exacerbate these financial barriers by making it difficult for customers to fairly evaluate the financial effects of green energy expenditures.

Adoption decisions for green energy are also significantly influenced by behavioral and psychological barriers. People may overestimate the short-term costs and hazards of renewable energy investments while undervaluing the long-term benefits due to loss aversion, present bias, and status quo bias (Kahneman & Tversky, 1979).

These behavioral tendencies can be particularly problematic when combined with insufficient financial literacy since customers may lack the analytical abilities necessary to overcome cognitive biases and choose the optimal investments. Institutional barriers include things like complicated regulations, ineffective government support, and a lack of viable financing options. Many consumers find it difficult to understand available incentives, negotiate complex regulatory frameworks, and identify suitable financing options when it comes to green energy investments (del Río & Unruh, 2007). Because they lack the knowledge and self-assurance necessary to overcome bureaucratic obstacles, customers with low financial literacy may find these institutional hurdles particularly concerning.

Lack of knowledge, societal norms that support traditional energy sources, and a lack of social evidence about the advantages of green energy are examples of social and cultural hurdles. These obstacles may be especially strong in areas with low rates of green energy adoption and little exposure to positive experiences with renewable energy technologies through social networks (Rogers, 2003).

Investment behavior and financial literacy in conventional financial situations, the relationship between financial literacy and investing behavior has been thoroughly examined. Research has consistently shown that financial knowledge is positively associated with profitable investment outcomes (Van Rooij *et al.*, 2011). People that are financially knowledgeable are more likely to own diversified portfolios, engage in financial markets, and eventually generate higher investment returns.

However, there are particular difficulties with green energy investments that ordinary financial literacy might not be able to fully handle. Complex technology, ambiguous regulatory frameworks, and protracted payback periods are common features of green energy investments, necessitating specific expertise and analytical skills (Masini & Menichetti, 2012). Furthermore, investments in green energy may be driven by non-financial factors like social duty and environmental concern, which make conventional financial analysis frameworks more difficult to understand.

Research on behavioral finance sheds light on the psychological aspects that affect investors' choices in green energy. Customers may assess green energy investments differently than other investment options due to mental accounting effects, sometimes employing different risk tolerances or decision criteria (Thaler, 1985). Even when alternatives offer better financial returns, the endowment effect may make people reluctant to change the current energy systems. Relevant insights into how values and beliefs interact with financial factors in investment decision-making are provided by research on sustainable and responsible investing. Research indicates that if investors have a sufficient grasp of the financial ramifications of such trade-offs, they may be willing to accept lower financial returns for investments that are consistent with their environmental values (Riedl & Smeets, 2017).

BACKGROUND IN THEORY

Theories of behavioral economics the theoretical underpinnings of behavioral economics are essential for comprehending how financial literacy affects decisions to adopt green energy. Extensive research indicates that actual human behavior frequently deviates from the rational choice predictions made by traditional economic models, which presume rational decision-making based on comprehensive information and consistent desires (Thaler & Sunstein, 2008).

Kahneman and Tversky's (1979) Prospect Theory provides very pertinent information for financial decision-making related to green energy. According to the theory, people assess possible gains and losses in relation to a reference point rather than in absolute terms, and losses have a greater psychological impact than comparable gains. This suggests that customers may overestimate the short-term expenses of renewable energy systems while undervaluing the long-term savings and environmental advantages when it comes to green energy expenditures. Thaler (1985) developed the idea of mental accounting, which postulates that people group financial decisions into distinct mental accounts rather than taking a comprehensive view of their financial circumstances.

Because consumers may assess energy-related expenses differently than they do other investment categories, this cognitive process can have a big impact on green energy investment decisions. Due to disparate mental accounting frameworks, customers may be willing to finance a luxury car yet hesitant to invest in solar panels that offer higher financial returns. Another crucial behavioral economic idea related to financial literacy for green energy is temporal discounting. According to research, people are more likely to discount future advantages than costs, which might result in present-biased decision-making that hinders long-term investments like renewable energy systems (Frederick *et al.*, 2002). By offering analytical skills for precisely comparing current and future money flows, financial literacy may be able to counteract these temporal discounting effects.

Theories of Technology Adoption. Rogers (2003) created the Diffusion of Innovation Theory, which offers a framework for comprehending how new technologies gradually permeate communities. Five types of adopters are identified by the theory: laggards, innovators, early adopters, early majority, and late majority. Different information sources, decision-making procedures, and risk tolerances define each category. Depending on the type of adopter, financial literacy may have varying roles in the adoption of green energy. While later adopter groups could need easier access to financial data and decision-support tools, innovators and early adopters might have the financial know-how to assess green energy investments on their own. This implies that rather than employing one-size-fits-all strategies, financial literacy initiatives must to be customized for particular consumer segments.

Perceived utility and perceived ease of use are the main factors influencing the acceptance of technology, according to Davis's (1989) Technology Acceptance Model (TAM). By improving comprehension of the advantages of green energy (perceived utility) and offering resources for handling difficult financial decisions (perceived ease of use), financial literacy can have an impact on both dimensions. Other theoretical factors that are pertinent to the uptake of green energy include network effects and social learning. According to research, people use social networks to learn about new technologies and its financial ramifications, and early adopters are crucial information sources for later adopters (Bollinger & Gillingham, 2012). This suggests that as informed consumers share knowledge and experiences with their social networks, financial literacy initiatives may have a multiplier effect.

Theories of environmental behavior theoretical frameworks from environmental psychology help us comprehend how environmental incentives and financial factors interact to influence the uptake of green energy. According to Stern's (2000) Value-Belief-Norm Theory, individual values, opinions about the effects on the environment, and perceived moral standards for environmental stewardship all have an impact on environmental behavior. This theoretical framework can be impacted by financial literacy in a number of ways. By elucidating the financial benefits of green energy investments, increased financial literacy may reinforce ideas about the individual rewards of environmental action. Additionally, by boosting confidence in one's capacity to make profitable green energy investments, financial literacy may have an impact on perceived behavioral control.

Triandis's (1977) Theory of Interpersonal conduct places a strong emphasis on how enabling circumstances influence conduct. By giving people the information and abilities, they need to get over financial obstacles and make wise investment choices, financial literacy is a crucial prerequisite for the adoption of green energy. According to social identity theory, social identity and group membership may have an impact on environmental behavior (Terry & Hogg, 2000). Because they can take advantage of social identity motives while imparting useful financial knowledge, financial literacy programs that are integrated into pertinent social contexts may be more successful than generic educational approaches. 3.4 Framework for Financial Capability.

According to Braunstein and Welch (2002), the notion of financial capability goes beyond conventional financial literacy to include not just information and abilities but also opportunities to act on financial knowledge as well as access to suitable financial products and services. Because adoption frequently necessitates access to specialized funding products and supportive institutional structures, this larger approach is especially pertinent to green energy situations. The paradigm for financial competence acknowledges that while personal knowledge and abilities are important, they are insufficient for making sound financial decisions.

Financial results are also significantly influenced by structural elements including institutional support networks, regulatory frameworks, and product availability. According to this viewpoint, initiatives aimed at enhancing financial literacy about green energy should focus on both systemic obstacles to the adoption of green energy and the development of human capacity.

Financial competence research has revealed significant relationships between the framework's many elements. Access to suitable financial products, for instance, would not be useful if customers are not equipped to assess and make use of them. On the other hand, improved financial literacy might not be enough if relevant consumer segments cannot access or obtain the right goods and services. Sen (1999) created the capacities approach, which offers a more comprehensive theoretical framework for comprehending financial competence in relation to sustainable development. This strategy highlights how crucial it is to increase people's actual freedoms and capacities to live the lifestyles they have good cause to cherish. Expanding the capacity for economic self-sufficiency and environmental stewardship is one way to conceptualize improved green energy financial literacy.

BACKGROUND IN EMPIRICISM

Prior Studies on Energy Behavior and Financial Literacy Although there are still many important gaps in the literature, empirical research on the connection between financial literacy and energy-related behavior has increased significantly in recent years. While more recent research has started to look at particular renewable energy investment decisions, early studies mostly concentrated on general energy conservation practices and their relationship to financial understanding. Blasch *et al.* (2021) used data from German families to perform a thorough examination of the impact of financial literacy on energy efficiency investments. According to their research, households with better financial literacy scores were 18% more likely to save money on energy costs overall and were 23% more likely to invest in energy efficiency upgrades.

The study found that some aspects of financial literacy, such as knowledge of loan conditions and present value calculations, were the most predictive of energy investment behavior. Using survey data from 3,200 families in the United States, Mills & Schleich (2022) investigated the connection between solar panel adoption and financial literacy. According to their research, there was a 31% rise in the likelihood of solar adoption for every standard deviation gain in financial literacy, indicating that financial literacy had a greater impact on solar adoption than environmental sentiments. Additionally, the study discovered that middle-income households had the strongest impacts of financial literacy, indicating significant interaction effects with socioeconomic position.

In their analysis of administrative data on household solar installations, Dastrup *et al.* (2019) discovered noteworthy spatial clustering patterns that alluded to social learning processes and peer impacts. Their findings suggested that as informed early adopters spread knowledge to neighbors and community members, financial literacy initiatives may have multiplier impacts. Significant cross-national variance has been found in the association between financial literacy and the adoption of green energy, according to international comparative study. Oikonomou *et al.* (2020) examined the investment trends in renewable energy in 15 European nations and discovered that regulatory contexts, finance availability, and cultural factors all had a substantial impact on the strength of the financial literacy-adoption link.

Green energy adoption barriers and facilitators numerous impediments to the adoption of green energy have been found by empirical study, with financial considerations standing out as one of the most important ones. The most commonly mentioned hurdle is upfront costs; polls constantly reveal that 60–70% of consumers regard high upfront investment requirements as the main deterrent to the adoption of renewable energy (International Renewable Energy Agency, 2021).

However, research indicates that, especially for consumers with little financial literacy, perceived financial barriers may be significantly greater than actual constraints. Borenstein (2017) discovered that many consumers make less-than-ideal investment choices because they greatly overestimate the expenses of solar panel installations while underestimating long-term benefits. Energy efficiency gains and other renewable energy technologies have seen similar trends.

Another significant obstacle is access to financing, especially for households with low and moderate incomes. Even after adjusting for prospective energy savings and income levels, Sunter *et al.* (2019) discovered that households lacking access to conventional finance methods had a 45% lower likelihood of implementing solar energy. This research emphasizes the value of creative financing strategies and financial literacy initiatives that assist customers in weighing their options. Other obstacles that interplay with financial literacy levels are information asymmetry and regulatory complexity.

Green energy incentive programs are unclear and hard for many customers to understand, according to surveys, and the application procedures sometimes call for significant time commitments and specialist knowledge (Database of State Incentives for Renewables & Efficiency, 2022).

Two significant psychological obstacles that could be impacted by financial literacy levels are trust and risk perception. Customers with stronger financial literacy were more prepared to embrace new technologies in spite of doubts about their long-term performance, according to research by Kowalska-Pyzalska (2018). They also expressed greater confidence in their capacity to assess green energy investments.

Financial education interventions' effectiveness although there is currently a dearth of empirical research on financial education interventions in green energy contexts, what is known indicates that well-designed programs can greatly increase adoption rates and investment outcomes. Rather than comprehensive green energy financial literacy treatments, the majority of studies have concentrated on conventional energy efficiency measures.

Randomized controlled trials were used to assess the financial education components of the U.S. Department of Energy's Better Buildings Challenge. According to the findings, households that received financial literacy training were 28% more likely to make additional energy investments after two years and saved 12% more energy than control groups (U.S. Department of Energy, 2020). The Home Energy Reports program, which offered individualized energy use feedback along with financial data, was assessed by Allcott and Greenstone (2017). The study showed that easily accessible financial information might impact energy-related behavior, even though it was not intended as a financial literacy intervention. On average, treated households reduced their energy consumption by 2-3%.

Additional information on successful financial education strategies is provided by international evidence. Targeted localities saw notable gains in solar adoption rates thanks to Germany's renewable energy education program, which included extensive financial literacy components (Federal Ministry for Economic Affairs and Energy, 2021). The program's combination of financial training, technical education, and continuing support services was credited with its success. According to research on the features of program design, successful interventions ought to be customized for certain local settings and demographic groupings. While focused treatments that address particular barriers and knowledge gaps yield better results, generic financial education programs are not very effective (Fernandes *et al.*, 2014).

SOCIOECONOMIC INEQUALITIES IN ACCESS TO GREEN ENERGY

Significant social gaps in the adoption of green energy have been shown by empirical study, and these discrepancies may be partially explained by variations in financial literacy. Numerous obstacles affect low-income households, such as restricted access to cash, low credit scores, rental housing circumstances, and a lack of opportunity for financial education. Even after adjusting for pertinent variables like homeownership rates and electricity prices, O'Shaughnessy *et al.* (2021) examined demographic trends in residential solar adoption and discovered that adoption rates were negatively connected with neighborhood poverty levels. According to the study, these inequities may be exacerbated by financial literacy gaps, which make it more difficult for low-income households to understand financing choices and assess long-term benefits.

Studies on community solar initiatives shed light on strategies for resolving socioeconomic inequality. These initiatives have the potential to lower complexity and financial obstacles by enabling people to buy or lease parts of larger solar projects. To promote equal access, financial literacy programs might be required, as participation rates continue to be biased toward higher-income households (Hoen *et al.*, 2019). Low-income households spend a disproportionate amount of their income on energy expenditures, according to energy burden research, which suggests that they stand to gain a great deal from green energy investments that lower long-term costs. Nonetheless, the initial expenses and intricacy of these investments pose obstacles that could be resolved with improved financial literacy and focused assistance initiatives (Drehobl & Ross, 2016).

METHODS OF RESEARCH

Research methods and design in order to provide thorough insights on green energy financial literacy and its relationship to sustainable development outcomes, this study uses a mixed-methods research methodology that blends quantitative and qualitative approaches. In order to capture the depth of understanding through in-depth qualitative investigation and the breadth of the phenomena through statistical analysis, the mixed-methods technique was chosen. According to Creswell and Plano Clark (2017), the study design is an explanatory sequential mixed-methods model in which the collection and analysis of quantitative data is followed by the gathering of qualitative data intended to elucidate and elaborate on the quantitative findings. This method improves the validity and dependability of conclusions by enabling triangulation of findings across various data sources and methodologies.

In order to evaluate the financial literacy of green energy across a range of demographic groups and investigate correlations with adoption intentions and actions, the quantitative component employs a cross-sectional survey approach. Semi-structured interviews with important stakeholders, such as consumers, business leaders, legislators, and financial service providers, are used in the qualitative component to acquire a deeper understanding of the obstacles, enablers, and possibilities for action. The Institutional Review Board approved the study, which was carried out in compliance with ethical standards for research involving human beings. Every participant gave their informed consent, and participant identity and confidentiality were guaranteed via data gathering methods.

Gathering quantitative information, a complete survey instrument was administered to a nationally representative sample of 2,500 persons who were 18 years of age or older as part of the study's quantitative component. To guarantee sufficient representation across pertinent population segments, the sample was stratified by important demographic factors such as age, income, education level, geographic location, and homeownership status. A seasoned survey research company used a mix of address-based sampling, telephone recruitment, and web panels to find survey respondents. In order to reduce selection bias and guarantee representation of communities with different levels of technology access and desire, the multi-mode approach was created.

In order to capture domain-specific knowledge and attitudes, the survey instrument was created by adding original items to pre-existing financial literacy scales that had been modified for use in green energy contexts. Demographics, general financial literacy, green energy financial knowledge, green energy attitudes and intents, and green energy adoption behaviors were the five primary components of the test.

The Financial Industry Regulatory Authority (FINRA) Investor Education Foundation surveys and the Organization for Economic Co-operation and Development

(OECD) financial literacy framework were among the established scales used to measure general financial literacy. These topics addressed fundamental financial ideas such risk assessment, diversification, inflation, and interest rates.

25 unique items covering five domains—the costs and benefits of renewable technology, financing options and mechanisms, available incentives and subsidies, long-term financial analysis skills, and confidence in green energy financial decision-making—were used to measure green energy financial literacy. In order to guarantee content quality and suitable difficulty levels, items were created after expert consultation and pilot testing. Through in-depth inquiries concerning prior investments in renewable energy technology, energy efficiency enhancements, and green financial products, green energy adoption behaviors were evaluated. Established behavioral intention questionnaires modified for use in green energy contexts were used to gauge adoption intentions.

Participants were given the option to select their preferred method of data collection, which included web-based platforms and computer-assisted telephone interviewing (CATI). The overall response rate was 47%, which is deemed appropriate for this kind of research, and the average survey completion time was 28 minutes. 5.3 Gathering Qualitative Information 45 semi-structured interviews were conducted with stakeholders in four different groups as part of the qualitative component: financial service providers (n = 7), industry experts (n = 10), consumers (n = 20), and policymakers (n = 8). Purposive sampling was used to choose interview subjects in order to guarantee that a range of viewpoints and experiences pertinent to financial literacy for green energy were represented.

Consumer interviews were conducted with people who had recently decided to invest in green energy or were currently thinking about doing so. Referrals from community organizations, environmental organizations, and installers of renewable energy were used to find participants. To get a range of viewpoints on obstacles and enablers, the consumer sample comprised people with different financial literacy, income, and green energy expertise levels. Installers of renewable energy, experts in clean energy financing, environmental consultants, and members of renewable energy trade associations were among the industry experts interviewed. Professional viewpoints on consumer financial literacy requirements, prevalent misunderstandings, and successful teaching strategies were examined in these interviews.

State energy offices, public utility commissions, and local government organizations that support renewable energy were among the policymakers interviewed. These interviews looked at policy viewpoints regarding prospects for policy interventions and obstacles to financial literacy. Interviews with financial service providers included financial advisors with expertise in renewable energy investments, loan officers from banks and credit unions that provide green energy financing, and representatives from specialized green energy financing firms. Through these interviews, industry viewpoints on consumer financial literacy requirements and product development prospects were investigated.

With participant consent, all interviews were performed over the phone or via video conference, recorded, and then verbatim transcribed for analysis. Semi-structured interview techniques ensured that important research topics were covered while facilitating the investigation of emerging themes. The average interview lasted 62 minutes, with a range of 45 to 90 minutes. 5.4 Techniques for Data Analysis Descriptive statistics, correlation analysis, and multivariate regression modeling were used in quantitative data analysis to investigate the connections among financial literacy, demographic traits, and green energy outcomes. The statistical program STATA 17 was used for all studies.

All-important variables' means, standard deviations, and frequency distributions were calculated as part of the descriptive analysis. The percentage of right answers in each domain was used to calculate financial literacy scores, and composite scores were produced by averaging results from several areas. Bivariate associations between financial literacy metrics and green energy outcomes, such as adoption behaviors, intentions, and attitudes, were investigated using correlation analysis. For continuous variables, Pearson correlation coefficients were computed; for associations involving dichotomous variables, point-biserial correlations were employed.

After adjusting for pertinent demographic and socioeconomic factors, multivariate regression analysis was used to investigate the connections between financial literacy and green energy outcomes. Whereas logistic regression was utilized for binary outcomes, such adoption decisions, linear regression was utilized for continuous outcomes. Using structural equation modeling, mediation analysis was done to see if perceived behavioral control, risk perception, or trust in green technology mediated the effects of financial literacy on the uptake of green energy. Bootstrapping techniques were added to the Baron and Kenny (1986) methodology to test indirect effects.

The framework method was used to analyze qualitative data using a thematic analysis approach (Ritchie & Spencer, 1994). Two researchers independently classified interview transcripts using the qualitative analysis program NVivo 12. Prior to deductive coding based on theoretical frameworks and research topics, first coding was inductive, allowing themes to arise from the data. Several steps were involved in the qualitative analysis process: getting to know the data by reading the transcripts several times, identifying preliminary themes and patterns, creating a coding framework, systematically coding all of the transcripts, charting the data within themes, and interpreting the results. Consistency in coding and theme development was assured by the researchers' regular meetings.

Joint displays, meta-inferences, and mixed-methods analysis matrices that pinpointed areas of complementarity, divergence, and convergence between the two data sources were used to integrate quantitative and qualitative findings (Fetters *et al.*, 2013).

Findings and interpretation feature of the sample after eliminating cases with significant missing data, 2,347 survey respondents made up the final analytical sample. In line with standard survey research patterns, the sample showed good representation across major demographic parameters, with a small overrepresentation of higher education levels.

30% of respondents were 60 years of age or older, 24% were 30-44, 28% were 45-59, and 18% were between the ages of 18 and 29. The distribution of genders was 48% male and 52% female. 15% of respondents had only completed high school, 31% had some college, 34% had a bachelor's degree, and 20% had a graduate degree. According to the income distribution, 22% of households made less than $40,000 per year, 28% made between $40,000 and $74,999, 31% made between $75,000 and $124,999, and 19% made more than $125,000. 36% of respondents were from cities, 42% were from suburbs, and 22% were from rural areas.

The sample's 73% homeownership percentage is in line with the national average. The features of energy bills revealed that the average monthly cost of power was $127, with a significant range of $45 to $340. 34% of respondents said they were actively considering investments in renewable energy, 12% had made significant energy efficiency improvements in the last three years, and about 8% had already installed solar panels.

ANALYSIS OF RESULTS

Evaluation of green energy financial literacy the evaluation of financial literacy related to green energy found notable gaps in knowledge in every area examined. The majority of respondents showed a weak comprehension of financial concepts connected to investments in renewable energy, as seen by the total mean score of 32.4% accurate replies (SD = 18.7%).

Performance in each of the five evaluated domains varied significantly. With 41.2% of accurate answers, renewable technology costs and benefits had the best average performance, indicating that consumers are somewhat aware of the fundamental benefits of green energy. There is still much space for growth, though, as roughly 60% of replies were wrong even in this comparatively strong domain. At 24.1% accurate answers, financing mechanisms and options showed the lowest average performance. This finding is especially troubling because access to funding is crucial for the development of renewable energy. The availability of specialist financing products was underestimated, interest rates for green energy loans were overestimated, and the lease vs buy choices for solar installations were not well understood.

On average, the available incentives and subsidies yielded 28.7% correct answers. Federal tax credits for investments in renewable energy were unknown to many respondents, and even fewer were aware of state and municipal incentive programs. It's interesting to note that respondents from states with more substantial incentive programs did not exhibit noticeably higher awareness levels, indicating that attempts to promote and educate the program may not be sufficient. On average, long-term financial analytical skills produced 31.8% accurate answers. This area evaluated the capacity to compare total cost of ownership among energy options, compute payback periods, and comprehend concepts related to net present value. Poor performance in this area raises the possibility that many consumers may choose to invest in energy in less-than-ideal ways because they are unable to undertake the necessary financial analysis.

With 38.9% of respondents suggesting adequate confidence levels, confidence in green energy financial decision-making performed the second best. Nevertheless, additional investigation showed alarming trends of overconfidence among those with low objective knowledge scores, indicating that self-reported knowledge might not be a reliable indicator of true financial literacy.

financial literacy demographic trends significant differences in green energy financial literacy were found among population groupings based on demographic pattern analysis. The design of focused intervention programs and fair access to the advantages of renewable energy are significantly impacted by these inequities.

Financial literacy scores were most strongly correlated with educational achievement. The average number of accurate answers was 47.3% for respondents with doctoral degrees and 19.8% for those with only a high school diploma ($F(3,2343) = 127.4$, $p < 0.001$). For areas requiring numerical analysis, like long-term financial assessment and financing systems, the education disparity was very noticeable.

Financial literacy and income level showed a strong positive correlation, but one that was marginally smaller than the impacts of education. Respondents with high incomes (>$125,000 annually) had an average of 43.1% right answers, whereas those with low incomes (<$40,000 annually) had an average of 25.7% ($F(3,2343) = 89.2$, $p < 0.001$). Even after adjusting for educational attainment, this income disparity persisted, indicating separate influences of financial resources on financial literacy on green energy.

A curvilinear association between age and financial literacy was found; respondents between the ages of 45 and 59 had the greatest scores (37.2% correct), while those between the ages of 18 and 29 and 60 and older had the lowest scores (29.1% and 28.8%, respectively) ($F(3,2343) = 34.7$, $p < 0.001$). The junction of life stage, technology familiarity, and financial experience that affects interest in and learning about green energy may be shown in this pattern.

Geographic differences revealed that respondents from urban and suburban areas performed much better than those from rural areas (35.1% and 33.8% correct, respectively) ($F(2,2344) = 52.1$, $p < 0.001$). Due to fewer local renewable energy installations that offer social proof and educational opportunities, as well as restricted access to information resources, rural individuals performed especially poorly in domains pertaining to available incentives and financing alternatives.

Male respondents averaged 34.1% correct replies, whereas female respondents averaged 30.9% ($t(2345) = 3.2$, $p < 0.01$), indicating moderate but statistically significant gender disparities. Whereas gender disparities were negligible in areas pertaining to incentives and fundamental cost-benefit analysis, these inequalities were most noticeable in technical areas like funding mechanisms and long-term financial analysis. Strong correlations were found between financial literacy and homeownership status, with homeowners' average 35.8% accurate answers compared to renters' 24.1% ($t(2345) = 8.7$, $p < 0.001$). This discrepancy most likely stems from homeowners' increased exposure to home-related financial decisions that foster important analytical abilities as well as the rising significance of green energy investments for them.

The connection between adoption of green energy and financial literacy strong positive correlations were found across a number of outcome measures when the relationship between green energy financial literacy and actual adoption behaviors was examined. These results provide credence to the theoretical idea that financial literacy is a crucial component of investment choices in renewable energy. Adoption of solar panels was most strongly correlated with financial literacy. On the financial literacy test, respondents with solar panels gave an average of 52.3% of the right answers, while those without solar panels gave an average of 30.1% ($t(2345) = 12.4$, $p < 0.001$). Even after adjusting for property status, income, and education, the difference of 22.2 percentage points remained significant and implies a considerable impact size.

Similar trends were seen in energy efficiency investments, with households that had significantly improved their efficiency averaging 41.7% accurate answers compared to 29.8% for those that had not ($t(2345) = 8.9$, $p < 0.001$). While smaller upgrades like replacing LED lighting had weaker connections with financial literacy, the relationship was especially high for more complicated efficiency investments like heat pump installs and whole-house retrofits. Intentions to invest in green energy were significantly correlated with financial literacy in every domain that was measured.

The average number of right answers was 39.6% for respondents who strongly intended to invest in renewable energy within the next two years, compared to 28.4% for those who did not ($t(2345) = 9.1$, $p < 0.001$). This association implies that financial literacy affects future investment planning in addition to adoption decisions made in the past.

The likelihood of adopting green energy as a function of financial literacy was investigated using logistic regression analysis, which also controlled for pertinent economic and demographic variables. The probabilities of adopting solar panels increased by 28% for every 10 percentage points that financial literacy scores increased (OR = 1.28, 95% CI: 1.19-1.38, $p < 0.001$). Investments in energy efficiency and other renewable energy technologies showed comparable effect sizes. The correlation between adoption and financial literacy differed considerably throughout knowledge domains.

The most powerful predictor of adoption decisions was long-term financial analysis ability (OR = 1.45 for a 10-point increase, $p < 0.001$), which was followed by understanding of funding processes (OR = 1.33, $p < 0.001$). Remarkably, adoption was less strongly correlated with awareness of available incentives (OR = 1.18, $p < 0.05$), indicating that the capacity to perform thorough financial analysis may be more important than knowledge of financial assistance programs.

Pathways and mediating factors potential mediation mechanisms through which financial literacy affects decisions to use green energy were investigated using structural equation modeling analysis. Several significant indirect effects that contribute to the explanation of the mechanisms behind the literacy-adoption association were identified by this investigation.

Higher financial literacy was linked to reduced perceived risks of green energy investments ($\beta = -0.31, p < 0.001$), which in turn predicted higher adoption likelihood ($\beta = 0.24, p < 0.001$). Risk perception was found to be a significant mediator. About 18% of the overall impact on adoption decisions was attributable to the indirect effect of financial literacy through risk perception. According to this research, people who possess financial literacy are better able to evaluate investment risks and get over their extreme risk aversion. A further crucial mediating mechanism was perceived behavioral control. Financial literacy significantly impacted adoption intentions and actions ($\beta = 0.35, p < 0.001$) and was a strong predictor of confidence in one's ability to make wise green energy investment decisions ($\beta = 0.42, p < 0.001$).

The significance of self-efficacy in energy investment decisions is highlighted by the indirect effect, which accounted for 23% of the overall financial literacy effect through perceived behavioral control. The mediating effects of trust in green energy technology were not as strong as anticipated. The association between trust and adoption was weak ($\beta = 0.14, p < 0.05$), although financial literacy was favorably correlated with trust levels ($\beta = 0.19, p < 0.01$). This implies that when it comes to making judgments about green energy investments, trust may not be as important as risk perception and self-efficacy.

Peer adoption and social norms are examples of social impact elements that have demonstrated intriguing interaction effects with financial literacy. Social influence variables did not significantly affect adoption decisions among respondents who were highly financially literate. On the other hand, neighborhood peer adoption significantly predicted the likelihood of individual adoption among those with low financial literacy ($\beta = 0.38$, $p < 0.001$). According to this pattern, social evidence can be especially significant for customers who are unsure of their own aptitude for financial analysis.

Obstacles to the adoption of green energy self-reported obstacles to the adoption of green energy were analyzed, and significant correlations with financial literacy were found. Comprehending these connections offers valuable perspectives on how financial education could tackle particular barriers to investment in green energy. Sixty-eight percent of respondents who had not adopted green energy solutions cited high upfront costs as the main obstacle. However, there were considerable differences in the link between real expenses and cost worries depending on the amount of financial literacy. The costs of installing solar panels were significantly underestimated by respondents with limited financial literacy; on average, their predictions were 47% higher than the going rates.

On the other hand, cost predictions from respondents who were very financially literate were, on average, within 12% of real market costs. The second most commonly mentioned obstacle, according to 54% of non-adopters, was uncertainty regarding financial rewards. Strong negative associations were seen between this barrier and financial literacy levels ($r = -0.43$, $p < 0.001$), indicating that increased financial awareness could considerably lessen concerns about return uncertainty.

According to qualitative interviews, there was a great deal of ambiguity regarding investment returns because many customers were not aware with the techniques used to calculate payback periods and net present value. 41% of non-adopters reported lacking confidence in the dependability of technology, and this factor had a slight connection with financial literacy ($r = -0.12$, $p < 0.05$). This implies that while financial literacy may aid consumers in assessing warranty terms and performance guarantees, technical expertise may be more closely linked to worries about technological trust.

With strong negative connections to financial literacy ($r = -0.38$, $p < 0.001$), 39% of non-adopters reported difficulty obtaining sufficient funding. Low financial literacy respondents overestimated the difficulties of being eligible for favorable loan terms and frequently were not aware of specialist green energy financing options.

Additionally, a lot of people were not aware of financing options that required no down payment, like power purchase agreements and solar leases. Thirty-three percent of non-adopters cited incentive program complexity as a barrier. This barrier had the highest correlation with financial literacy ($r = -0.51$, $p < 0.001$), indicating that improved financial education and streamlined program design might significantly lessen this barrier. Numerous respondents voiced their dissatisfaction with the intricate application procedures and their ambiguity on the prerequisites for eligibility for incentives.

qualitative results regarding customer experiences the qualitative interview component revealed nuanced viewpoints that extended and supplemented the quantitative findings, offering deep insights into consumer experiences with green energy financial decisions. Consumer interviews brought to light the psychological and emotional aspects of green energy investment choices that financial literacy metrics fall short of capturing. Numerous participants expressed feeling overpowered by the intricacy of the possibilities at their disposal and doubting their capacity to make the best choices.

According to a participant: "I know solar panels can save money, but there are so many companies, financing options, and technical specifications that I don't know where to start." A crucial element that surfaced was the need of trustworthy advisors, as customers indicated a significant preference for financial advice from reliable, objective sources. Due to potential conflicts of interest, many respondents expressed distrust regarding the information supplied by solar installation.

Preferred information sources were found to be community-based educational initiatives and reliable financial advisors, however many consumers had restricted access to these options. Many interviewees expressed dissatisfaction with contradicting information from many sources, indicating that information overload was a frequent experience. Many consumers were unable to make decisions because of the intricate financial arrangements and technological complexity of renewable energy systems. Due to uncertainty and information overload, a number of respondents said they had put off making financial decisions for years. Throughout the interviews, the significance of social proof and peer learning became clear.

Talking with friends, family, or neighbors was frequently mentioned by consumers who had implemented green energy solutions as a deciding factor. The potential of peer-to-peer education systems is highlighted by the fact that early adopters often reported acting as informal consultants for others in their social networks. During the interviews, deficiencies in financial literacy showed up in a number of distinct ways. The present value calculations required to compare lease and buy choices were difficult for many customers. Others found it challenging to comprehend the ramifications of various loan terms and how they affected the overall project expenditures. A number of participants conveyed uncertainty over the impact of tax incentives on their own tax circumstances.

Views from industry and policy from a professional standpoint, interviews with legislators and industry experts shed light on systemic obstacles and chances to improve financial literacy for green energy. Installers of renewable energy have often noted that one of the biggest obstacles to their business operations is consumer financial literacy. Many said they spent a lot of time teaching clients about incentive schemes, payback estimates, and financing possibilities. According to one installer: Financial education makes up half of our sales process. Consumers wish to switch to solar power but are unsure of their possibilities. A few of installations had created internal training materials and procedures to fill the common gaps in financial literacy.

Financial service providers recognized that many loan officers lack the specialist knowledge needed for green energy loans. Employees of traditional banks may not be aware with performance-based underwriting techniques and frequently find it difficult to assess green energy investments. For employees engaged in green energy financing, a number of organizations reported putting in place specialist training programs. Interviews with policymakers showed that while they were aware of the obstacles to financial literacy, there were few systematic ways to overcome them.

The majority of incentive schemes prioritize eligibility requirements and reward amounts over customer support and education. Although they mentioned resource limitations and a lack of clarity regarding practical strategies, a number of state energy offices indicated interest in creating financial literacy elements for their renewable energy initiatives.

Standardized information and decision-support tools are essential, according to industry experts, as they can help consumers make better decisions and lessen uncertainty. One major obstacle that may be overcome with improved coordination and standardization initiatives is the proliferation of various finance products and program forms. There has been much discussion about how technological solutions might help remove obstacles to financial literacy. Online calculators, decision-support software, and educational materials, according to a number of experts, could assist customers in making difficult financial decisions. They did point out, though, that these tools need to be thoughtfully created to be used by people with different degrees of financial expertise.

DISCUSSION

Analysis of the main results the results of this study offer compelling empirical evidence in favor of the theoretical claim that financial literacy is a key component of the adoption of green energy and environmentally sustainable development. With effect sizes equivalent to those discovered for income and education effects in comparable situations, the observed association between financial knowledge and investment behavior in renewable energy is both statistically significant and practically useful. A major obstacle to sustainable energy transitions is highlighted by the study's overall low level of green energy finance literacy (32.4% average accurate responses). Given the complexity of green energy investment decisions and the significant financial benefits that many customers might obtain by adopting renewable energy, this conclusion is very worrisome.

These deficiencies in financial literacy could account for a portion of the discrepancy between the prospective advantages and the actual adoption rates. Important information about chances for focused intervention can be gleaned from the differences in financial literacy performance across several knowledge categories. The comparatively high performance in fundamental cost-benefit analysis indicates that although consumers are somewhat aware of the benefits of green energy, they lack the technical financial know-how required to assess particular investment opportunities. This pattern implies that rather than concentrating on raising broad awareness, educational interventions should emphasize the development of practical financial analytical skills.

Even after adjusting for education and income, there is still a high correlation between adoption behavior and financial literacy, suggesting that financial literacy influences green energy investment choices independently. This research shows that focused financial education may have significant effects on adoption rates across a range of demographic groups, which runs counter to viewpoints that see financial literacy as merely a stand-in for socioeconomic position. The ways that financial literacy affects adoption decisions can be explained by the mediating roles that risk perception and perceived behavioral control play. These results support well-established theories of planned behavior and imply that interventions in financial education should target both subjective confidence in one's ability to make decisions and objective information.

Consequences for research and theory the importance of domain-specific financial literacy in intricate investment decisions is highlighted in this study, which advances theoretical knowledge of technology adoption processes. The financial complexity that many green energy systems have may not be sufficiently addressed by traditional technology adoption models, which place a greater emphasis on perceived utility and usability. The results are consistent with behavioral economic theories, which highlight the role of cognitive talents in overcoming heuristics and biases that can obstruct the best possible decision-making. Improved analytical skills enable customers to make more accurate risk assessments and steer clear of excessive loss aversion, according to the association between financial literacy and decreased risk perception.

The observed interplay between social influence elements and financial literacy sheds light on how diverse technology adoption processes are. One-size-fits-all methods to technology promotion may be challenged by the reduced significance of peer effects among financially educated customers, which raises the possibility that various adopter segments may need different kinds of information and support. Important concerns regarding equitable access to the advantages of green energy are brought up by the notable demographic differences in financial literacy levels. Communities with low financial literacy may find it more difficult to take advantage of the financial advantages of sustainable energy technology, which could exacerbate environmental justice issues that have been brought up in relation to renewable energy transitions.

By highlighting the applicability of financial capability frameworks in environmental situations, the research also advances knowledge of these frameworks. The discovery that institutional elements like program design and product availability interact with financial knowledge points to the necessity for all-encompassing strategies for developing financial competence.

Implications for policy and suggestions the results of the study have important ramifications for the formulation and application of policies at various governmental levels and among institutional actors. These ramifications go beyond conventional energy policy to include social justice, financial regulation, and education.

For renewable energy incentive schemes to be as effective as possible at the federal level, financial literacy elements should be included. Financial incentives are the main emphasis of current programs, but nothing is done to guarantee that customers know how to access and maximize these advantages. Program results could be significantly enhanced by making little changes such mandating uniform disclosure forms, offering educational resources, and providing decision-support tools.

In order to supplement current energy programs, state and municipal governments ought to think about creating comprehensive financial literacy projects related to green energy. Public education campaigns, community seminars, internet resources, and collaboration programs with financial institutions and neighborhood organizations are a few examples of such projects. Given the noted regional differences in financial literacy, it may be necessary to provide tailored outreach and assistance to underserved and rural groups.

Green energy financial literacy should be incorporated into current financial education courses at educational institutions. Given the close connection between financial literacy and education, formal educational channels may be useful for fostering the development of pertinent information and abilities. From consumer education classes in high school to continuing education courses for adults, integration can take place at several levels. Financial regulatory bodies ought to investigate whether the distinctive features of green energy financial products are sufficiently addressed by the consumer protection frameworks in place. To guarantee informed decision-making, the intricacy of solar leases, power purchase agreements, and specialty financing products may necessitate stricter disclosure regulations and consumer protection safeguards.

Financial literacy should be taken into account by utility regulatory agencies when assessing renewable energy initiatives and pricing schemes. Demand response programs, net metering regulations, and time-of-use pricing all necessitate that customers comprehend intricate financial linkages, which can be difficult for those with low financial literacy.

Consequences for industry and financial institutions the study's conclusions point to a number of chances for financial institutions to provide cutting-edge goods and services that fill the knowledge gaps in green energy finance while opening up new markets. Traditional banks, credit unions, fintech firms, and specialized green energy lenders are all affected.

To accommodate customers with different levels of financial sophistication, product design should place a high priority on simplicity and openness. Many customers, especially those in underprivileged areas who stand to gain the most from green energy investments, may find complex financial instruments with several options and variable terms unsuitable. It's possible that standardized products with comparable, unambiguous words work better than highly personalized ones. For financial institutions, educational assistance services are a significant difference and business opportunity.

By investing in client education and decision-support systems, banks and credit unions can gain a competitive edge while advancing larger societal objectives. These services might consist of online calculators, educational seminars, financial counseling, and collaboration initiatives with installers of renewable energy.

It might be necessary to modify risk assessment and underwriting procedures to take into consideration the particulars of green energy projects. The advantages of energy cost savings for borrower repayment capacity may not be sufficiently reflected in traditional credit rating. Access could be increased while maintaining good risk management through performance-based underwriting that takes prospective energy savings into account. Financial institutions and other players in the green energy ecosystem can form partnerships. While sharing expenses and knowledge, partnerships between utility companies, government organizations, community groups, and installers of renewable energy could result in more extensive customer support systems.

Opportunities to scale financial literacy support while cutting expenses are presented by technological solutions. Tools for decision-support and individualized financial advice may be offered by automated advisory services, internet platforms, and mobile apps. These solutions must, however, be made to be usable by people with different degrees of financial sophistication and technological comfort.

Restrictions and prospects for further study there are a number of limitations to this study that should be recognized and taken into consideration for subsequent research. The capacity to show causal correlations between adoption behaviors and financial literacy is limited by the cross-sectional methodology. Although the theoretical framework and empirical patterns indicate that adoption decisions are influenced by financial literacy, it is impossible to rule out other possible explanations such reverse causality or omitted variable bias.

The practical knowledge and abilities required for actual green energy investment decisions may not be well captured by the survey-based financial literacy assessment. Assessments conducted in laboratories or in the field, where real-world decision-making processes are observed, may yield more reliable indicators of functional financial literacy. Furthermore, rather than using confirmed installation data, the study relied on self-reported adoption habits, which could increase measurement error. Despite its size and ethnic diversity, the sample was restricted to American adults who spoke English.

The generalizability of findings across various policy environments, cultural contexts, and economic systems would be established with the use of cross-cultural and international research. Countries with diverse funding systems, incentive structures, or cultural perspectives on environmental preservation may have quite varied relationships between financial literacy and the adoption of green energy.

To more clearly define causal links and investigate the long-term effects of financial literacy treatments on adoption habits, future studies should use longitudinal designs. Stronger proof on the efficacy of interventions and the best features of program design would come from randomized controlled trials of financial education initiatives. Studies on particular subgroups may give light on focused intervention tactics. More efficient and fair program design may result from studies that concentrate on low-income neighborhoods, rural communities, senior citizens, or other groups that face unique obstacles to the adoption of green energy.

More research is necessary to fully understand the connection between various forms of green energy technologies and financial literacy. Although the adoption of solar energy was the main emphasis of this study, the criteria and obstacles for financial literacy may differ for other technologies, such as wind, geothermal, and energy storage systems. Policy and program design may benefit from an examination of the institutional and contextual elements that regulate the relationship between financial literacy and adoption. The way that financial literacy is translated into adoption behaviors may be influenced by a number of factors, including market structures, regulatory settings, available financing alternatives, and social norms.

CONCLUSIONS

Strong evidence that green energy finance literacy is a crucial component of sustainable energy transitions and environmental development outcomes is provided by this extensive study. The results of the survey show that, despite a comparatively high level of public awareness of the advantages of green energy, all demographic groups still lack the precise financial knowledge required to assess and pursue investments in renewable energy.

According to the study, there is a significant positive association ($r = 0.67$, $p < 0.001$) between the actual adoption of renewable energy technology and the degree of financial literacy regarding green energy. The study also finds socioeconomic differences in access to financial information about green energy, with lower-income groups exhibiting considerably lower literacy rates even though they may stand to gain more from lower energy costs.

The research ends with thorough policy recommendations for improving financial literacy in the context of green energy through inclusive funding arrangements, streamlined financial tools, and focused educational initiatives. These results have important ramifications for accomplishing environmental sustainability objectives and quickening the world's shift to sustainable energy sources. Even after adjusting for income, education, and other pertinent variables, the study shows a strong correlation between financial literacy levels and green energy adoption habits. These connections function in a number of ways, such as decreased perception of danger, more confidence in one's capacity to make decisions, and better proficiency with intricate incentive and funding structures.

The study's effect sizes imply that focused financial literacy initiatives may have a significant influence on the rates of adoption of renewable energy sources and environmental consequences. Important equity concerns regarding access to sustainable energy advantages are raised by the notable demographic differences in green energy financial literacy. Comprehensive financial literacy programs, in conjunction with supportive institutional and legislative changes, can help overcome the various obstacles faced by underprivileged groups, including lower-income areas and rural populations. These differences underline the necessity of inclusive strategies for sustainable energy transitions that guarantee fair access to the advantages of the economy and the environment.

The study pinpoints particular areas of knowledge where educational interventions may have the most effects. Targeted education could greatly enhance consumer decision-making in three critical areas: financing systems, long-term financial analysis skills, and knowledge of available incentives. The observation that confidence levels frequently surpass objective knowledge implies that practical skills should be the main focus of interventions rather than raising general understanding.

The viewpoints of industry and policy stakeholders highlight structural obstacles that cannot be overcome by personal financial education. Structural issues that need for coordinated solutions from various stakeholders include complicated and inconsistent financing products, insufficient consumer protection regimes, and a lack of institutional support for the development of financial literacy.

The research's ramifications go beyond the use of green energy to more general issues of sustainable development, environmental justice, and financial capacity. Making sure that every demographic has the information and resources needed to engage in sustainable energy markets is becoming more and more crucial as societies around the world struggle with the pressing demands of energy transitions and climate action. The results of the study provide credence to a multifaceted strategy that combines business innovation, institutional capacity building, policy reform, and individual education to improve green energy financial literacy. Educational interventions ought to be provided through reputable community channels and customized for particular demographic groups.

Financial institutions ought to create more user-friendly products and better customer service. Programs and regulatory frameworks pertaining to renewable energy should incorporate financial literacy considerations. While addressing the limitations of this study, future research should keep improving our understanding of successful intervention techniques. The evidence foundation for developing policies and programs would be strengthened by cross-cultural analysis, randomized controlled trials of educational initiatives, and longitudinal research methods. Furthermore, studies looking into how financial literacy relates to other facets of sustainable development may shed light on how these findings might be used more broadly.

Making the switch to sustainable energy systems is both economically and environmentally necessary. This study shows that obstacles to financial literacy could be preventing the achievement of economic and environmental objectives. Societies may hasten sustainable energy transitions while guaranteeing that the advantages are widely distributed throughout all demographic groups by tackling these obstacles with all-encompassing, empirically supported strategies.

Investing in green energy financial literacy appears to be a high-leverage way to promote environmentally sustainable development, according to the data in this study. Enhancing financial literacy is a desirable policy choice due to the comparatively low costs of educational interventions in comparison to the possible advantages of greater adoption of renewable energy. Furthermore, other aspects of financial decision-making and environmental behavior may benefit from the abilities and information acquired through green energy financial literacy programs.

Continued focus on the demands for financial literacy will be crucial as the renewable energy sector develops with new technology, funding sources, and legislative frameworks. In order to ensure widespread participation in sustainable energy transitions, consumer education and support services are even more crucial because the complexity of green energy markets may increase before they diminish.

Although the study discussed here offers a starting point for comprehending the part financial literacy plays in the adoption of sustainable energy, much more has to be done to convert these findings into practical interventions and legislative changes. This translation is essential given the pressing need to address climate change and the potential for green energy solutions to yield both financial and environmental advantages. While guaranteeing that the advantages of clean energy transitions are distributed fairly throughout society, success in raising green energy financial literacy could hasten the achievement of sustainable development goals.

REFERENCES

Ajzen, I. (1991). The theory of planned behavior. *Organizational Behavior and Human Decision Processes, 50*(2), 179–211. DOI: 10.1016/0749-5978(91)90020-T

Allcott, H., & Greenstone, M. (2017). Measuring the welfare effects of residential energy efficiency programs. *Journal of Public Economics, 145*, 40–53. DOI: 10.1016/j.jpubeco.2016.11.009

Bandura, A. (1986). *Social foundations of thought and action: A social cognitive theory*. Prentice-Hall.

Baron, R. M., & Kenny, D. A. (1986). The moderator-mediator variable distinction in social psychological research: Conceptual, strategic, and statistical considerations. *Journal of Personality and Social Psychology, 51*(6), 1173–1182. DOI: 10.1037/0022-3514.51.6.1173 PMID: 3806354

Blasch, J., Boogen, N., Datta, S., & Filippini, M. (2021). The role of energy literacy for residential energy demand. *Energy Economics, 96*, 105155. DOI: 10.1016/j.eneco.2021.105155

Bollinger, B., & Gillingham, K. (2012). Peer effects in the diffusion of solar photovoltaic panels. *Marketing Science, 31*(6), 900–912. DOI: 10.1287/mksc.1120.0727

Borenstein, S. (2017). Private net benefits of residential solar PV: The role of electricity tariffs, tax incentives, and rebates. *Journal of the Association of Environmental and Resource Economists, 4*(S1), S85–S122. DOI: 10.1086/691978

Braunstein, S., & Welch, C. (2002). Financial literacy: An overview of practice, research, and policy. *Federal Reserve Bulletin, 88*(11), 445–457. DOI: 10.17016/bulletin.2002.88-11

Brundtland Commission. (1987). *Our common future: Report of the World Commission on Environment and Development*. Oxford University Press.

Carfì, D., Donato, A., & Gambarelli, G. (2018). Economically sustainable development and green economy: A game-theoretic model. *Decisions in Economics and Finance, 41*(1), 25–43. DOI: 10.1007/s10203-018-0213-7

Chen, H., & Volpe, R. P. (2019). An analysis of personal financial literacy among college students. *Financial Services Review, 7*(2), 107–128. DOI: 10.1016/S1057-0810(99)80006-7

Creswell, J. W., & Plano Clark, V. L. (2017). *Designing and conducting mixed methods research* (3rd ed.). SAGE Publications.

Dastrup, S. R., Graff Zivin, J., Costa, D. L., & Kahn, M. E. (2019). Understanding the solar home price premium: Electricity generation and "green" social status. *European Economic Review*, *56*(5), 961–973. DOI: 10.1016/j.euroecorev.2012.02.006

Database of State Incentives for Renewables & Efficiency. (2022). *Annual report on renewable energy incentive programs*. North Carolina State University.

Davis, F. D. (1989). Perceived usefulness, perceived ease of use, and user acceptance of information technology. *Management Information Systems Quarterly*, *13*(3), 319–340. DOI: 10.2307/249008

del Río, P., & Unruh, G. (2007). Overcoming the lock-out of renewable energy technologies in Spain: The cases of wind and solar electricity. *Renewable & Sustainable Energy Reviews*, *11*(7), 1498–1513. DOI: 10.1016/j.rser.2005.12.003

Drehobl, A., & Ross, L. (2016). *Lifting the high energy burden in America's largest cities: How energy efficiency can improve low income and underserved communities*. American Council for an Energy-Efficient Economy.

Federal Ministry for Economic Affairs and Energy. (2021). *Renewable energy education initiative: Final evaluation report*. German Federal Government.

Fernandes, D., Lynch, J. G.Jr, & Netemeyer, R. G. (2014). Financial literacy, financial education, and downstream financial behaviors. *Management Science*, *60*(8), 1861–1883. DOI: 10.1287/mnsc.2013.1849

Fetters, M. D., Curry, L. A., & Creswell, J. W. (2013). Achieving integration in mixed methods designs—Principles and practices. *Health Services Research*, *48*(6pt2), 2134–2156. DOI: 10.1111/1475-6773.12117 PMID: 24279835

Frederick, S., Loewenstein, G., & O'Donoghue, T. (2002). Time discounting and time preference: A critical review. *Journal of Economic Literature*, *40*(2), 351–401. DOI: 10.1257/jel.40.2.351

Hoen, B., Rand, J., Elmallah, S., & Paulos, B. (2019). *Sharing the sun: Community solar markets, rates, and participation*. Lawrence Berkeley National Laboratory.

International Energy Agency. (2023). *World energy outlook 2023*. International Energy Agency.

International Renewable Energy Agency. (2021). *Renewable energy statistics 2021*. International Renewable Energy Agency.

Kahneman, D., & Tversky, A. (1979). Prospect theory: An analysis of decision under risk. *Econometrica*, *47*(2), 263–291. DOI: 10.2307/1914185

Klapper, L., Lusardi, A., & Van Oudheusden, P. (2015). *Financial literacy around the world: Insights from the Standard & Poor's ratings services global financial literacy survey*. World Bank Group.

Kowalska-Pyzalska, A. (2018). What makes consumers adopt to innovative energy services in the energy market? A review of incentives and barriers. *Renewable & Sustainable Energy Reviews*, *82*, 3570–3581. DOI: 10.1016/j.rser.2017.10.103

Lusardi, A., & Mitchell, O. S. (2014). The economic importance of financial literacy: Theory and evidence. *Journal of Economic Literature*, *52*(1), 5–44. DOI: 10.1257/jel.52.1.5 PMID: 28579637

Masini, A., & Menichetti, E. (2012). The impact of behavioural factors in the renewable energy investment decision making process: Conceptual framework and empirical findings. *Energy Policy*, *40*, 28–38. DOI: 10.1016/j.enpol.2010.06.062

Mills, B., & Schleich, J. (2022). Solar panel adoption and peer effects: Evidence from residential energy consumption data. *Energy Economics*, *107*, 105825. DOI: 10.1016/j.eneco.2022.105825

Nassiry, D. (2018). The role of fintech in unlocking green finance: Policy insights for developing countries. *ADBI Working Paper Series*, No. 883. Asian Development Bank Institute.

O'Shaughnessy, E., Barbose, G., Wiser, R., Forrester, S., & Darghouth, N. (2021). The impact of policies and business models on income equity in rooftop solar adoption. *Nature Energy*, *6*(1), 84–91. DOI: 10.1038/s41560-020-00724-2

Oikonomou, V., Becchis, F., Steg, L., & Russolillo, D. (2020). Energy saving and energy efficiency concepts for policy making. *Energy Policy*, *37*(11), 4787–4796. DOI: 10.1016/j.enpol.2009.06.035

Painuly, J. P. (2001). Barriers to renewable energy penetration; a framework for analysis. *Renewable Energy*, *24*(1), 73–89. DOI: 10.1016/S0960-1481(00)00186-5

Riedl, A., & Smeets, P. (2017). Why do investors hold socially responsible mutual funds? *The Journal of Finance*, *72*(6), 2505–2550. DOI: 10.1111/jofi.12547

Ritchie, J., & Spencer, L. (1994). Qualitative data analysis for applied policy research. In Bryman, A., & Burgess, R. G. (Eds.), *Analyzing qualitative data* (pp. 173–194). Routledge. DOI: 10.4324/9780203413081_chapter_9

Rogers, E. M. (2003). *Diffusion of innovations* (5th ed.). Free Press.

Sen, A. (1999). *Development as freedom*. Oxford University Press.

Sovacool, B. K. (2009). The cultural barriers to renewable energy and energy efficiency in the United States. *Technology in Society*, *31*(4), 365–373. DOI: 10.1016/j.techsoc.2009.10.009

Stern, P. C. (2000). New environmental theories: Toward a coherent theory of environmentally significant behavior. *The Journal of Social Issues*, *56*(3), 407–424. DOI: 10.1111/0022-4537.00175

Sunter, D. A., Castellanos, S., & Kammen, D. M. (2019). Disparities in rooftop photovoltaics deployment in the United States by race and ethnicity. *Nature Sustainability*, *2*(1), 71–76. DOI: 10.1038/s41893-018-0204-z

Terry, D. J., & Hogg, M. A. (2000). *Attitudes, behavior, and social context: The role of norms and group membership*. Lawrence Erlbaum Associates.

Thaler, R. (1985). Mental accounting and consumer choice. *Marketing Science*, *4*(3), 199–214. DOI: 10.1287/mksc.4.3.199

Thaler, R. H., & Sunstein, C. R. (2008). *Nudge: Improving decisions about health, wealth, and happiness*. Yale University Press.

Triandis, H. C. (1977). *Interpersonal behavior*. Brooks/Cole Publishing Company.

United Nations Framework Convention on Climate Change. (2021). *Nationally determined contributions under the Paris Agreement: Synthesis report by the secretariat*. United Nations.

U.S. Department of Energy. (2020). *Better Buildings Challenge: Impact evaluation of residential energy efficiency programs*. U.S. Department of Energy.

Van Rooij, M., Lusardi, A., & Alessie, R. (2011). Financial literacy and stock market participation. *Journal of Financial Economics*, *101*(2), 449–472. DOI: 10.1016/j.jfineco.2011.03.006

Compilation of References

Abera, Y. A. (2024). Sustainable building materials: A comprehensive study on eco-friendly alternatives for construction. *Composites and Advanced Materials*, *33*, 26349833241255957. Advance online publication. DOI: 10.1177/26349833241255957

Abid, I., Fuad, S. M. Z. A., Chowdhury, M. J. M., Chowdhury, M. S., & Ferdous, M. S. (2024). A Systematic Literature Review on the Use of Blockchain Technology in Transition to a Circular Economy. *arXiv (Cornell University)*. https://doi.org// arxiv.2408.11664DOI: 10.48550

Abideen, A. Z., Sundram, V. P. K., Pyeman, J., Othman, A. K., & Sorooshian, S. (2021). Food Supply Chain Transformation through Technology and Future Research Directions—A Systematic Review. *Logistics (Basel)*, *5*(4), 83. DOI: 10.3390/ logistics5040083

Adenan, F. A., Ariff, A., & Adnan, S. M. (2024). ESG in Small and Medium-Sized Enterprises: Malaysian Perspectives. *E3S Web of Conferences, 574*, 3001. DOI: 10.1051/e3sconf/202457403001

Agarwal, T., Chiesa, I., Costantini, M., Lopamarda, A., Tirelli, M. C., Borra, O. P., Varshapally, S. V. S., Kumar, Y. A. V., Koteswara Reddy, G., De Maria, C., Zhang, L. G., & Maiti, T. K. (2023). Chitosan and its derivatives in 3D/4D (bio) printing for tissue engineering and drug delivery applications. *International Journal of Biological Macromolecules*, *246*, 125669. DOI: 10.1016/j.ijbiomac.2023.125669

Agostinelli, S., Cumo, F., Guidi, G., & Tomazzoli, C. (2021). Cyber-Physical Systems Improving Building Energy Management: Digital Twin and Artificial Intelligence. *Energies*, *14*(8), 2338. DOI: 10.3390/en14082338

Agrawal, A., Bhaskar, A., & Jain, R. (2022). Comparative life cycle assessment of additive and conventional manufacturing of aerospace components. *Journal of Manufacturing Processes*, *75*, 167–177. DOI: 10.1016/j.jmapro.2022.01.011

Agrawal, K., Goktas, P., Holtkemper, M., Beecks, C., & Kumar, N. (2025). AI-driven transformation in food manufacturing: A pathway to sustainable efficiency and quality assurance. *Frontiers in Nutrition*, *12*, 1553942. Advance online publication. DOI: 10.3389/fnut.2025.1553942

Ahmed, A., Arya, S., Gupta, V., Furukawa, H., & Khosla, A. (2021a). 4D printing: Fundamentals, materials, applications and challenges. *Polymer*, *228*, 123926. DOI: 10.1016/j.polymer.2021.123926

Ahmed, F., Ahmed, M., Kabir, M. A., & Islam, M. M. (2025). Revolutionizing Business Analytics. *The Impact of Artificial Intelligence and Machine Learning.*, *1*(1), 147–173. DOI: 10.63125/f7yjxw69

Ahmed, W., & Zaneldin, E. (2020). The Innovation of Using 3D Printing Technology in Mechanical and Manufacturing Engineering. *IOP Conference Series. Materials Science and Engineering*, *938*(1), 12002. DOI: 10.1088/1757-899X/938/1/012002

Ahmed, Y., & Wang, T. (2019). Challenges in grid integration of solar PV systems. *Energy Reports*, *5*, 101–113. DOI: 10.1016/j.egyr.2019.01.012

Ahn, C. R., Lee, S. H., Peña-Mora, F., & AbouRizk, S. (2010). Toward Environmentally Sustainable Construction Processes: The U.S. and Canada's Perspective on Energy Consumption and GHG/CAP Emissions. *Sustainability (Basel)*, *2*(1), 354–370. DOI: 10.3390/su2010354

Ajzen, I. (1991). The theory of planned behavior. *Organizational Behavior and Human Decision Processes*, *50*(2), 179–211. DOI: 10.1016/0749-5978(91)90020-T

Akadiri, P. O., Olomolaiye, P., & Chinyio, E. (2012). Multi-criteria evaluation model for selecting sustainable materials for building projects. *Automation in Construction*, *30*, 113–125. DOI: 10.1016/j.autcon.2012.10.004

Akbarnezhad, A., Ong, K. C. G., & Chandra, L. R. (2013). Economic and environmental assessment of deconstruction strategies using building information modeling. *Automation in Construction*, *37*, 131–144. DOI: 10.1016/j.autcon.2013.10.017

Akhtar, I., & Zhang, T. (2021). Evaluating ML models for demand-side energy prediction. *Energy Modelling and Forecasting*, *6*(1), 58–71.

Alam, M., Arifuzzaman, M., Islam, M. K., Alfuhaid, A. F., & Al-Mamun, A. (2022). Sustainable Solution for Deteriorated and Aged RCC Structures: A Review of Buildings, Bridges and Pavements [Review of Sustainable Solution for Deteriorated and Aged RCC Structures: A Review of Buildings, Bridges and Pavements]. IOP Conference Series Earth and Environmental Science, 1026(1), 12009. IOP Publishing. DOI: 10.1088/1755-1315/1026/1/012009

Alghieth, M. (2025). Sustain AI: A Multi-Modal Deep Learning Framework for Carbon Footprint Reduction in Industrial Manufacturing. *Sustainability (Basel), 17*(9), 4134. DOI: 10.3390/su17094134

Ali, H., Chen, T., & Hao, Y. (2021). Sustainable Manufacturing Practices, Competitive Capabilities, and Sustainable Performance: Moderating Role of Environmental Regulations. *Sustainability (Basel), 13*(18), 10051. DOI: 10.3390/su131810051

Ali, M. S., & McKeown, C. (2020). Blockchain and Sustainability: Ethical and Environmental Considerations. *Sustainability, 12*(11), 3572.

Allcott, H., & Greenstone, M. (2017). Measuring the welfare effects of residential energy efficiency programs. *Journal of Public Economics, 145*, 40–53. DOI: 10.1016/j.jpubeco.2016.11.009

Alli, Y. A., Anuar, H., Manshor, M. R., Okafor, C. E., Kamarulzaman, A. F., Akçakale, N., Mohd Nazeri, F. N., Bodaghi, M., Suhr, J., & Mohd Nasir, N. A. (2024). Optimization of 4D/3D printing via machine learning: A systematic review. *Hybrid Advances, 6*, 100242. DOI: 10.1016/j.hybadv.2024.100242

Allwood, J. M., Ashby, M. F., Gutowski, T. G., & Worrell, E. (2011). Material efficiency: A white paper. *Resources, Conservation and Recycling, 55*(3), 362–381. DOI: 10.1016/j.resconrec.2010.11.002

Almusaed, A., Yitmen, İ., Myhren, J. A., & Almssad, A. (2024). Assessing the Impact of Recycled Building Materials on Environmental Sustainability and Energy Efficiency: A Comprehensive Framework for Reducing Greenhouse Gas Emissions. *Buildings (Basel, Switzerland), 14*(6), 1566. DOI: 10.3390/buildings14061566

Al-Raeei, M. (2024). Artificial intelligence for climate resilience: Advancing sustainable goals in SDGs 11 and 13 and its relationship to pandemics. *Discover Sustainability, 5*(1), 513. Advance online publication. DOI: 10.1007/s43621-024-00775-5

Alshebly, Y. S., Nafea, M., Mohamed Ali, M. S., & Almurib, H. A. F. (2021). Review on recent advances in 4D printing of shape memory polymers. *European Polymer Journal, 159*, 110708. https://doi.org/https://doi.org/10.1016/j.eurpolymj.2021.110708. DOI: 10.1016/j.eurpolymj.2021.110708

Alting, D. L., & Jøgensen, D. J. (1993). The Life Cycle Concept as a Basis for Sustainable Industrial Production. *CIRP Annals, 42*(1), 163–167. DOI: 10.1016/S0007-8506(07)62417-2

Ameh, B. (2024). Digital tools and AI: Using technology to monitor carbon emissions and waste at each stage of the supply chain, enabling real-time adjustments for sustainability improvements. *International Journal of Science and Research Archive*, *13*(1), 2741–2757. DOI: 10.30574/ijsra.2024.13.1.1995

Ameta, K. L., Solanki, V. S., Singh, V., Devi, A. P., Chundawat, R. S., & Haque, S. (2022). Critical appraisal and systematic review of 3D & 4D printing in sustainable and environment-friendly smart manufacturing technologies. *Sustainable Materials and Technologies*, *34*, e00481. DOI: 10.1016/j.susmat.2022.e00481

Artificial intelligence in manufacturing – systematic literature review. (2025). *Scientific Papers of Silesian University of Technology Organization and Management Series*, (215). Advance online publication. DOI: 10.29119/1641-3466.2025.215.39

Artificial Intelligence: Towards Sustainable Intelligence. (2023). In *Communications in computer and information science*. Springer Science+Business Media. DOI: 10.1007/978-3-031-47997-7

Ashby, M. F. (2012). *Materials and sustainable development*. Butterworth-Heinemann.

Atapattu, A. J., Perera, L., Nuwarapaksha, T. D., Udumann, S. S., & Dissanayaka, N. S. (2024). *Challenges in Achieving Artificial Intelligence in Agriculture*., DOI: 10.1007/978-981-97-5878-4_2

Athukoralalage, S. S., Balu, R., Dutta, N. K., & Choudhury, N. R. (2019). 3D bioprinted nanocellulose-based hydrogels for tissue engineering applications: A brief review. In *Polymers* (Vol. 11, Issue 5). MDPI AG. DOI: 10.3390/polym11050898

Attah, R. U., Garba, B. M. P., Gil-Ozoudeh, I., & Iwuanyanwu, O. (2024). Digital transformation in the energy sector: Comprehensive review of sustainability impacts and economic benefits. *International Journal of Advanced Economics*, *6*(12), 760–776. DOI: 10.51594/ijae.v6i12.1751

Attar, H., Calin, M., Zhang, L. C., Scudino, S., & Eckert, J. (2014). Manufacture by selective laser melting and mechanical behavior of commercially pure titanium. *Materials Science and Engineering A*, *593*, 170–177. DOI: 10.1016/j.msea.2013.11.038

Attiya, E. M. E., Shebl, M. A., & Nasser, M. M.Eman M. E. AttiyaMohamed A. SheblMaii M. Nasser. (2020). A Comparative Analysis of LEED and GPRS for the Applicability in Egyptian Office Buildings. *International Journal of Engineering Research & Technology (Ahmedabad)*, *3*(3). Advance online publication. DOI: 10.17577/IJERTV9IS030299

Aufa, A. N., Ismail, Z., & Zaki Hassan, M. (2023). Emerging trends in 4d printing of hydrogels in the biomedical field: A review. *Materials Today: Proceedings*. https://doi.org/https://doi.org/10.1016/j.matpr.2023.01.101

Ayadi, R., Forouheshfar, Y., & Moghadas, O. (2025). Enhancing system resilience to climate change through artificial intelligence: A systematic literature review. *Frontiers in Climate*, *7*, 1585331. Advance online publication. DOI: 10.3389/fclim.2025.1585331

Azadi, S., Kasraian, D., Nourian, P., & Van Wesemael, P. (2025). What have urban digital twins contributed to urban planning and decision making? From a systematic literature review toward a Socio-Technical research and Development agenda. *Smart Cities*, *8*(1), 32. DOI: 10.3390/smartcities8010032

Badmus, O., Rajput, S., Arogundade, J. B., & Williams, M. H. (2024). AI-driven business analytics and decision making. *World Journal of Advanced Research and Reviews*, *24*(1), 616–633. DOI: 10.30574/wjarr.2024.24.1.3093

Baharuddin, D., Zainordin, N., & Jusoh, Z. M. (2025). The Institutional ESG Disclosure Framework for Malaysian Consultancy Quantity Surveying Practices: A Systematic Literature Review. *Journal of Ecohumanism*, *4*(2). Advance online publication. DOI: 10.62754/joe.v4i2.6600

Baig, A., Barse, R., Paryekar, A., & Jagtap, V. (2024). Evolution of pharmaceuticals using 3D and 4D printing. *Intelligent Pharmacy*. https://doi.org/https://doi.org/10.1016/j.ipha.2024.05.008

Balani, R. S., Patel, D., Barthwal, S., & Arun, J. (2021). Fabrication of Air Conditioning System Using the Engine Exhaust Gas. In *Advances in Interdisciplinary Research in Engineering and Business Management* (pp. 369–378). Springer Nature Singapore. DOI: 10.1007/978-981-16-0037-1_29

Baldassarre, B., Konietzko, J., Brown, P., Calabretta, G., Bocken, N., Karpen, I. O., & Hultink, E. J. (2020). Addressing the design-implementation gap of sustainable business models by prototyping: A tool for planning and executing small-scale pilots. *Journal of Cleaner Production*, *255*, 120295. DOI: 10.1016/j.jclepro.2020.120295

Bamgbade, J. A., Kamaruddeen, A. M., & Nawi, M. N. M. (2017). Towards environmental sustainability adoption in construction firms: An empirical analysis of market orientation and organizational innovativeness impacts. *Sustainable Cities and Society*, *32*, 486–495. DOI: 10.1016/j.scs.2017.04.015

Bamgbade, J. A., Mohammed, K. A., & Nawi, M. N. M. (2016). Assessing the sustainable construction of large construction companies in Malaysia. *AIP Conference Proceedings*, *1761*, 20027. DOI: 10.1063/1.4960867

Bandara, L. V., & Buics, L. (2024). Digital Twins in Sustainable Supply Chains: A Comprehensive Review of Current Applications and Enablers for Successful Adoption. *Engineering Proceedings*, *79*(1), 64. DOI: 10.3390/engproc2024079064

Bandura, A. (1986). *Social foundations of thought and action: A social cognitive theory*. Prentice-Hall.

Baron, R. M., & Kenny, D. A. (1986). The moderator-mediator variable distinction in social psychological research: Conceptual, strategic, and statistical considerations. *Journal of Personality and Social Psychology*, *51*(6), 1173–1182. DOI: 10.1037/0022-3514.51.6.1173 PMID: 3806354

Batty, M., Axhausen, K. W., Giannotti, F., Pozdnoukhov, A., Bazzani, A., Wachowicz, M., Ouzounis, G., & Portugali, Y. (2012). Smart cities of the future. *The European Physical Journal. Special Topics*, *214*(1), 481–518. DOI: 10.1140/epjst/e2012-01703-3

Baufeld, B., Biest, O., & Gault, R. (2011). Additive manufacturing of Ti–6Al–4V components by shaped metal deposition : Microstructure and mechanical properties. *Materials & Design*, *31*(4), 199–206. DOI: 10.1016/j.matdes.2009.05.021

Bazaz, S. M., Penttilä, S., Ollikainen, M., Ratava, J., & Varis, J. (2021). Industry 4.0 readiness of manufacturing sector in the Baltic Region. *IOP Conference Series. Materials Science and Engineering*, *1140*(1), 12019. DOI: 10.1088/1757-899X/1140/1/012019

Bécue, A., Maia, E., Feeken, L., Borchers, P., & Praça, I. (2020). A New Concept of Digital Twin Supporting Optimization and Resilience of Factories of the Future. *Applied Sciences (Basel, Switzerland)*, *10*(13), 4482. DOI: 10.3390/app10134482

Behl, M., & Lendlein, A. (2007). Shape-memory polymers. *Materials Today*, *10*(4), 20–28. https://doi.org/https://doi.org/10.1016/S1369-7021(07)70047-0. DOI: 10.1016/S1369-7021(07)70047-0

Beier, G., Ullrich, A., Niehoff, S., Reißig, M., & Habich, M. (2020). Industry 4.0: How it is defined from a sociotechnical perspective and how much sustainability it includes – A literature review—. *Journal of Cleaner Production*, *259*, 120856. DOI: 10.1016/j.jclepro.2020.120856

Benachio, G. L. F., Freitas, M. do C. D., & Tavares, S. F. (2019). Green Supply Chain Management in the Construction Industry: A literature review [Review of Green Supply Chain Management in the Construction Industry: A literature review]. IOP Conference Series Earth and Environmental Science, 225, 12011. IOP Publishing. DOI: 10.1088/1755-1315/225/1/012011

Berman, B. (2012). 3-D printing : The new industrial revolution. *Business Horizons*, *55*(2), 155–162. DOI: 10.1016/j.bushor.2011.11.003

Besigomwe, K. (2025). Closed-Loop Manufacturing with AI-Enabled Digital Twin Systems. *Cognizance Journal of Multidisciplinary Studies*, *5*(1), 18–38. DOI: 10.47760/cognizance.2025.v05i01.002

Bezerra, R. R. R., Martins, V. W. B., & Macêdo, A. N. (2024). Validation of Challenges for Implementing ESG in the Construction Industry Considering the Context of an Emerging Economy Country. *Applied Sciences (Basel, Switzerland)*, *14*(14), 6024. DOI: 10.3390/app14146024

Bhuiyan, M. A. H. (2024). Carbon Footprint Measurement and Mitigation Using AI. SSRN *Electronic Journal*. DOI: 10.2139/ssrn.4746446

Bibri, S. E., Huang, J., Jagatheesaperumal, S. K., & Krogstie, J. (2024). The synergistic interplay of artificial intelligence and digital twin in environmentally planning sustainable smart cities: A comprehensive systematic review. *Environmental Science and Ecotechnology*, *20*, 100433. DOI: 10.1016/j.ese.2024.100433 PMID: 38831974

Binnemans, K., Jones, P. T., Blanpain, B., Van Gerven, T., Yang, Y., Walton, A., & Buchert, M. (2015). Recycling of rare earths: A critical review. *Journal of Cleaner Production*, *104*, 1–22.

Blasch, J., Boogen, N., Datta, S., & Filippini, M. (2021). The role of energy literacy for residential energy demand. *Energy Economics*, *96*, 105155. DOI: 10.1016/j.eneco.2021.105155

Bocken, N. M. P.. (2016). Product Design and Sustainability: A Circular Economy Perspective. *Journal of Industrial Ecology*, *20*(6), 1059–1070.

Bocken, N. M. P., Short, S. W., Rana, P., & Evans, S. (2016). A literature and practice review to develop sustainable business model archetypes. *Journal of Cleaner Production*, *65*, 42–56. DOI: 10.1016/j.jclepro.2013.11.039

Bocken, N. M., de Pauw, I., Bakker, C., & van der Grinten, B. (2016). Product design and business model strategies for a circular economy. *Journal of Industrial and Production Engineering*, *33*(5), 308–320. DOI: 10.1080/21681015.2016.1172124

Bollinger, B., & Gillingham, K. (2012). Peer effects in the diffusion of solar photovoltaic panels. *Marketing Science*, *31*(6), 900–912. DOI: 10.1287/mksc.1120.0727

Bond, G., Mahjoubnia, A., Zhao, W., King, S. D., Chen, S., & Lin, J. (2023). 4D printing of biocompatible, hierarchically porous shape memory polymeric structures. *Biomaterials Advances*, *153*, 213575. https://doi.org/https://doi.org/10.1016/j.bioadv.2023.213575. DOI: 10.1016/j.bioadv.2023.213575

Bonenberg, W., & Wei, X. (2015). Green BIM in Sustainable Infrastructure. *Procedia Manufacturing*, *3*, 1654–1659. DOI: 10.1016/j.promfg.2015.07.483

Bonetti, L., Natali, D., Pandini, S., Messori, M., Toselli, M., & Scalet, G. (2024). 4D printing of semi-crystalline crosslinked polymer networks with two-way shape-memory effect. *Materials & Design*, *238*, 112725. https://doi.org/https://doi.org/10.1016/j.matdes.2024.112725. DOI: 10.1016/j.matdes.2024.112725

Borenstein, S. (2017). Private net benefits of residential solar PV: The role of electricity tariffs, tax incentives, and rebates. *Journal of the Association of Environmental and Resource Economists*, *4*(S1), S85–S122. DOI: 10.1086/691978

Bortolini, R., Rodrigues, R., Alavi, H., Vecchia, L. F. D., & Forcada, N. (2022). Digital Twins' Applications for building Energy Efficiency: A review. *Energies*, *15*(19), 7002. DOI: 10.3390/en15197002

Bourell, D. L., Leu, M. C., & Rosen, D. W. (2009). A brief history of additive manufacturing and the 2009 roadmap for additive manufacturing. *Proceedings of the 16th Solid Freeform Fabrication Symposium*, 1-12.

Bovo, M., Santolini, E., & Barbaresi, A. (2025). An Alternative Modular Wooden System for Fast Assembly/Disassembly of Buildings. *Buildings (Basel, Switzerland)*, *15*(7), 1196. DOI: 10.3390/buildings15071196

Braunstein, S., & Welch, C. (2002). Financial literacy: An overview of practice, research, and policy. *Federal Reserve Bulletin*, *88*(11), 445–457. DOI: 10.17016/bulletin.2002.88-11

Brossier, T., Habib, M., Benkhaled, B. T., Volpi, G., Lapinte, V., & Blanquer, S. (2024). 4D printing of hydrogels based on poly(oxazoline) and poly(acrylamide) copolymers by stereolithography. *Materials Advances*, *5*(7), 2750–2758. DOI: 10.1039/D3MA00665D

Brown, K. L. (2021). *Machine learning applications in sustainable energy systems*. Wiley.

Brundtland Commission. (1987). *Our common future: Report of the World Commission on Environment and Development*. Oxford University Press.

Brynjolfsson, E., & McAfee, A. (2014). *The Second Machine Age: Work, Progress, and Prosperity in a Time of Brilliant Technologies*. W. W. Norton & Company.

Burke, H., Zhang, A., & Wang, J. X. (2021). Integrating product design and supply chain management for a circular economy. *Production Planning and Control*, *34*(11), 1097–1113. DOI: 10.1080/09537287.2021.1983063

Cacace, S., Boccadoro, M., & Semeraro, Q. (2024). Investigation on the effect of the gas-to-metal ratio on powder properties and PBF-LB/M processability. *Progress in Additive Manufacturing*, *9*(4), 889–904. DOI: 10.1007/s40964-023-00490-z

Campbell, T. A., Tibbits, S., & Garrett, B. (2014). *THE NEXT WAVE: 4D PRINTING PROGRAMMING THE MATERIAL WORLD*.

Campilho, R. D. S. G., & Silva, F. J. G. (2023). Industrial Process Improvement by Automation and Robotics. *Machines (Basel)*, *11*(11), 1011. DOI: 10.3390/machines11111011

CarbonChain. (2022). Understand your cobalt emissions. *CarbonChain Blog*. https://www.carbonchain.com/blog/understand-your-cobalt-emissions

Carfì, D., Donato, A., & Gambarelli, G. (2018). Economically sustainable development and green economy: A game-theoretic model. *Decisions in Economics and Finance*, *41*(1), 25–43. DOI: 10.1007/s10203-018-0213-7

Carter, L. N., Martin, C., Withers, P. J., & Attallah, M. M. (2014). The influence of the laser scan strategy on grain structure and cracking behaviour in SLM powder-bed fabricated nickel superalloy. *Journal of Alloys and Compounds*, *615*, 338–347. DOI: 10.1016/j.jallcom.2014.06.172

Castro, N. J., Meinert, C., Levett, P., & Hutmacher, D. W. (2017). Current developments in multifunctional smart materials for 3D/4D bioprinting. *Current Opinion in Biomedical Engineering*, *2*, 67–75. https://doi.org/https://doi.org/10.1016/j.cobme.2017.04.002. DOI: 10.1016/j.cobme.2017.04.002

Cevallos, M. G. O., Jaramillo, G. A. L., Vélez, J. L. B., Zambrano, M. I. U., Montesdeoca, L. D. Z., & Palomeque, M. A. B.Miguel Giancarlo Ormaza CevallosGustavo Alberto Lozano JaramilloJosé Luis Bernardo VélezMaritza Irinuska Ureta Zambrano-Lady Diana Zambrano MontesdeocaManuel Augusto Bermúdez Palomeque. (2024). Implementation of Artificial Intelligence in Quality Management in SMEs: Benefits and Challenges. *Evolutionary Studies in Imaginative Culture*, *1489*, 1489–1500. Advance online publication. DOI: 10.70082/esiculture.vi.1590

Chakraborty, S., & Adhikari, S. (2020). Machine learning based digital twin for dynamical systems with multiple time-scales. *Computers & Structures*, *243*, 106410. DOI: 10.1016/j.compstruc.2020.106410

Challouf, K., Alhloul, A., & Németh, N. (2025). Mapping the role of industry 4.0 technologies in green supply chain management: A bibliometric and structured text analysis. *Discover Sustainability*, *6*(1), 949. Advance online publication. DOI: 10.1007/s43621-025-01827-0

Chandra, P. (2023). Role of AI in managing renewable energy demand. *Journal of Clean Energy Technology*, *11*(1), 25–38.

Chaudhary, A., & Akhtar, A. (2022). Template for Evaluating Cradle-to-Site Environmental Life Cycle Impacts of Buildings in India. *ACS Environmental Au*, *3*(2), 94–104. DOI: 10.1021/acsenvironau.2c00035

Chauhan, M., & Sahoo, D. R. (2024). Towards a Greener Tomorrow: Exploring the Potential of AI, Blockchain, and IoT in Sustainable Development. *Nature Environment and Pollution Technology*, *23*(2), 1105–1113. DOI: 10.46488/NEPT.2024.v23i02.044

Chen. (2020). Enhanced creep performance in a polycrystalline superalloy driven by atomic-scale phase transformation. *Acta Materialia Communications*. https://arxiv.org/abs/2007.01676

Chen, G., Chen, J., Zhao, Z., & Ruan, X. (2005). An object-oriented hierarchical case representation of automotive panels in a computer-aided process planning system. *International Journal of Advanced Manufacturing Technology*, *26*(11-12), 1323–1330. Advance online publication. DOI: 10.1007/s00170-004-2105-8

Chen, H., & Volpe, R. P. (2019). An analysis of personal financial literacy among college students. *Financial Services Review*, *7*(2), 107–128. DOI: 10.1016/S1057-0810(99)80006-7

Chen, J., Virrueta, C., Zhang, S., Mao, C., & Wang, J. (2024). 4D printing: The spotlight for 3D printed smart materials. *Materials Today*, *77*, 66–91. https://doi.org/https://doi.org/10.1016/j.mattod.2024.06.004. DOI: 10.1016/j.mattod.2024.06.004

Chen, W., Pan, W., & Wang, J. (2020). Leveraging IoT and AI for innovative and sustainable manufacturing. *Sustainability*, *12*(14), 5731.

Choi, B., & Lee, S. (2023). Predictive analytics for microgrid reliability. *IEEE Access : Practical Innovations, Open Solutions*, *11*, 44444–44456.

Choi, S. B., Lee, H., & Kang, S. W. (2020). Cross-disciplinary collaboration in sustainable product development : The role of environmental expertise. *Sustainability, 12*(5). Advance online publication. DOI: 10.3390/su12052049

Choi, S., Kim, H., & Kim, Y. (2020). Artificial Intelligence and Machine Learning: The Future of Sustainable Product Design. *Sustainability*, *12*(10), 3150.

Chopra, V., Fuentes-Velasco, V., Nacif-Lopez, S. R., Melendez-Malpicca, J., Mendez-Hernandez, A. S., Ramos-Mendez-Iris, L. F., Arroyo-Jimenez, D. A., Reyes-Segura, D. G., Gonzalez-Y-Mendoza, P., Sanchez-Hernandez, K. A., Spinola-Corona, E., Vazquez-del-Mercado-Pardiño, J. A., & Chauhan, G. (2024, October). Gonzalez-Y-Mendoza, P., Sanchez-Hernandez, K. A., Spinola-Corona, E., Vazquez-del-Mercado-Pardiño, J. A., & Chauhan, G. (2024). Advancements in 3D-4D printing of hydroxyapatite composites for bone tissue engineering. *Ceramics International*, *50*(20), 38819–38840. Advance online publication. DOI: 10.1016/j.ceramint.2024.07.266

Choudhary, S., & Kumar, N. (2021). AI-based energy forecasting for smart agriculture. *Agritech and Renewable Integration*, *6*(2), 80–93.

Chow, T. C., Zailani, S., Rahman, M. K., Zhang, Q., Bhuiyan, M. A., & Patwary, A. K. (2021). Impact of sustainable project management on project plan and project success of the manufacturing firm: Structural model assessment. *PLoS One*, *16*(11), e0259819. Advance online publication. DOI: 10.1371/journal.pone.0259819

Cioffi, R., Travaglioni, M., Piscitelli, G., Petrillo, A., & Parmentola, A. (2020). Smart Manufacturing Systems and Applied Industrial Technologies for a Sustainable Industry: A Systematic Literature Review. *Applied Sciences (Basel, Switzerland)*, *10*(8), 2897. DOI: 10.3390/app10082897

Cipriano, B. H., Banik, S. J., Sharma, R., Rumore, D., Hwang, W., Briber, R. M., & Raghavan, S. R. (2014). Superabsorbent Hydrogels That Are Robust and Highly Stretchable. *Macromolecules*, *47*(13), 4445–4452. DOI: 10.1021/ma500882n

Creswell, J. W., & Plano Clark, V. L. (2017). *Designing and conducting mixed methods research* (3rd ed.). SAGE Publications.

D'Souza, M., & Fernandes, A. (2021). Role of data analytics in microgrid energy management. *Energy Data Science*, *4*(2), 77–91.

Daios, A., Kladovasilakis, N., Kelemis, A., & Kostavelis, I. (2025). AI Applications in Supply Chain Management: A Survey. *Applied Sciences (Basel, Switzerland)*, *15*(5), 2775. DOI: 10.3390/app15052775

Darabă, D., Pop, F. S., & Daraba, C. (2024). Digital Twin Used in Real-Time Monitoring of Operations Performed on CNC Technological Equipment. *Applied Sciences (Basel, Switzerland)*, *14*(22), 10088. DOI: 10.3390/app142210088

Das, M., & Reddy, V. (2022). Machine learning-based load forecasting models. *International Journal of Power Systems*, *29*(3), 210–225.

Das, S. K., Bressanelli, G., & Saccani, N. (2024). Clustering the Research at the Intersection of Industry 4.0 Technologies, Environmental Sustainability and Circular Economy: Evidence from Literature and Future Research Directions. *Circular Economy and Sustainability*, *4*(4), 2473–2504. Advance online publication. DOI: 10.1007/s43615-024-00393-3

Dastrup, S. R., Graff Zivin, J., Costa, D. L., & Kahn, M. E. (2019). Understanding the solar home price premium: Electricity generation and "green" social status. *European Economic Review*, *56*(5), 961–973. DOI: 10.1016/j.euroecorev.2012.02.006

Database of State Incentives for Renewables & Efficiency. (2022). *Annual report on renewable energy incentive programs*. North Carolina State University.

Davis, F. D. (1989). Perceived usefulness, perceived ease of use, and user acceptance of information technology. *Management Information Systems Quarterly*, *13*(3), 319–340. DOI: 10.2307/249008

DebRoy, T., Wei, H. L., Zuback, J. S., Mukherjee, T., Elmer, J. W., Milewski, J. O., & Zhang, W. (2018). Additive manufacturing of metallic components – Process, structure and properties. *Progress in Materials Science*, *92*, 112–224. DOI: 10.1016/j.pmatsci.2017.10.001

del Río, P., & Unruh, G. (2007). Overcoming the lock-out of renewable energy technologies in Spain: The cases of wind and solar electricity. *Renewable & Sustainable Energy Reviews*, *11*(7), 1498–1513. DOI: 10.1016/j.rser.2005.12.003

Deng, C., Qu, J., Dong, J., Guo, Y., Wu, X., Fang, Y., Sun, X., Wei, Y., & Li, Z. (2024). 4D printing of magnetic smart structures based on light-cured magnetic hydrogel. *Chemical Engineering Journal*, *494*, 152992. https://doi.org/https://doi.org/10.1016/j.cej.2024.152992. DOI: 10.1016/j.cej.2024.152992

Despeisse, M., Ball, P., Evans, S., & Levers, A. (2012). Industrial ecology at factory level – a conceptual model. *Journal of Cleaner Production*, *31*, 30–39. DOI: 10.1016/j.jclepro.2012.02.027

Digital Twin: enabling technologies, challenges and open research. (2020). IEEE Journals & Magazine | IEEE Xplore. https://ieeexplore.ieee.org/abstract/document/9103025

Dowaidar, M. (2024). Guidelines for biomimetic 3D/4D printing in drug delivery. *Materials Chemistry and Physics*, *325*, 129692. DOI: 10.1016/j.matchemphys.2024.129692

Drehobl, A., & Ross, L. (2016). *Lifting the high energy burden in America's largest cities: How energy efficiency can improve low income and underserved communities*. American Council for an Energy-Efficient Economy.

Duraivelu, K. (2022). Digital transformation in manufacturing industry – A comprehensive insight. *Materials Today: Proceedings*, *68*, 2396–2401. https://doi.org/10.1016/j.matpr.2022.07.409

Egger, J., & Masood, T. (2019). Augmented reality in support of intelligent manufacturing – A systematic literature review. *Computers & Industrial Engineering*, *140*, 106195. DOI: 10.1016/j.cie.2019.106195

El-Agamy, R. F., Sayed, H. A., Akhatatneh, A. M. A., Aljohani, M., & Elhosseini, M. A. (2024). Comprehensive analysis of digital twins in smart cities: A 4200-paper bibliometric study. *Artificial Intelligence Review*, *57*(6), 154. Advance online publication. DOI: 10.1007/s10462-024-10781-8

Ellen MacArthur Foundation. (2013). Towards the circular economy: Economic and business rationale for an accelerated transition. Retrieved from https://ellenmacarthurfoundation.org

Emani, V. V. (2025). Robotic bricklaying and 3d printing robots in construction: automating masonry work and building structures layer by layer, and robotic welding in construction (enhancing precision and efficiency in structural steel work), International journal of Busness quantitative DOI: 10.2139/ssrn.5297498

European Commission. (2020). *Circular Economy Action Plan: For a cleaner and more competitive Europe*. https://europa.eu

European Commission. (2022) The Barcelona Urban Platform and integrated smart city operations. https://smart-cities-marketplace.ec.europa.eu

Faludi, J., Hoffenson, S., Kwok, S. Y., Saidani, M., Hallstedt, S. I., Telenko, C., & Martinez, V. (2020). A Research Roadmap for Sustainable Design Methods and Tools. *Sustainability (Basel)*, *12*(19), 8174. DOI: 10.3390/su12198174

Fatorachian, H. (2023). The Significance of Industry 5.0 in the Globalization of Supply Chain Management. [EEL]. *European Economic Letters*, *13*(5), 872–885. DOI: 10.52783/eel.v13i5.843

Federal Ministry for Economic Affairs and Energy. (2021). *Renewable energy education initiative: Final evaluation report*. German Federal Government.

Fera, M., Fruggiero, F., Lambiase, A., Macchiaroli, R., & Papa, A. (2016). The evaluation of the additive manufacturing process to improve the performances of the supply chain. *Procedia CIRP*, *47*, 82–87. DOI: 10.1016/j.procir.2016.03.050

Fernandes, D., Lynch, J. G.Jr, & Netemeyer, R. G. (2014). Financial literacy, financial education, and downstream financial behaviors. *Management Science*, *60*(8), 1861–1883. DOI: 10.1287/mnsc.2013.1849

Ferrari, F., Bandini, M., Lorusso, M., & Pavese, M. (2021). Sustainability assessment of powder reuse strategies in metal additive manufacturing. *Sustainability*, *13*(3), 1173. DOI: 10.3390/su13031173

Fetters, M. D., Curry, L. A., & Creswell, J. W. (2013). Achieving integration in mixed methods designs—Principles and practices. *Health Services Research*, *48*(6pt2), 2134–2156. DOI: 10.1111/1475-6773.12117 PMID: 24279835

Firoozi, A. A., Firoozi, A. A., Oyejobi, D. O., Avudaiappan, S., & Flores, E. S. (2024). Emerging Trends in Sustainable Building Materials: Technological Innovations, Enhanced Performance, and Future Directions. *Results in Engineering*, *24*, 103521. DOI: 10.1016/j.rineng.2024.103521

Florido-Benítez, L. (2024). The use of digital twins to address smart tourist destinations' future challenges. *Platforms*, *2*(4), 234–254. DOI: 10.3390/platforms2040016

Ford, S., & Despeisse, M. (2016). Additive manufacturing and sustainability: An exploratory study of the advantages and challenges. *Journal of Cleaner Production*, *137*, 1573–1587. DOI: 10.1016/j.jclepro.2016.04.150

Frank, A. G., Dalenogare, L. S., & Ayala, N. F. (2019). Industry 4.0 technologies: Implementation patterns in manufacturing companies. *International Journal of Production Economics*, *210*, 15–26. DOI: 10.1016/j.ijpe.2019.01.004

Frank, T., Duflou, J. R., & Dewulf, W. (2019). Automation in manufacturing: An overview of trends and challenges. *Procedia CIRP*, *80*, 537–542.

Franzè, C., Paolucci, E., & Pessot, E. (2023). Sustained value creation driven by digital connectivity: A multiple case study in the mechanical components industry. *Technovation*, *129*, 102918. DOI: 10.1016/j.technovation.2023.102918

Frazier, W. E. (2014). Metal additive manufacturing : A review. *Journal of Materials Engineering and Performance*, *23*(6), 1917–1928. DOI: 10.1007/s11665-014-0958-z

Frederick, S., Loewenstein, G., & O'Donoghue, T. (2002). Time discounting and time preference: A critical review. *Journal of Economic Literature*, *40*(2), 351–401. DOI: 10.1257/jel.40.2.351

Früchtl, M., Leis, M., & Wertheim, R. (2020). A comprehensive and interdisciplinary perspective on sustainable manufacturing towards sustainable life cycles. *Procedia Manufacturing*, *43*, 383–390. DOI: 10.1016/j.promfg.2020.02.197

Fujigaya, T., Morimoto, T., Niidome, Y., & Nakashima, N. (2008). NIR Laser-Driven Reversible Volume Phase Transition of Single-Walled Carbon Nanotube/Poly(N-isopropylacrylamide) Composite Gels. *Advanced Materials*, *20*(19), 3610–3614. https://doi.org/https://doi.org/10.1002/adma.200800494. DOI: 10.1002/adma.200800494

Gallab, M., Nardo, M. D., & Naciri, L. (2024). Navigating contemporary challenges and future prospects in digital industry evolution. *Discover Applied Sciences*, *6*(5), 259. Advance online publication. DOI: 10.1007/s42452-024-05913-2

Gandía, J. A. G., Ancillo, A. de L., & Núñez, M. T. del V. (2025). The Role of Artificial Intelligence and Knowledge in Enhancing Corporate Sustainability. *Journal of Innovation & Knowledge*, *10*(5), 100792. DOI: 10.1016/j.jik.2025.100792

Gao, R. X., Wang, L., Helu, M., & Teti, R. (2020). Big data analytics for smart factories of the future. *CIRP Annals*, *69*(2), 668–692. DOI: 10.1016/j.cirp.2020.05.002

Gao, Y., Xu, S., Wu, R., Wang, J., & Wei, J. (2008). Preparation and characteristic of electric stimuli responsive hydrogel composed of polyvinyl alcohol/poly (sodium maleate-co-sodium acrylate). *Journal of Applied Polymer Science*, *107*(1), 391–395. https://doi.org/https://doi.org/10.1002/app.26375. DOI: 10.1002/app.26375

Garcia, R. R., Alcock, J. R., & Maier, S. (2021). Energy and environmental performance of additive manufacturing technologies : A review of literature and current practice. *Sustainable Materials and Technologies*, *28*, 00260. DOI: 10.1016/j.susmat.2021.e00260

Gardner, P., Borgo, M. D., Ruffini, V., Hughes, A. J., Zhu, Y., & Wagg, D. (2020). Towards the Development of an Operational Digital Twin. *Vibration*, *3*(3), 235–265. DOI: 10.3390/vibration3030018

Garetti, M., Mummolo, G., & Taisch, M. (2012). Special issue on "sustainable manufacturing.". *Production Planning and Control*, *23*(2-3), 79–82. DOI: 10.1080/09537287.2011.591617

Garetti, M., & Taisch, M. (2012). Sustainable manufacturing: Trends and research challenges. *Production Planning and Control*, *23*(2–3), 83–104. DOI: 10.1080/09537287.2011.591619

Garmendia, I., Coello, J., Fernández, E., & Martinez, A. (2020). Digital twins for predictive maintenance in additive manufacturing : State of the art. *Procedia Manufacturing*, *45*, 272–279. DOI: 10.1016/j.promfg.2020.04.028

Gebhardt, A., & Hötter, J.-S. (2016). *Additive manufacturing : 3D printing for prototyping and manufacturing*. Carl Hanser Verlag GmbH Co KG.

Gebler, M., Schoot Uiterkamp, A. J. M., & Visser, C. (2014). A global sustainability perspective on 3D printing technologies. *Energy Policy*, *74*, 158–167. DOI: 10.1016/j.enpol.2014.08.033

Geissdoerfer, M., Savaget, P., Bocken, N. M. P., & Hultink, E. J. (2017). The circular economy – A new sustainability paradigm? *Journal of Cleaner Production*, *143*, 757–768. DOI: 10.1016/j.jclepro.2016.12.048

Ge, Q., Dunn, C. K., Qi, H. J., & Dunn, M. L. (2014). Active origami by 4D printing. *Smart Materials and Structures*, *23*(9), 94007. DOI: 10.1088/0964-1726/23/9/094007

Ge, Q., Sakhaei, A. H., Lee, H., Dunn, C. K., Fang, N. X., & Dunn, M. L. (2016a). Multimaterial 4D Printing with Tailorable Shape Memory Polymers. *Scientific Reports*, *6*(1), 31110. Advance online publication. DOI: 10.1038/srep31110

Ghadge, A., Shastri, Y., & Shankar, R. (2020). Sustainable additive manufacturing : An exploration of the emerging trends and challenges. *Journal of Manufacturing Technology Management*, *31*(2), 401–427. DOI: 10.1108/JMTM-07-2019-0254

Ghahremani-Nahr, J., Aliahmadi, A., & Nozari, H. (2022). An IoT-based sustainable supply chain framework and blockchain. *International Journal of Innovation in Engineering*, *2*(1), 12–21. DOI: 10.59615/ijie.2.1.12

Ghisellini, P., Cialani, C., & Ulgiati, S. (2016). A review on circular economy: The expected transition to a balanced interplay of environmental and economic systems. *Journal of Cleaner Production*, *114*, 11–32. DOI: 10.1016/j.jclepro.2015.09.007

Ghoreishi, M., Franconi, A., Mate, O.-A., Godsell, J., & Pynnönen, M. (2024). Supply Chain 4.0 and the transition towards a circular economy. In *Edward Elgar Publishing eBooks* (p. 238). Edward Elgar Publishing. DOI: 10.4337/9781803920207.00018

Ghosh, S. (2017a). *Predictive modeling of solidification during laser additive manufacturing of nickel superalloys : Recent developments, future directions*. DOI: 10.48550/arXiv.1707.09292

Giannakopoulos, G., Shaikh, K. M., Perez, M. A., & Adegbenro, P. (2025). Sustainable Practices in Green Engineering: A Pathway to a Greener Future using. *AI*, *1*(1), 1–9. DOI: 10.21276/jccci/2025.v1.i1.1

Gibson, I., Rosen, D. W., & Stucker, B. (2021). *Additive manufacturing technologies : 3D printing, rapid prototyping, and direct digital manufacturing* (3rd éd.). Springer. DOI: 10.1007/978-3-030-56127-7

Gibson, I., Rosen, D. W., & Stucker, B. (2015). *Additive manufacturing technologies: 3D Printing, rapid prototyping, and direct digital manufacturing*. Springer. DOI: 10.1007/978-1-4939-2113-3

Giudice, F., Barbagallo, R., & Fargione, G. (2020). A Design for Additive Manufacturing approach based on process energy efficiency: Electron beam melted components. *Journal of Cleaner Production*, *290*, 125185. DOI: 10.1016/j.jclepro.2020.125185

Glogar, M., Petrak, S., & Naglić, M. M. (2025). Digital Technologies in the Sustainable Design and Development of Textiles and Clothing—A Literature Review. *Sustainability (Basel)*, *17*(4), 1371. DOI: 10.3390/su17041371

Gonzalez, R., & Kim, H. (2021). Deep learning in renewable energy fault prediction. *Energy AI*, *2*(1), 45–59.

Goswami, S. S., Mondal, S., Sarkar, S., Gupta, K. K., Sahoo, S. K., & Halder, R. (2024). Artificial Intelligence-Enabled Supply Chain Management: Unlocking New Opportunities and Challenges. *Artificial Intelligence and Applications (Commerce, Calif.)*, *3*(1), 110–121. Advance online publication. DOI: 10.47852/bonviewAIA42021814

Goubran, S. (2019). On the Role of Construction in Achieving the SDGs. *Journal of Sustainability Research*, *1*(2). Advance online publication. DOI: 10.20900/jsr20190020

Govindan, K., Soleimani, H., & Kannan, D. (2015). Reverse logistics and closed-loop supply chain: A comprehensive review to explore the future. *European Journal of Operational Research*, *240*(3), 603–626. DOI: 10.1016/j.ejor.2014.07.012

Goyal, L. K., & Singh, H. (2020). BIM Approach for Sustainable Design of Flat Slab Buildings: A Review [Review of BIM Approach for Sustainable Design of Flat Slab Buildings: A Review]. IOP Conference Series Materials Science and Engineering, 955(1), 12012. IOP Publishing. DOI: 10.1088/1757-899X/955/1/012012

Green, M., & Hall, S. (2023). Data standards and interoperability in AM software. *Software, Practice & Experience*, *53*(5), 1124–1137. DOI: 10.1002/spe.3125

Grira, S., Khalifeh, H. A., Alkhedher, M., & Ramadan, M. (2023). 3D printing algae-based materials: Pathway towards 4D bioprinting. *Bioprinting (Amsterdam, Netherlands)*, *33*, e00291. DOI: 10.1016/j.bprint.2023.e00291

Gross, B. C., Erkal, J. L., Lockwood, S. Y., Chen, C., & Spence, D. M. (2014). Evaluation of 3D printing and its potential impact on biotechnology and the chemical sciences. *Analytical Chemistry*, *86*(7), 3240–3253. DOI: 10.1021/ac403397r

Gu, D., Meiners, W., Wissenbach, K., & Poprawe, R. (2012). Laser additive manufacturing of metallic components : Materials, processes and mechanisms. *International Materials Reviews*, *57*(3), 133–164. DOI: 10.1179/1743280411Y.0000000014

Guinee, J. B. (2016). *Handbook on life cycle assessment: Operational guide to the ISO standards*. Springer Science & Business Media.

Gunawardena, D., Mendis, P., Ngo, D., Aye, L., & Alfano, J. (2014). Sustainable Prefabricated Modular Buildings. DOI: 10.13140/2.1.4847.3920

Guo, N., & Leu, M. C. (2013). Additive manufacturing : Technology, applications and research needs. *Frontiers of Mechanical Engineering*, *8*(3), 215–243. DOI: 10.1007/s11465-013-0248-8

Guo, Y. B., Klink, A., Bártolo, P., & Guo, W. (2023). Digital twins for electro-physical, chemical, and photonic processes. *CIRP Annals*, *72*(2), 593–619. DOI: 10.1016/j.cirp.2023.05.007

Gupta, P., Vermani, K., & Garg, S. (2002). Hydrogels: From controlled release to pH-responsive drug delivery. *Drug Discovery Today*, *7*(10), 569–579. https://doi.org/https://doi.org/10.1016/S1359-6446(02)02255-9. DOI: 10.1016/S1359-6446(02)02255-9

Gupta, S., Dangayach, G. S., & Singh, A. K. (2015). Key Determinants of Sustainable Product Design and Manufacturing. *Procedia CIRP*, *26*, 99–102. DOI: 10.1016/j.procir.2014.07.166

Gurova, O., Merritt, T., Papachristos, E., & Vaajakari, J. (2020). Sustainable Solutions for Wearable Technologies: Mapping the Product Development Life Cycle. *Sustainability (Basel)*, *12*(20), 8444. DOI: 10.3390/su12208444

Hamedani, S. S., Aslam, S., & Hamedani, S. S. (2025). AI in business operations: Driving urban growth and societal sustainability. *Frontiers in Artificial Intelligence*, *8*, 1568210. DOI: 10.3389/frai.2025.1568210

Hanafi, A., Moawed, M., & Abdellatif, O. (2024). Advancing Sustainable Energy Management: A Comprehensive Review of Artificial Intelligence Techniques in Building [Review of Advancing Sustainable Energy Management: A Comprehensive Review of Artificial Intelligence Techniques in Building]. *Deleted Journal*, *53*(2), 26–46. DOI: 10.21608/erjsh.2023.226854.1196

Harun, S. N., Kamaruddin, N., & Choo, J. (2025). INTERIOR DESIGN AND RENOVATION WORKS FOR 2 LEVEL FACTORY OFFICE AT ROVSKI, USJ.

HARUN, S. N., Kamaruddin, N., & Choo, J. (2025). *INTERIOR DESIGN AND RENOVATION WORKS FOR 2 LEVEL FACTORY OFFICE AT ROVSKI, USJ.*

Hasan, M. R., Liu, Z., & Rahman, A. (2023). *Energy consumption modeling for DED-based hybrid additive manufacturing*. /arXiv.2305.09104DOI: 10.21203/rs.3.rs-2905463/v1

Hasanzadeh, R., Mihankhah, P., Azdast, T., Rasouli, A., Shamkhali, M., & Park, C. B. (2023). Biocompatible tissue-engineered scaffold polymers for 3D printing and its application for 4D printing. *Chemical Engineering Journal*, *476*, 146616. DOI: 10.1016/j.cej.2023.146616

Hatami, H., Almahmeed, W., Kesharwani, P., & Sahebkar, A. (2024a). Exploring the potential of 3D and 4D printing in advancing stent manufacturing for cardiovascular diseases. *European Polymer Journal*, *212*, 113035. https://doi.org/https://doi.org/10.1016/j.eurpolymj.2024.113035. DOI: 10.1016/j.eurpolymj.2024.113035

Hatledal, L. I., Skulstad, R., Li, G., Styve, A., & Zhang, H. (2020). Co-simulation as a Fundamental Technology for Twin Ships. *Modeling Identification and Control A Norwegian*. *Modeling, Identification and Control*, *41*(4), 297–311. DOI: 10.4173/mic.2020.4.2

Hauck, S., Greif, L., Benner, N., & Ovtcharova, J. (2025). Advancing Sustainable Additive Manufacturing: Analyzing Parameter Influences and Machine Learning Approaches for CO2 Prediction. *Sustainability (Basel)*, *17*(9), 3804. DOI: 10.3390/su17093804

Hauke, B., Kuhnhenne, M., Lawson, M., Veljković, M., & Siebers, R. (2016). What does 'sustainable construction' mean? An overview (p. 1). DOI: 10.1002/9781118740828.ch1

Hauschild, M. Z., Rosenbaum, R. K., & Olsen, S. I. (2018). *Life cycle assessment: Theory and practice*. Springer. DOI: 10.1007/978-3-319-56475-3

Hazrat Ali, M., Abilgaziyev, A., & Adair, D. (2019). 4D printing: A critical review of current developments, and future prospects. *International Journal of Advanced Manufacturing Technology*, *105*(1-4), 701–717. DOI: 10.1007/s00170-019-04258-0

Hernandez, L., & Patel, K. (2022). AI-assisted demand response for residential energy systems. *IEEE Internet of Things Journal*, *9*(7), 5600–5612.

Herrmann, C., Thiede, S., & Kara, S. (2020). Sustainability in additive manufacturing : A review and future perspectives. *Journal of Cleaner Production*, *261*, 121234. DOI: 10.1016/j.jclepro.2020.121234

Herzog, D., Seyda, V., Wycisk, E., & Emmelmann, C. (2016). Additive manufacturing of metals. *Acta Materialia*, *117*, 371–392. DOI: 10.1016/j.actamat.2016.07.019

Hirai, T., Nemoto, H., Suzuki, T., Hayashi, S., & Hirai, M. (1993). Actuation of Poly(vinyl alcohol) Gel by Electric Field. *Journal of Intelligent Material Systems and Structures*, *4*(2), 277–279. DOI: 10.1177/1045389X9300400219

Hoen, B., Rand, J., Elmallah, S., & Paulos, B. (2019). *Sharing the sun: Community solar markets, rates, and participation*. Lawrence Berkeley National Laboratory.

Ho, F. H., Hanim, A.-R. S., Ariffin, R. G. R., Rahman, M. N. A., Hassan, M. F., Arifin, A. M. T., Haq, R. H. A., Ahmad, S., & Abdullah, H. (2019). Analytical analysis for material efficiency practice: An exploratory study of Malaysia manufacturing cases. *Journal of Physics: Conference Series*, *1150*, 12006. DOI: 10.1088/1742-6596/1150/1/012006

Holmström, J., Liotta, G., & Chaudhuri, A. (2017). Sustainability outcomes through direct digital manufacturing-based operational practices: A design theory approach. *Journal of Cleaner Production*, *167*, 951–961. DOI: 10.1016/j.jclepro.2017.03.092

Holmström, J., Partanen, J., Tuomi, J., & Walter, M. (2010). Rapid manufacturing in the spare parts supply chain. *Journal of Manufacturing Technology Management*, *21*(6), 687–697. DOI: 10.1108/17410381011063996

Hua, J., Wang, R., Hu, Y. C., Chen, Z., Chen, L., Osman, A. I., Farghali, M., Huang, L., Feng, J., Wang, H., Zhang, X., Zhou, X., & Yap, P. (2025). Artificial intelligence for calculating and predicting building carbon emissions: a review [Review of *Artificial intelligence for calculating and predicting building carbon emissions: a review*]. *Environmental Chemistry Letters*. Springer Science+Business Media. DOI: 10.1007/s10311-024-01799-z

Huang, J., Irfan, M., Fatima, S. S., & Shahid, R. M. (2023). The role of lean six sigma in driving sustainable manufacturing practices: An analysis of the relationship between lean six sigma principles, data-driven decision making, and environmental performance. *Frontiers in Environmental Science*, *11*, 1184488. Advance online publication. DOI: 10.3389/fenvs.2023.1184488

Hunhevicz, J., Motie, M., & Hall, D. (2021). Digital building twins and blockchain for performance-based (smart) contracts. *Automation in Construction*, *133*, 103981. DOI: 10.1016/j.autcon.2021.103981

Hussain, M., & George, T. (2022). AI in predictive maintenance for smart grids. *Electric Power Systems Research*, *208*, 107886.

Hussain, M., Shah, F., & Kiani, M. (2019). Role of IoT in Sustainable Product Lifecycle Management. *International Journal of Advanced Manufacturing Technology*, *101*(5), 1445–1460.

Hussin, J. M., Rahman, I. A., & Memon, A. H. (2013). The Way Forward in Sustainable Construction: Issues and Challenges. *International Journal of Advances in Applied Sciences*, *2*(1). Advance online publication. DOI: 10.11591/ijaas.v2i1.1321

Iluyomade, T. D., & Okwandu, A. C.Tosin Daniel IluyomadeAzubuike Chukwudi Okwandu. (2024). Innovative materials in sustainable construction: A review [Review of Innovative materials in sustainable construction: A review]. *International Journal of Science and Research Archive*, *12*(1), 2435–2447. DOI: 10.30574/ijsra.2024.12.1.1048

International Energy Agency. (2023). *World energy outlook 2023*. International Energy Agency.

International Renewable Energy Agency. (2021). *Renewable energy statistics 2021*. International Renewable Energy Agency.

Ishrat, M., Khan, W., Faisal, S. M., Farsi, M. M. A., & Farooqui, N. A. (2025). Digital Twins Future prospects in smart cities. In *CRC Press eBooks* (pp. 53–68). https://doi.org/DOI: 10.1201/9781003510338-3

Islam, F., & Chowdhury, A. (2021). Machine learning approaches for solar irradiance prediction. *Solar Computing and Forecasting*, *8*(2), 75–89.

Ismaeil, M. K. L., & Lalla, A. F.ISMAEIL. (2024). The Role and Impact of Artificial Intelligence on Supply Chain Management: Efficiency, Challenges, and Strategic Implementation. *Journal of Ecohumanism*, *3*(4), 89–106. DOI: 10.62754/joe.v3i4.3461

Jackson, T. (2005). *Motivating sustainable consumption: A review of evidence on consumer behaviour and behavioural change.*

Jacobson, M. Z. (2022). 100% renewable energy systems for cities. *Renewable Energy Reviews, 109*, 235–250.

Jamaludin, A. F., Husain, M. F. H., & Razali, M. N. (2025). Assessing environmental, social and governance of property- listed companies in Malaysia.pdf.

Javaid, M., & Haleem, A. (2019). 4D printing applications in medical field: A brief review. *Clinical Epidemiology and Global Health*, *7*(3), 317–321. https://doi.org/https://doi.org/10.1016/j.cegh.2018.09.007. DOI: 10.1016/j.cegh.2018.09.007

Jeong, B., & Gutowska, A. (2002). Lessons from nature: Stimuli-responsive polymers and their biomedical applications. *Trends in Biotechnology*, *20*(7), 305–311. https://doi.org/https://doi.org/10.1016/S0167-7799(02)01962-5. DOI: 10.1016/S0167-7799(02)01962-5

Jiang, F., Ma, L., Broyd, T., Chen, W., & Luo, H. (2021). Digital twin enabled sustainable urban road planning. *Sustainable Cities and Society*, *78*, 103645. DOI: 10.1016/j.scs.2021.103645

Jiang, J., Xiong, Y., Zhang, Z., & Rosen, D. (2022). Machine learning integrated design for additive manufacturing. *Journal of Intelligent Manufacturing*, *33*(4), 1073–1086. DOI: 10.1007/s10845-020-01715-6

Jinoop, A., Paul, C. P., & Bindra, K. S. (2019). Laser-assisted directed energy deposition of nickel superalloys : A review. *International Journal of Advanced Manufacturing Technology*, *102*, 2321–2340. DOI: 10.1177/1464420719852658

Jin, S., Gu, J., Shi, Y., Shao, K., Yu, X., & Yue, G. (2013). Preparation and electrical sensitive behavior of poly (N-vinylpyrrolidone-co-acrylic acid) hydrogel with flexible chain nature. *European Polymer Journal*, *49*(7), 1871–1880. https://doi.org/https://doi.org/10.1016/j.eurpolymj.2013.04.022. DOI: 10.1016/j.eurpolymj.2013.04.022

Johnson, D., & Lee, C. (2021). Benchmarking sustainability in metal additive manufacturing. *Materials Today Sustainability*, *12*, 100087. DOI: 10.1016/j.mtsust.2021.100087

Jolaiy, S., Yousefi, A., Hosseini, M., Zolfagharian, A., Demoly, F., & Bodaghi, M. (2024). Limpet-inspired design and 3D/4D printing of sustainable sandwich panels: Pioneering supreme resiliency, recoverability and repairability. *Applied Materials Today*, *38*, 102243. DOI: 10.1016/j.apmt.2024.102243

Jones, S. R., & Seepersad, C. C. (2016). Integrating materials science and software engineering for advanced manufacturing. *Additive Manufacturing*, *12*, 272–279. DOI: 10.1016/j.addma.2016.05.005

Joshi, M., & Verghese, A. (2022). Real-time energy analytics for smart buildings. *Intelligent Energy Systems*, *7*(3), 120–133.

Kahneman, D., & Tversky, A. (1979). Prospect theory: An analysis of decision under risk. *Econometrica*, *47*(2), 263–291. DOI: 10.2307/1914185

Kamali, M. H., & Hewage, K. (2016). Development of performance criteria for sustainability evaluation of modular versus conventional construction methods. *Journal of Cleaner Production*, *142*, 3592–3606. DOI: 10.1016/j.jclepro.2016.10.108

Kamble, S. S., Gunasekaran, A., & Gawankar, S. A. (2020). Achieving sustainable performance in a data-driven agriculture supply chain: A review for research and applications. *Computers and Electronics in Agriculture*, *168*, 105204.

Kantaros, A., & Ganetsos, T. (2024). Integration of Cyber-Physical Systems, Digital Twins and 3D Printing in Advanced Manufacturing: A Synergistic Approach. *American Journal of Engineering and Applied Sciences*, *17*(1), 1–22. DOI: 10.3844/ajeassp.2024.1.22

Kapoor, P., & Jain, D. (2023). Optimizing battery storage using reinforcement learning. *Energy Informatics and Control Systems*, *9*(1), 40–53.

Karadağ, E., Saraydın, D., & Güven, O. (2001). Radiation Induced Superabsorbent Hydrogels. Acrylamide/Itaconic Acid Copolymers. *Macromolecular Materials and Engineering*, *286*(1), 34–42. https://doi.org/https://doi.org/10.1002/1439-2054(20010101)286:1<34:AID-MAME34>3.0.CO;2-J. DOI: 10.1002/1439-2054(20010101)286:1<34::AID-MAME34>3.0.CO;2-J

Kashem, M. A., Shamsuddoha, M., Nasir, T., & Chowdhury, A. A. (2023). Supply Chain Disruption versus Optimization: A Review on Artificial Intelligence and Blockchain [Review of Supply Chain Disruption versus Optimization: A Review on Artificial Intelligence and Blockchain]. *Knowledge (Beverly Hills, Calif.)*, *3*(1), 80–96. DOI: 10.3390/knowledge3010007

Keles, C., Rios, F. C., & Hoque, S. (2025). Digital Technologies and Circular Economy in the Construction Sector: A Review of Lifecycle Applications, Integrations, Potential, and Limitations [Review of *Digital Technologies and Circular Economy in the Construction Sector: A Review of Lifecycle Applications, Integrations, Potential, and Limitations*]. *Buildings, 15*(4), 553. Multidisciplinary Digital Publishing Institute. DOI: 10.3390/buildings15040553

Kellens, K., Baumers, M., Gutowski, T. G., Flanagan, W., Lifset, R., & Duflou, J. R. (2017). Environmental dimensions of additive manufacturing: Mapping application domains and their environmental implications. *Journal of Industrial Ecology*, *21*(S1), S49–S68.

Kellens, K., Mertens, R., Paraskevas, D., Dewulf, W., & Duflou, J. R. (2017). Environmental impact of additive manufacturing processes : Does AM contribute to a more sustainable way of part manufacturing? *Procedia CIRP*, *61*, 582–587. DOI: 10.1016/j.procir.2016.11.153

Kelly, A. (2024). Impact of Artificial Intelligence on Supply Chain Optimization. *Journal of Technology and Systems*, *6*(6), 15–27. DOI: 10.47941/jts.2153

Khair, M. A., & Sandu, A. K. (2023). *Blockchain-Optimized Supply Chain Traceability System for Transparent Logistics*. https://hal.science/hal-04543221

Khalid, M. Y., Arif, Z. U., Noroozi, R., Hossain, M., Ramakrishna, S., & Umer, R. (2023). 3D/4D printing of cellulose nanocrystals-based biomaterials: Additives for sustainable applications. *International Journal of Biological Macromolecules*, *251*, 126287. DOI: 10.1016/j.ijbiomac.2023.126287

Kharat, M. G., Kapoor, S., Parhi, S., Kharat, M. G., & Pandey, S. (2024). Operationalizing sustainability in pharmaceuticals: Green supply chain metrics for circular economy. *Sustainable Futures : An Applied Journal of Technology, Environment and Society*, *9*, 100413. DOI: 10.1016/j.sftr.2024.100413

Khattra, S. K., Singh, H., & Singh, J. (2020). Leveraging the Potential of BIM towards Sustainable Construction. *IOP Conference Series. Materials Science and Engineering*, *955*(1), 12011. DOI: 10.1088/1757-899X/955/1/012011

Kholkhoev, B. Ch., Nikishina, A. N., Bardakova, K. N., Matveev, Z. A., Sapozhnikov, D. A., Efremov, Y. M., Timashev, P. S., & Burdukovskii, V. F. (2024). 4D-printing of high-temperature shape-memory polymers based on polyimide, N,N-dimethylacrylamide and photoactive cross-linkers. *Polymer*, *299*, 126978. https://doi.org/https://doi.org/10.1016/j.polymer.2024.126978. DOI: 10.1016/j.polymer.2024.126978

Kim, B. J., Hong, S., & Jeong, H. E. (2024). Preface for Advanced Manufacturing for ESG. *International Journal of Precision Engineering and Manufacturing-Green Technology*, *11*(3), 707–708. DOI: 10.1007/s40684-024-00613-z

Kim, S. J., Yoon, S. G., Lee, Y. H., & Kim, S. I. (2004). Bending behavior of hydrogels composed of poly(methacrylic acid) and alginate by electrical stimulus. *Polymer International*, *53*(10), 1456–1460. https://doi.org/https://doi.org/10.1002/pi.1560. DOI: 10.1002/pi.1560

King, W. E., Anderson, A. T., Ferencz, R. M., Hodge, N. E., Kamath, C., Khairallah, S. A., & Rubenchik, A. M. (2015). Laser powder bed fusion additive manufacturing of metals; physics, computational, and materials challenges. *Applied Physics Reviews*, *2*(4), 041304. DOI: 10.1063/1.4937809

King, W. E., Bajaj, P., & Blue, C. (2021). Effects of multiple reuse on oxygen and nitrogen levels in metal powders. *Additive Manufacturing*, *39*, 101845. DOI: 10.1016/j.addma.2021.101845

Klapper, L., Lusardi, A., & Van Oudheusden, P. (2015). *Financial literacy around the world: Insights from the Standard & Poor's ratings services global financial literacy survey*. World Bank Group.

Kloxin, A. M., Kasko, A. M., Salinas, C. N., & Anseth, K. S. (2009). Photodegradable Hydrogels for Dynamic Tuning of Physical and Chemical Properties. *Science*, *324*(5923), 59–63. DOI: 10.1126/science.1169494

Kokare, A. M., Ghosh, D., & Venkatesh, V. C. (2023). Data quality in LCA studies of additive manufacturing : Challenges and the way forward. *Journal of Industrial Ecology*, *27*(2), 483–498. DOI: 10.1111/jiec.13302

Kokare, A. M., Godina, R., & Oliveira, J. P. (2023). Life cycle assessment of additive manufacturing processes : A review. *Journal of Manufacturing Systems*, *68*, 536–559. DOI: 10.1016/j.jmsy.2023.05.007

Kong, Z., Kong, L., & Bai, K. (2024). Research on the Development Path of E-commerce Logistics in the Context of Carbon Neutrality. *Journal of Physics: Conference Series*, *2735*(1), 12001. DOI: 10.1088/1742-6596/2735/1/012001

Korhonen, J., Honkasalo, A., & Seppälä, J. (2018). Circular economy: The concept and its limitations. *Ecological Economics*, *143*, 37–46. DOI: 10.1016/j.ecolecon.2017.06.041

Körner, C. (2016). Additive manufacturing of metallic components by selective electron beam melting – A review. *International Materials Reviews*, *61*(5), 361–377. DOI: 10.1080/09506608.2016.1176289

Kowalska-Pyzalska, A. (2018). What makes consumers adopt to innovative energy services in the energy market? A review of incentives and barriers. *Renewable & Sustainable Energy Reviews*, *82*, 3570–3581. DOI: 10.1016/j.rser.2017.10.103

Kozlovská, M., & Spišáková, M. (2013). Contruction waste generation across construction project life-cycle. *Organization Technology and Management in Construction An International Journal*, *5*(1), 687–695. DOI: 10.5592/otmcj.2013.1.5

KPMG. (2021). *Sustainability Reporting and the Circular Economy: Compliance with environmental standards.*

Krupp, T., & Seifi, M. (2018). Strategies for residual stress reduction in SLM: Scan patterns and preheat. *Additive Manufacturing*, *23*, 454–462. DOI: 10.1016/j.addma.2018.06.012

Kruth, J. P., Mercelis, P., Vaerenbergh, J., Froyen, L., & Rombouts, M. (2005). Binding mechanisms in selective laser sintering and selective laser melting. *Rapid Prototyping Journal*, *11*(1), 26–36. DOI: 10.1108/13552540510573365

Kshirsagar, M., Ambike, S. D., Jaya Prakash, N., Kandasubramanian, B., & Deshpande, P. (2023). Origami engineering: Creating dynamic functional materials through folded structures. *Hybrid Advances*, *4*, 100092. DOI: 10.1016/j.hybadv.2023.100092

Kubec, V., Lukáš, O., Vavřík, J., & Brázda, M. (2023). Possibilities of application of virtual prototyping and 3D printing methods in practice. In *Proceedings of the 61st International Conference of Machine Design Departments (ICMD 2020)* (pp. 316–325). Atlantis Press. https://doi.org/DOI: 10.2991/978-94-6463-182-1_33

Kukreja, A. K. (2025). AI Adoption in SMEs: Integrating Supply Chain and Financial Strategies for Competitive Advantage. *International Journal For Multidisciplinary Research*, *7*(1), 35821. Advance online publication. DOI: 10.36948/ijfmr.2025.v07i01.35821

Kulkarni, A. (2018). *Additive manufacturing of nickel-based superalloy*. DOI: 10.48550/arXiv.1805.11664

Kumar, R., Aggarwal, V., & Surinder, M. G. (2021). Sustainable materials and techniques in affordable high-rise buildings - A case study. E3S Web of Conferences, 309, 1080. DOI: 10.1051/e3sconf/202130901080

Kumar, J., Kumar, G., Mehdi, H., & Kumar, M. (2024). Optimization of FSW parameters on mechanical properties of different aluminum alloys of AA6082 and AA7050 by response surface methodology. [IJIDeM]. *International Journal on Interactive Design and Manufacturing*, *18*(3), 1359–1371. DOI: 10.1007/s12008-023-01425-2

Kumar, P. (2024). AI for Optimizing Supply Chain Management. *International Journal for Research in Applied Science and Engineering Technology*, *12*(11), 549–560. DOI: 10.22214/ijraset.2024.65059

Kumar, P., Suryavanshi, P., Kumar Dwivedy, S., & Banerjee, S. (2024). Stimuli-responsive materials for 4D Printing: Mechanical, Manufacturing, and Biomedical Applications. *Journal of Molecular Liquids*, *410*, 125553. https://doi.org/https://doi.org/10.1016/j.molliq.2024.125553. DOI: 10.1016/j.molliq.2024.125553

Kumar, S., & Rao, P. (2022). Performance analysis of solar water heaters in India. *Journal of Thermal Science*, *47*(2), 201–215.

Kumar, V., & Shahin, K. (2025). Artificial Intelligence and Machine Learning for Sustainable Manufacturing: Current Trends and Future Prospects. *Intelligent and Sustainable Manufacturing*, *2*(1), 10002. DOI: 10.70322/ism.2025.10002

Kusiak, A. (2018). Smart manufacturing. *International Journal of Production Research*, *56*(1-2), 508–517. DOI: 10.1080/00207543.2017.1351644

Kwon, Y. S., & Yun, H. (2021). Generative design for sustainable lightweight structures using topology optimization and additive manufacturing. *Sustainable Materials and Technologies*, *29*, 00308. DOI: 10.1016/j.susmat.2021.e00308

Lacy, P., & Rutqvist, J. (2015). *Waste to wealth : The circular economy advantage*. Palgrave Macmillan.

Lacy, P., & Rutqvist, J. (2015). *Waste to Wealth: The Circular Economy Advantage*. Palgrave Macmillan.

Lai, J., Ye, X., Liu, J., Wang, C., Li, J., Wang, X., Ma, M., & Wang, M. (2021). 4D printing of highly printable and shape morphing hydrogels composed of alginate and methylcellulose. *Materials & Design*, *205*, 109699. https://doi.org/https://doi.org/10.1016/j.matdes.2021.109699. DOI: 10.1016/j.matdes.2021.109699

Lalegani Dezaki, M., & Bodaghi, M. (2024). 4D printing and programming of continuous fibre-reinforced shape memory polymer composites. *European Polymer Journal*, *210*, 112988. https://doi.org/https://doi.org/10.1016/j.eurpolymj.2024.112988. DOI: 10.1016/j.eurpolymj.2024.112988

Larimian, T., & Borkar, T. (2019). Additive manufacturing of emerging materials. In *Emerging Materials for Additive Manufacturing*. Springer.

Lauff, C., Kotys-Schwartz, D., & Rentschler, M. E. (2018). What is a Prototype? What are the Roles of Prototypes in Companies? *Journal of Mechanical Design*, *140*(6), 061102. Advance online publication. DOI: 10.1115/1.4039340

Laverne, F., Marquardt, R., Segonds, F., Koutiri, I., & Perry, N. (2019). Improving resources consumption of additive manufacturing use during early design stages: A case study. *International Journal of Sustainable Engineering*, *12*(6), 365–375. DOI: 10.1080/19397038.2019.1620897

Lee, C., Kim, S., Kim, H., & Ahn, S.-H. (2007). Measurement of anisotropic compressive strength of rapid prototyping parts. *Journal of Materials Processing Technology*, *187*, 627–630. DOI: 10.1016/j.jmatprotec.2006.11.095

Lee, J., Bagheri, B., & Kao, H. A. (2019). A Cyber-Physical Systems architecture for Industry 4.0-based manufacturing systems. *Manufacturing Letters*, *3*, 18–23. DOI: 10.1016/j.mfglet.2014.12.001

Lee, S., & Kim, J. (2021). Reinforcement learning for energy scheduling in smart homes. *Journal of Intelligent Systems*, *12*(2), 210–224.

Lehtola, V. V., Koeva, M., Elberink, S. O., Raposo, P., Virtanen, J., Vahdatikhaki, F., & Borsci, S. (2022). Digital twin of a city: Review of technology serving city needs. *International Journal of Applied Earth Observation and Geoinformation*, *114*, 102915. DOI: 10.1016/j.jag.2022.102915

Leicht, A., Åkerfeldt, P., & Pederson, R. (2018). Influence of build orientation on microstructure and mechanical properties of Inconel 718 produced by Electron Beam Melting. *Metals*, *8*(9), 682. DOI: 10.3390/met8090682

Leist, S. K., & Zhou, J. (2016). Current status of 4D printing technology and the potential of light-reactive smart materials as 4D printable materials. *Virtual and Physical Prototyping*, *11*(4), 249–262. DOI: 10.1080/17452759.2016.1198630

Leong, T. G., Randall, C. L., Benson, B. R., Bassik, N., Stern, G. M., & Gracias, D. H. (2024). *Tetherless thermobiochemically actuated microgrippers*.

Leong, T. G., Randall, C. L., Benson, B. R., Bassik, N., Stern, G. M., & Gracias, D. H. (2009). Tetherless thermobiochemically actuated microgrippers. *Proceedings of the National Academy of Sciences of the United States of America*, *106*(3), 703–708. DOI: 10.1073/pnas.0807698106

Liang, J., Feng, R., & Li, Q. (2018). Residual stress in metal additive manufacturing. *Procedia CIRP*, *71*, 348–353. DOI: 10.1016/j.procir.2018.05.039

Liao, E., Zhang, Y., Su, Y., Zhang, C., Geng, C., Li, C., Liu, X., Liu, Y., & Lu, A. (2024). Active mixing of two-component viscoelastic silicone ink at molecular level for spatiotemporally controlled 3D/4D printing of cellular silicones. *Polymer*, *302*, 127053. DOI: 10.1016/j.polymer.2024.127053

Li, B., He, P., Wang, J., Pan, X., Wang, Y., Wang, Z., & Baniotopoulos, C. C. (2024). Mechanical characteristic and stress-strain modelling of friction stir welded 6061-T6 aluminium alloy butt joints. *Thin-walled Structures*, *198*, 111645. DOI: 10.1016/j.tws.2024.111645

Li, J., & Zhang, X. (2023). Effect of hot isostatic pressing on the stress-rupture properties of second-generation single crystal nickel superalloys processed by laser powder bed fusion. *Materials & Design*, *232*, 111555. DOI: 10.1016/j.jmrt.2024.07.064

Li, L., Aslam, S., Wileman, A., & Perinpanayagam, S. (2021). Digital Twin in Aerospace Industry: A Gentle Introduction. *IEEE Access : Practical Innovations, Open Solutions, 10*, 9543–9562. DOI: 10.1109/ACCESS.2021.3136458

Li, L., & Hsieh, Y.-L. (2005). Ultra-fine polyelectrolyte hydrogel fibres from poly(acrylic acid)/poly(vinyl alcohol). *Nanotechnology, 16*(12), 2852–2860. DOI: 10.1088/0957-4484/16/12/020

Lin, K.-Y., & Wei, S. (2023). Advancing the Industrial Circular Economy: The Integrative Role of Machine Learning in Resource Optimization. *Journal of Green Economy and Low-Carbon Development, 2*(3), 122–136. DOI: 10.56578/jgelcd020302

Liu, G., Zhang, C., Zhao, M., Guo, W., & Luo, Q. (2020). Comparison of Nanomaterials with Other Unconventional Materials Used as Additives for Soil Improvement in the Context of Sustainable Development. Nanomaterials, 11(1), 15. Multidisciplinary Digital Publishing Institute. DOI: 10.3390/nano11010015

Liu, S. H., Rahmawati, Y., & Zawawi, N. A. W. A. (2019). Critical success factors of collaborative approach in delivering sustainable construction. MATEC Web of Conferences, 270, 5003. DOI: 10.1051/matecconf/201927005003

Liu, S., Chen, X., & Zhang, Y. (2020). Chapter 14 - Hydrogels and hydrogel composites for 3D and 4D printing applications. In K. K. Sadasivuni, K. Deshmukh, & M. A. Almaadeed (Eds.), *3D and 4D Printing of Polymer Nanocomposite Materials* (pp. 427–465). Elsevier. https://doi.org/https://doi.org/10.1016/B978-0-12-816805-9.00014-4

Liu, F., & Urban, M. W. (2010). Recent advances and challenges in designing stimuli-responsive polymers. *Progress in Polymer Science, 35*(1), 3–23. https://doi.org/https://doi.org/10.1016/j.progpolymsci.2009.10.002. DOI: 10.1016/j.progpolymsci.2009.10.002

Liu, P., Ma, L., & Zhang, Y. (2020). Energy consumption analysis of additive manufacturing : A case study of SLM process. *Journal of Cleaner Production, 245*, 118878. DOI: 10.1016/j.jclepro.2019.118878

Liu, W., Lv, Y., Wang, Q., Sun, B., & Han, D. (2024). A Systematic Review of the Digital Twin Technology in Buildings, Landscape and Urban Environment from 2018 to 2024. *Buildings (Basel, Switzerland), 14*(11), 3475. DOI: 10.3390/buildings14113475

Liu, Y., & Xie, B. (2021). Smart appliance coordination for home energy efficiency. *Home Automation & AI, 9*(3), 93–107.

Liu, Z. S., & Rempel, G. L. (1997). Preparation of superabsorbent polymers by crosslinking acrylic acid and acrylamide copolymers. *Journal of Applied Polymer Science*, *64*(7), 1345–1353. https://doi.org/https://doi.org/10.1002/(SICI)1097-4628(19970516)64:7<1345:AID-APP14>3.0.CO;2-W. DOI: 10.1002/(SICI)1097-4628(19970516)64:7<1345::AID-APP14>3.0.CO;2-W

Liu, Z., Wang, M., Zhang, Y., & Tang, Y. (2023). Artificial intelligence-enabled design and quality monitoring in additive manufacturing : A review. *Additive Manufacturing*, *55*, 102936. DOI: 10.1016/j.addma.2022.102936

Li, X., Wang, B., Peng, T., & Xu, X. (2021). Greentelligence: Smart Manufacturing for a Greener Future. *Chinese Journal of Mechanical Engineering*, *34*(1), 116. Advance online publication. DOI: 10.1186/s10033-021-00656-2

Li, Y.-C., Zhang, Y. S., Akpek, A., Shin, S. R., & Khademhosseini, A. (2016). 4D bioprinting: The next-generation technology for biofabrication enabled by stimuli-responsive materials. *Biofabrication*, *9*(1), 12001. DOI: 10.1088/1758-5090/9/1/012001

Li, Y., Zheng, W., Li, B., Dong, J., Gao, G., & Jiang, Z. (2022). Double-layer temperature-sensitive hydrogel fabricated by 4D printing with fast shape deformation. *Colloids and Surfaces. A, Physicochemical and Engineering Aspects*, *648*, 129307. https://doi.org/https://doi.org/10.1016/j.colsurfa.2022.129307. DOI: 10.1016/j.colsurfa.2022.129307

Longo, F., Mirabelli, G., Solina, V., Alberto, U., Paola, G. D., Giordano, L., & Ziparo, M. (2022). A Simulation-Based Framework for Manufacturing Design and Resilience Assessment: A Case Study in the Wood Sector. *Applied Sciences (Basel, Switzerland)*, *12*(15), 7614. DOI: 10.3390/app12157614

Lozano-Ramírez, N. E., & Garcia-Lopez, N. P. (2025). Mapping Construction 4.0 and Sustainability for Innovation in the Construction Industry of Developing Countries. 1(1). DOI: 10.7771/3067-4883.1512

Lu, X., Liu, G., & Lu, J. (2024). Development of Ceramic 3D/4D Printing in China. *Additive Manufacturing Frontiers*, 200158. https://doi.org/https://doi.org/10.1016/j.amf.2024.200158

Luo, J., Liu, P., Xu, W., Zhao, T., & Biljecki, F. (2024). A perception-powered urban digital twin to support human-centered urban planning and sustainable city development. *Cities (London, England)*, *156*, 105473. DOI: 10.1016/j.cities.2024.105473

Lusardi, A., & Mitchell, O. S. (2014). The economic importance of financial literacy: Theory and evidence. *Journal of Economic Literature*, *52*(1), 5–44. DOI: 10.1257/jel.52.1.5 PMID: 28579637

Mahato, R., & Dixit, A. (2020). Comparative lifecycle emissions of solar vs. *Coal. Environmental Energy Metrics*, *14*(1), 100–114.

Mahmoudi, M., & Kovacevic, R. (2015). An investigation on the influence of direct laser melting process parameters on microstructure and mechanical properties of Inconel 625 build parts. *Journal of Materials Processing Technology*, *214*(11), 2623–2630. DOI: 10.1016/j.jmatprotec.2014.06.001

Ma, J., Li, X., & Bao, Y. (2015). Advances in cellulose-based superabsorbent hydrogels. *RSC Advances*, *5*(73), 59745–59757. DOI: 10.1039/C5RA08522E

Mallakpour, S., Tabesh, F., & Hussain, C. M. (2024). Entering a new era in pharmaceutics through 3D/4D printing technologies. In *Medical Additive Manufacturing* (pp. 267–280). Elsevier., DOI: 10.1016/B978-0-323-95383-2.00006-8

Mann, S., Potdar, V., Gajavilli, R. S., & Chandan, A. (2018). Blockchain Technology for Supply Chain Traceability. *Transparency and Data Provenance*, *22*, 22–26. Advance online publication. DOI: 10.1145/3301403.3301408

Manu, B. A. (2024). Innovative Construction Materials: Advancing Sustainability, Durability, Efficiency, and Cost-effectiveness in Modern Infrastructure. *International Journal of Research Publication and Reviews*, *5*(12), 4987–4999. DOI: 10.55248/gengpi.5.1224.0215

Mariana, Y., Suryawinata, B. A., & Peranginangin, E. (2022). Sustainability in West Java traditional house: A case study from Julang Ngapak House. *IOP Conference Series. Earth and Environmental Science*, *998*(1), 12040. DOI: 10.1088/1755-1315/998/1/012040

Marrucci, L., Daddi, T., & Iraldo, F. (2023). Creating environmental performance indicators to assess corporate sustainability and reward employees. *Ecological Indicators*, *159*, 111667. DOI: 10.1016/j.ecolind.2023.111489

Martelli, P. A., Bassini, E., & Ugues, D. (2024). Effect of hot isostatic pressing pressure and solution annealing cooling rate on CM247 LC nickel-based superalloy processed by laser-based powder bed fusion. *Progress in Additive Manufacturing*, *10*(1), 605–618. DOI: 10.1007/s40964-024-00645-6

Martínez-Gómez, J., & Morán, P. (2022). Integration of social life cycle assessment into sustainability decisions in additive manufacturing. *The International Journal of Life Cycle Assessment*, *27*(8), 1580–1592. DOI: 10.1016/B978-0-323-90386-8.00057-7

Martin, K. E., Shilton, K., & Smith, J. R. (2020). Ethical Implications of Smart Consumer Products: Privacy, Consent, and Control. *Journal of Business Ethics, 168*(1), 39–55.

Masini, A., & Menichetti, E. (2012). The impact of behavioural factors in the renewable energy investment decision making process: Conceptual framework and empirical findings. *Energy Policy, 40*, 28–38. DOI: 10.1016/j.enpol.2010.06.062

Mazzetto, S. (2024). A review of urban Digital Twins integration, challenges, and future directions in smart city development. *Sustainability (Basel), 16*(19), 8337. DOI: 10.3390/su16198337

McKee, S. (2020). *What is a digital twin?* DOI: 10.61557/ETSQ9929

McKinnon, A., Browne, M., Whiteing, A., & Piecyk, M. (2015). *Green logistics: Improving the environmental sustainability of logistics*. Kogan Page Publishers.

McKinsey & Company. (2020). *Smart manufacturing: The next wave of industrial transformation.*

McKinsey. (2020). *The Next Horizon for Product Sustainability: Technology-enabled solutions to unlock value.*

Mehta, P. (2020). Role of policy in promoting green energy in developing nations. *Global Energy Policy Review, 8*(2), 70–83.

Meier, C., Weissbach, R., Weinberg, J., Wall, W. A., & Hart, A. J. (2018). *Critical influences of particle size and adhesion on the powder layer uniformity in metal additive manufacturing*. DOI: 10.1016/j.jmatprotec.2018.10.037

Meloni, M., Cai, J., Zhang, Q., Sang-Hoon Lee, D., Li, M., Ma, R., Parashkevov, T. E., & Feng, J. (2021). Engineering Origami: A Comprehensive Review of Recent Applications, Design Methods, and Tools. *Advanced Science (Weinheim, Baden-Wurttemberg, Germany), 8*(13), 2000636. Advance online publication. DOI: 10.1002/advs.202000636

Mertens, R., Kullmann, J., Treitz, M., & Wycisk, E. (2017). Laser additive manufacturing processes : Thermodynamic and ecological evaluation. *Procedia CIRP, 61*, 143–148. DOI: 10.1016/j.procir.2016.11.202

Miao, Z., Du, J., Dong, F., Liu, Y., & Wang, X. (2020). Identifying technology evolution pathways using topic variation detection based on patent data: A case study of 3D printing. *Futures, 118*, 102530. https://doi.org/https://doi.org/10.1016/j.futures.2020.102530. DOI: 10.1016/j.futures.2020.102530

Mills, B., & Schleich, J. (2022). Solar panel adoption and peer effects: Evidence from residential energy consumption data. *Energy Economics*, *107*, 105825. DOI: 10.1016/j.eneco.2022.105825

Mi, S., Feng, Y., Zheng, H., Wang, Y., Gao, Y., & Tan, J. (2020). Prediction maintenance integrated decision-making approach supported by digital twin-driven cooperative awareness and interconnection framework. *Journal of Manufacturing Systems*, *58*, 329–345. DOI: 10.1016/j.jmsy.2020.08.001

Misnan, M. S., Ismail, M. Z., & Yan, T. J. (2024). Construction Project Management Issues and Development in Current for Future Construction Project: Challenges and Prospects in Sustainable Project Management. *International Journal of Research and Innovation in Social Science*, *1997*(II), 1997–2011. Advance online publication. DOI: 10.47772/IJRISS.2024.802141

Misopoulos, F., Manthou, V., & Michaelides, Z. (2019). Environmental and Social Sustainability in UK Construction Industry: A Systematic Literature Review. *European Journal of Economics and Business Studies*, *5*(1), 100. DOI: 10.26417/ejes.v5i1.p100-115

Moghrabi, I. A. R., Bhat, S. A., Szczuko, P., AlKhaled, R. A., & Dar, M. A. (2023). Digital Transformation and Its Influence on Sustainable Manufacturing and Business Practices. *Sustainability (Basel)*, *15*(4), 3010. DOI: 10.3390/su15043010

Mohammad Vaezi Srisit Chianrabutra, B. M., & Yang, S. (2013). Multiple material additive manufacturing – Part 1: A review. *Virtual and Physical Prototyping*, *8*(1), 19–50. DOI: 10.1080/17452759.2013.778175

Monek, G. D., & Fischer, S. (2024). Expert Twin: A Digital Twin with an Integrated Fuzzy-Based Decision-Making Module. *Decision Making Applications in Management and Engineering*, *8*(1), 1–21. DOI: 10.31181/dmame8120251181

Moradi, A., Ansari, R., Hassanzadeh-Aghdam, M. K., & Jang, S.-H. (2024). Thermomechanical behavior of carbon nanotube/graphene nanoplatelet-reinforced shape memory polymer nanocomposites: A micromechanics-based finite element approach. *European Journal of Mechanics. A, Solids*, *107*, 105360. https://doi.org/https://doi.org/10.1016/j.euromechsol.2024.105360. DOI: 10.1016/j.euromechsol.2024.105360

Morrison, T., & Chan, K. (2021). Barriers to renewable integration in fossil-dependent grids. *Energy Transition Studies*, *6*(3), 245–259.

Motalebi, A., Khondoker, M. A. H., & Kabir, G. (2024). A systematic review of life cycle assessments of 3D concrete printing. *Sustainable Operations and Computers*, *5*, 14–25. DOI: 10.1016/j.susoc.2023.08.003

Mrad, M., Frikha, M. A., & Boujelbène, Y. (2025). *State-of-The-Art Review: Artificial Intelligence and Robotics for Reducing Carbon Emissions in Supply Chain Management*. DOI: 10.2139/ssrn.5274912

Mügge, J., Seegrün, A., Hoyer, T.-K., Riedelsheimer, T., & Lindow, K. (2024). Digital Twins within the Circular Economy: Literature Review and Concept Presentation. *Sustainability (Basel)*, *16*(7), 2748. DOI: 10.3390/su16072748

Murr, L. E., Gaytan, S. M., & Ramirez, D. A. (2012). Preheating process optimization in EBM for nickel superalloys. *Journal of Materials Science and Technology*, *28*(1), 1–14. DOI: 10.1016/S1005-0302(12)60016-4

Murtaza, H. A., Mukhangaliyeva, A., Golman, B., Perveen, A., & Talamona, D. (2023). Quantitative characterization of metal powder morphology, size distribution, and flowability for additive manufacturing. *Journal of Materials Engineering and Performance*, *33*(20), 10853–10867. DOI: 10.1007/s11665-023-08761-0

Nadagouda, M. N., Ginn, M., & Rastogi, V. (2020). A review of 3D printing techniques for environmental applications [Review of *A review of 3D printing techniques for environmental applications*]. *Current Opinion in Chemical Engineering, 28*, 173. Elsevier BV. DOI: 10.1016/j.coche.2020.08.002

Nadgorny, M., Xiao, Z., Chen, C., & Connal, L. A. (2016). Three-Dimensional Printing of pH-Responsive and Functional Polymers on an Affordable Desktop Printer. *ACS Applied Materials & Interfaces*, *8*(42), 28946–28954. DOI: 10.1021/acsami.6b07388

Napoleone, A., Macchi, M., & Pozzetti, A. (2020). A review on the characteristics of cyber-physical systems for the future smart factories [Review of *A review on the characteristics of cyber-physical systems for the future smart factories*]. *Journal of Manufacturing Systems, 54*, 305. Elsevier BV. DOI: 10.1016/j.jmsy.2020.01.007

Nassiry, D. (2018). The role of fintech in unlocking green finance: Policy insights for developing countries. *ADBI Working Paper Series*, No. 883. Asian Development Bank Institute.

Nayak, G. N.Gautam Nandkishore Nayak. (2025). Leveraging Artificial Intelligence for Sustainable Electronics Manufacturing Supply Chains. *International Journal of Scientific Research in Computer Science Engineering and Information Technology*, *11*(1), 210–221. DOI: 10.32628/CSEIT25111230

Nazeer, A. (2021). AI-Powered Predictive Analytics for Supply Chain Optimization: A Risk-Resilient Framework. *International Journal of Emerging Trends in Computer Science and Information Technology*, *2*(1), 12–18. DOI: 10.63282/3050-9246.IJETCSIT-V2I1P102

Neyestani, B. (2017). A Review on Sustainable Building (Green Building) [Review of A Review on Sustainable Building (Green Building)]. SSRN Electronic Journal. RELX Group (Netherlands). DOI: 10.2139/ssrn.2968885

Ngo, T. D., Kashani, A., Imbalzano, G., Nguyen, K. T., & Hui, D. (2018). Additive manufacturing (3D printing) : A review of materials, methods, applications and challenges. *Composites. Part B, Engineering*, *143*, 172–196. DOI: 10.1016/j.compositesb.2018.02.012

Nguyen, T., & Le, M. (2021). AI-enhanced reliability assessment in energy distribution networks. *Power Systems Analysis Journal*, *7*(4), 120–135.

Nidumolu, R., Prahalad, C. K., & Rangaswami, M. R. (2009). Why sustainability is now the key driver of innovation. *Harvard Business Review*, *87*(9), 56–64.

Niinimäki, K., & Hassi, L. (2011). Emerging Design Strategies in Sustainable Production and Consumption. *Journal of Fashion Marketing and Management*, *15*(4), 393–410.

Nikishina, A. N., Kholkhoev, B. Ch., Bardakova, K. N., Matveev, Z. A., Kurdanova, Z. I., Shakhmurzova, K. T., Zhansitov, A. A., Khashirova, S. Yu., Timashev, P. S., & Burdukovskii, V. F. (2024). Photopolymerization-based 4D-printing of shape-memory materials containing high-performance polymers. *Reactive & Functional Polymers*, *202*, 106001. https://doi.org/https://doi.org/10.1016/j.reactfunctpolym.2024.106001. DOI: 10.1016/j.reactfunctpolym.2024.106001

O'Shaughnessy, E., Barbose, G., Wiser, R., Forrester, S., & Darghouth, N. (2021). The impact of policies and business models on income equity in rooftop solar adoption. *Nature Energy*, *6*(1), 84–91. DOI: 10.1038/s41560-020-00724-2

Ochieng, E. G., Wynn, T. S., Zuofa, T., Ruan, X., & Price, A. D. F.. (2014). Integration of Sustainability Principles into Construction Project Delivery. *J Archit Eng Tech*, *3*(1), 116. DOI: 10.4172/2168-9717.1000116

Oh, S. E., Kim, J., Kim, J.-Y., & Ahn, J. H. (2025). Food Safety Distribution Systems Using Private Blockchain: Ensuring Traceability and Data Integrity Verification. *Foods*, *14*(8), 1405. DOI: 10.3390/foods14081405

Oikonomou, V., Becchis, F., Steg, L., & Russolillo, D. (2020). Energy saving and energy efficiency concepts for policy making. *Energy Policy*, *37*(11), 4787–4796. DOI: 10.1016/j.enpol.2009.06.035

Olawade, D. B., Wada, O. Z., David-Olawade, A. C., Fapohunda, O., Ige, A. O., & Ling, J. (2024). Artificial intelligence potential for net zero sustainability: Current evidence and prospects. *Next Sustainability*, *4*, 100041. DOI: 10.1016/j.nxsust.2024.100041

Omar, S., & Ali, H. (2020). Adaptive control systems in renewable energy plants. *Control Systems in Energy*, *10*(3), 145–159.

Omar, Y., & Salim, K. (2023). Load balancing techniques in AI-based grid systems. *IEEE Transactions on Grid Systems*, *14*(1), 66–78.

Onyeke, F. O., Odujobi, O., Adikwu, F. E., & Elete, T. Y. (2024). The role of data-driven insights in industrial control systems: Advancing predictive maintenance and operational efficiency in refinery processes. *Engineering Science & Technology Journal*, *5*(12), 3266–3277. DOI: 10.51594/estj.v5i12.1775

Orsi, A., Guillén-Guillamón, I., & Pellicer, E. (2020). Optimization of Green Building Design Processes: Case Studies within the European Union. *Sustainability (Basel)*, *12*(6), 2276. DOI: 10.3390/su12062276

Osei, K., & Zhang, L. (2019). Life cycle assessment of solar PV in Ghana. *Sustainable Energy Research*, *98*, 400–412.

Özsoy, T. (2023). THE ROLE OF ARTIFICIAL INTELLIGENCE IN FACILITATING THE TRANSITION TO A CIRCULAR ECONOMY. *Nişantaşı Üniversitesi Sosyal Bilimler Dergisi/Nişantaşı Üniversitesi Sosyal Bilimler Dergisi, 11*(2), 369. DOI: 10.52122/nisantasisbd.1345828

Painuly, J. P. (2001). Barriers to renewable energy penetration; a framework for analysis. *Renewable Energy*, *24*(1), 73–89. DOI: 10.1016/S0960-1481(00)00186-5

Pal, S. (2023). Integrating AI in Sustainable Supply Chain Management: A New Paradigm for Enhanced Transparency and Sustainability. *International Journal for Research in Applied Science and Engineering Technology*, *11*(6), 2979–2984. DOI: 10.22214/ijraset.2023.54139

Park, J. (2025). A Holistic Approach to Managing Built Assets'. *Life Cycle Carbon Footprint.*, *1*(1). Advance online publication. DOI: 10.7771/3067-4883.1376

Park, J., & Yang, B. (2020). GIS-Enabled Digital Twin system for sustainable evaluation of carbon emissions: A case study of Jeonju City, South Korea. *Sustainability (Basel), 12*(21), 9186. DOI: 10.3390/su12219186

Park, S., & Kim, J. (2022). Digital twins in sustainable manufacturing. *Computers & Industrial Engineering, 168*, 108030. DOI: 10.1016/j.cie.2022.108030

Patel, D. (2023). Exploring the Frontiers of Microfluidics : Challenges and Future Prospects. In *Advances in MEMS and Microfluidic Systems* (p. 11-31).

Patel, D., Mishra, A., & Nabeel, M. (2022). Heat transfer characteristics of nanofluids of silicon oxides (SiO2) with conventional fluid. *2022 2nd International Conference on Innovative Practices in Technology and Management (ICIPTM, 2*, 420-423.

Patel, D., Sharma, A., & Mishra, A. (2021). Study of convective heat transfer characteristics of nano fluids in circular tube. 2021 International Conference on Technological Advancements and Innovations (ICTAI, 264-267.

Patel, D. (2024). Sustainable Renewable Energy Sources and Emerging Technologies. In *Optimization Techniques for Hybrid Power Systems* (pp. 343–361). Renewable Energy, Electric Vehicles, and Smart Grid. DOI: 10.4018/979-8-3693-0492-1.ch015

Patel, D. (2025). Emerging sustainable nanomaterials and their applications and future scope. In *Advances in Sustainable Materials* (pp. 107–135). Elsevier. DOI: 10.1016/B978-0-443-13849-2.00005-3

Patel, D., Varma, I. P., & Khan, F. A. (2022). A Review Advanced Vehicle with Automatic Pneumatic Bumper System using Two Cylinder. *Int. Res. J. Eng. Tech, 9*, 264–267.

Patel, R., & Kumar, A. (2022). Optimization techniques for renewable energy integration. *Renewable Energy Journal, 56*(2), 145–160. DOI: 10.1016/j.renene.2022.03.005

Patel, U. R., Ghaffarianhoseini, A., Ghaffarianhoseini, A., & Burgess, A. (2024). Digital Twin Technology for sustainable urban development: A review of its potential impact on SDG 11 in New Zealand. *Cities (London, England), 155*, 105484. DOI: 10.1016/j.cities.2024.105484

Patil, D. (2025). *Artificial Intelligence-Driven Supply Chain Optimization: Enhancing Demand Forecasting And Cost Reduction*. DOI: 10.2139/ssrn.5057408

Pattnaik, S., Liew, N., Kures, A. O., Pinsky, E., & Park, K. (2024). *Catalyzing Supply Chain Evolution: A Comprehensive Examination of Artificial Intelligence Integration in Supply Chain Management*. 57. DOI: 10.3390/engproc2024068057

Pei, E., & Loh, G. H. (2018). Technological considerations for 4D printing: An overview. *Progress in Additive Manufacturing*, *3*(1–2), 95–107. DOI: 10.1007/s40964-018-0047-1

Peldon, D., Banihashemi, S., LeNguyen, K., & Derrible, S. (2024). Navigating urban complexity: The transformative role of digital twins in smart city development. *Sustainable Cities and Society*, *111*, 105583. DOI: 10.1016/j.scs.2024.105583

Peng, T., Kellens, K., Tang, R., Chen, C., & Chen, G. (2018). Powder efficiency and sustainability. *Additive Manufacturing*, *21*, 694–704. DOI: 10.1016/j.addma.2018.04.022

Peng, T., Tang, R., & Kellens, K. (2018). Sustainability assessment of powder reuse and recycling in additive manufacturing. *Journal of Cleaner Production*, *192*, 801–813. DOI: 10.1016/j.jclepro.2018.05.173

Pereira, N. I. A., Oliveira, M. dos S., Reis, B. C. C., Nascimento, B. L., Carneiro, C. R., Arruda, T. R., Vieira, E. N. R., & Leite Junior, B. R. C. (2024). Unconventional sourced proteins in 3D and 4D food printing: Is it the future of food processing? *Food Research International*, *192*, 114849. DOI: 10.1016/j.foodres.2024.114849

Philips. (2020). *Philips Sustainability Report 2020*. Retrieved from https://www.philips.com

Phillips, T. (2020). *Philips' Circular Economy Commitment: Accelerating Progress Towards a Circular Economy. Philips Circular Economy Solutions*. Retrieved from https://www.philips.com/circular-economy

Pigosso, D. C. A., McAloone, T. C., & Rozenfeld, H. (2015). Characterisation of the state-of-the-art and identification of main trends for Ecodesign Tools and Methods: Classifying three decades of research and implementation. *Journal of Cleaner Production*, *108*, 1232–1244.

Pilate, F., Toncheva, A., Dubois, P., & Raquez, J.-M. (2016). Shape-memory polymers for multiple applications in the materials world. *European Polymer Journal*, *80*, 268–294. https://doi.org/https://doi.org/10.1016/j.eurpolymj.2016.05.004. DOI: 10.1016/j.eurpolymj.2016.05.004

Pillai, V., & Bose, D. (2020). Life cycle optimization in solar battery systems. *Smart Grid Design and Analysis*, *11*(3), 145–159.

Pollock, T. M., & Tin, S. (2006). Nickel-based superalloys for advanced turbine engines : Chemistry, microstructure, and properties. *Journal of Propulsion and Power*, *22*(2), 361–374. DOI: 10.2514/1.18239

Polo, S., Rubio, E. M., Ayllón, J., & de Agustina, B. (2025). Emerging Advances in Sustainable Manufacturing. *Processes (Basel, Switzerland)*, *13*(5), 1549. DOI: 10.3390/pr13051549

Popescu, C., Dissanayake, H., Mansi, E., & Stancu, A. (2024). Eco Breakthroughs: Sustainable Materials Transforming the Future of Our Planet. *Sustainability (Basel)*, *16*(23), 10790. DOI: 10.3390/su162310790

Porter, M. E., & Heppelmann, J. E. (2015). How smart, connected products are transforming companies. *Harvard Business Review*, *93*(10), 96–114.

Prajwal, S., Kumar, V., & Patil, N. N. (2021). Implementation of sustainable materials in an existing residential building and comparison. *IOP Conference Series. Materials Science and Engineering*, *1166*(1), 12038. DOI: 10.1088/1757-899X/1166/1/012038

Priya, E., Vasanthi, P., Prabhu, B., & Murugesan, P. (2025). Sawdust as a Sustainable Additive: Comparative Insights into its Role in Concrete and Brick Applications. *Cleaner Waste Systems*, *100286*, 100286. Advance online publication. DOI: 10.1016/j.clwas.2025.100286

Qiu, R., Liu, X., Tian, H., Hu, Z., Wang, K., & Zhao, L. (2024). Exploration of bigel 4D printing with spontaneous colour change for monitoring bio-actives kinetic behaviour based on the dual-units 3D printer. *Journal of Food Engineering*, *367*, 111861. DOI: 10.1016/j.jfoodeng.2023.111861

Qiu, Y., & Park, K. (2001). Environment-sensitive hydrogels for drug delivery. *Advanced Drug Delivery Reviews*, *53*(3), 321–339. https://doi.org/https://doi.org/10.1016/S0169-409X(01)00203-4. DOI: 10.1016/S0169-409X(01)00203-4

Rahimi, N., Molin, D. G., Cleij, T. J., van Zandvoort, M. A., & Post, M. J. (2012). Electrosensitive Polyacrylic Acid/Fibrin Hydrogel Facilitates Cell Seeding and Alignment. *Biomacromolecules*, *13*(5), 1448–1457. DOI: 10.1021/bm300161r

Rahmani, R., Jesus, C., & Lopes, S. I. (2024). Implementations of Digital Transformation and Digital Twins: Exploring the Factory of the Future. *Processes (Basel, Switzerland)*, *12*(4), 787. DOI: 10.3390/pr12040787

Rahmatabadi, D., Khajepour, M., Bayati, A., Mirasadi, K., Amin Yousefi, M., Shegeft, A., Ghasemi, I., Baniassadi, M., Abrinia, K., Bodaghi, M., & Baghani, M. (2024). Advancing sustainable shape memory polymers through 4D printing of polylactic acid-polybutylene adipate terephthalate blends. *European Polymer Journal*, *216*, 113289. https://doi.org/https://doi.org/10.1016/j.eurpolymj.2024.113289. DOI: 10.1016/j.eurpolymj.2024.113289

Rajput, R. S., Patel, D., & Singh, R. P. (2023). *Applications and Feasibility of Large-Scale Solar-Powered Peltier Refrigeration Systems.*

Raj, S., & Mehra, T. (2022). AI-based anomaly detection in energy systems. *IEEE Transactions on Sustainable Computing*, *7*(2), 315–327.

Ramani, K., Ramanujan, D., Bernstein, W. Z., Zhao, F., Sutherland, J. W., Handwerker, C. A., Choi, J., Kim, H., & Thurston, D. (2010). Integrated Sustainable Life Cycle Design: A Review [Review of Integrated Sustainable Life Cycle Design: A Review]. Journal of Mechanical Design, 132(9). American Society of Mechanical Engineers. DOI: 10.1115/1.4002308

Ramsperger, M., Markl, M., Berroth, K., Körner, C., & Singer, R. F. (2019). Hot cracking mechanism in nickel-based superalloys processed by selective electron beam melting. *Acta Materialia*, *159*, 77–87. DOI: 10.1016/j.actamat.2018.08.002

Rana, S. K., Kim, H., Pani, S. K., Rana, S., Joo, M.-I., Rana, A. K., & Aich, S. (2021). Blockchain-Based Model to Improve the Performance of the Next-Generation Digital Supply Chain. *Sustainability (Basel)*, *13*(18), 10008. DOI: 10.3390/su131810008

Rana, S., & Patel, D. (2021). Impact of AI on sustainable agriculture and energy use. *Agricultural Informatics*, *13*(1), 35–47.

Rane, N. L., Rane, J., & Paramesha, M. (2024). *Artificial Intelligence and business intelligence to enhance Environmental, Social, and Governance (ESG) strategies: Internet of things, machine learning, and big data analytics in financial services and investment sectors*. DOI: 10.70593/978-81-981367-4-9_3

Rane, N. L. (2023). Integrating Leading-Edge Artificial Intelligence (AI), Internet of Things (IoT), and Big Data Technologies for Smart and Sustainable Architecture, Engineering and Construction (AEC) Industry: Challenges and Future Directions. *International Journal of Data Science and Big Data Analytics*, *3*(2), 73–95. DOI: 10.51483/IJDSBDA.3.2.2023.73-95

Raviv, D., Zhao, W., McKnelly, C., Papadopoulou, A., Kadambi, A., Shi, B., Hirsch, S., Dikovsky, D., Zyracki, M., Olguin, C., Raskar, R., & Tibbits, S. (2014). Active Printed Materials for Complex Self-Evolving Deformations. *Scientific Reports*, *4*(1), 7422. DOI: 10.1038/srep07422

Reddy, V., & Sharma, N. (2022). Internet of Things (IoT) for smart grid applications. *Journal of Energy Automation*, *15*(2), 88–101. DOI: 10.1016/jea.2022.02.007

Reed, R. C. (2006). *The superalloys : Fundamentals and applications*. Cambridge University Press. DOI: 10.1017/CBO9780511541285

Reeves, P., Tuck, C., & Hague, R. (2011). Additive manufacturing for mass customisation. In *Customization 2.0* (pp. 275–289). Springer., DOI: 10.1007/978-1-84996-489-0_13

Reith, T., Bichler, L., Kopnarski, M., & May, S. (2020). Life cycle inventory database for additive manufacturing : Bridging data gaps. *Journal of Cleaner Production*, *247*, 119142. DOI: 10.1016/j.jclepro.2019.119142

Rejeski, D., Zhao, F., & Huang, Y. (2018). Research needs and recommendations on environmental implications of additive manufacturing. *Additive Manufacturing*, *19*, 21–28. DOI: 10.1016/j.addma.2017.10.019

Resman, M., Herakovič, N., & Debevec, M. (2025). Integrating Digital Twin Technology to Achieve Higher Operational Efficiency and Sustainability in Manufacturing Systems. *Systems*, *13*(3), 180. DOI: 10.3390/systems13030180

Rezaei, Z., Vahidnia, M. H., Aghamohammadi, H., Azizi, Z., & Behzadi, S. (2023). Digital twins and 3D information modeling in a smart city for traffic controlling: A review [Review of Digital twins and 3D information modeling in a smart city for traffic controlling: A review]. *Journal of Geography and Cartography*, *6*(1), 1865. DOI: 10.24294/jgc.v6i1.1865

Ricciardi, G., & Callegari, G. (2023). Digital Twins for Climate-Neutral and resilient cities. State of the art and future development as tools to support urban Decision-Making. In *The œurban book series* (pp. 617–626). https://doi.org/DOI: 10.1007/978-3-031-29515-7_55

Rickenbacher, L., Spierings, A. B., & Wegener, K. (2013). High temperature application of SLM-processed IN738LC parts. *Rapid Prototyping Journal*, *19*(4), 282–290. DOI: 10.1108/13552541311323281

Riedl, A., & Smeets, P. (2017). Why do investors hold socially responsible mutual funds? *The Journal of Finance*, *72*(6), 2505–2550. DOI: 10.1111/jofi.12547

Ritchie, J., & Spencer, L. (1994). Qualitative data analysis for applied policy research. In Bryman, A., & Burgess, R. G. (Eds.), *Analyzing qualitative data* (pp. 173–194). Routledge. DOI: 10.4324/9780203413081_chapter_9

Roberts, H., Zhang, J., Bariach, B., Cowls, J., Gilburt, B., Juneja, P., Tsamados, A., Ziosi, M., Taddeo, M., & Floridi, L. (2022). Artificial intelligence in support of the circular economy: ethical considerations and a path forward. SSRN *Electronic Journal*. DOI: 10.2139/ssrn.4080909

Rogers, E. M. (2003). *Diffusion of innovations* (5th ed.). Free Press.

Rojek, I., Mikołajewski, D., & Dostatni, E. (2020). Digital Twins in Product Lifecycle for Sustainability in Manufacturing and Maintenance. *Applied Sciences (Basel, Switzerland)*, *11*(1), 31. DOI: 10.3390/app11010031

Roque, A. J., Paleologos, E. K., O'Kelly, B. C., Tang, A. M., Reddy, K. R., Vitone, C., Mohamed, A. O., Koda, E., Goli, V. S. N. S., Vieira, C. S., Fei, X., Sollecito, F., Vaverková, M. D., Plötze, M., Petti, R., Podlasek, A., Puzrin, A. M., Cotecchia, F., Osiński, P., & Singh, D. N. (2021). Sustainable environmental geotechnics practices for a green economy. *Environmental Geotechnics*, *9*(2), 68–84. DOI: 10.1680/jenge.21.00091

Roy, I., & Gupta, M. N. (2003). Smart Polymeric Materials: Emerging Biochemical Applications. *Chemistry & Biology*, *10*(12), 1161–1171. https://doi.org/https://doi.org/10.1016/j.chembiol.2003.12.004. DOI: 10.1016/j.chembiol.2003.12.004

Roy, P., & Banerjee, A. (2022). Circular economy integration in solar panel lifecycle. *Journal of Green Technology*, *13*(1), 38–51.

Saberi, S., Kouhizadeh, M., Sarkis, J., & Shen, L. (2019). Blockchain technology and its relationships to sustainable supply chain management. *International Journal of Production Research*, *57*(7), 2117–2135. DOI: 10.1080/00207543.2018.1533261

Sachs, J. D. (2015). *The Age of Sustainable Development*. Columbia University Press. DOI: 10.7312/sach17314

Saeed, S., Arshad, M. Y., & Ahmed, A. S. (2023). Advancing circular economy in industrial chemistry and environmental engineering: Principles, alignment with United Nations sustainable development goals, and pathways to implementation. *European Journal of Chemistry*, *14*(3), 414–428. DOI: 10.5155/eurjchem.14.3.414-428.2452

Salhi, A., Algarni, F. S., Alshamrani, R., Althbiti, A., Ismail, A., & Hassan, B. M. (2025). Leveraging artificial intelligence to enable sustainable urban development through the creation of smart and environmentally friendly carbon-free cities. *Scientific Reports*, *15*(1), 35791. Advance online publication. DOI: 10.1038/s41598-025-16801-z

Sames, W. J., List, F. A., Pannala, S., Dehoff, R. R., & Babu, S. S. (2016). The metallurgy and processing science of metal additive manufacturing. *International Materials Reviews*, *61*(5), 315–360. DOI: 10.1080/09506608.2015.1116649

Samuel, P., Saini, A., Poongodi, T., & Nancy, P. (2023). Artificial intelligence–driven digital twins in Industry 4.0. In *Elsevier eBooks* (pp. 59–88). https://doi.org/DOI: 10.1016/B978-0-323-99205-3.00002-X

Sarmad, S., Yenici, G., Gürkan, K., Keçeli, G., & Gürdağ, G. (2013). Electric field responsive chitosan–poly(N,N-dimethyl acrylamide) semi-IPN gel films and their dielectric, thermal and swelling characterization. *Smart Materials and Structures*, *22*(5), 55010. DOI: 10.1088/0964-1726/22/5/055010

Sartal, A., Bellas, R., Mejías, A. M., & Garcia, A. (2014). The sustainable manufacturing concept, evolution and opportunities within Industry 4.0: A literature review [Review of *The sustainable manufacturing concept, evolution and opportunities within Industry 4.0: A literature review*]. *Advances in Mechanical Engineering*, *12*(5). SAGE Publishing. DOI: 10.1177/1687814020925232

Savastano, M., Amendola, C., Bellini, F., & D'Ascenzo, F. (2019). Contextual Impacts on Industrial Processes Brought by the Digital Transformation of Manufacturing: A Systematic Review [Review of *Contextual Impacts on Industrial Processes Brought by the Digital Transformation of Manufacturing: A Systematic Review*]. *Sustainability, 11*(3), 891. Multidisciplinary Digital Publishing Institute. DOI: 10.3390/su11030891

Schmaljohann, D. (2006). Thermo- and pH-responsive polymers in drug delivery. *Advanced Drug Delivery Reviews*, *58*(15), 1655–1670. https://doi.org/https://doi .org/10.1016/j.addr.2006.09.020. DOI: 10.1016/j.addr.2006.09.020

Schmidt, O., Chadwick, E., & Tress, B. (2021). Life Cycle Costing in Additive Manufacturing : Unlocking economic insights. *Additive Manufacturing*, *45*, 102015. DOI: 10.1016/j.addma.2021.102015

Sen, A. (1999). *Development as freedom*. Oxford University Press.

Sengupta, S., & Dreyer, H. C. (2024). Digitalization and sustainability. In *Routledge eBooks* (p. 98). Informa., DOI: 10.4324/9781032693415-8

Seong, K., Jiao, J., Lewis, R. H., Farahi, A., Navrátil, P., Casebeer, N., Davis, B., Jones, J. B., & Niyogi, D. (2025). Towards a digital twin for smart resilient cities: Real-time fire and smoke tracking and prediction platform for community awareness (FireCom). *Computational Urban Science*, *5*(1), 49. Advance online publication. DOI: 10.1007/s43762-025-00212-x

Sershen, S. R., Mensing, G. A., Ng, M., Halas, N. J., Beebe, D. J., & West, J. L. (2005). Independent Optical Control of Microfluidic Valves Formed from Optomechanically Responsive Nanocomposite Hydrogels. *Advanced Materials*, *17*(11), 1366–1368. https://doi.org/https://doi.org/10.1002/adma.200401239. DOI: 10.1002/ adma.200401239

Shafiq, M. T., & Afzal, M. (2020). Potential of Virtual Design Construction Technologies to Improve Job-Site Safety in Gulf Corporation Council. *Sustainability (Basel)*, *12*(9), 3826. DOI: 10.3390/su12093826

Shahat, E., Hyun, C. T., & Yeom, C. (2021). City Digital Twin Potentials: A Review and Research Agenda [Review of *City Digital Twin Potentials: A Review and Research Agenda*]. *Sustainability, 13*(6), 3386. Multidisciplinary Digital Publishing Institute. DOI: 10.3390/su13063386

Shahbazi, M., Jäger, H., Ettelaie, R., Mohammadi, A., & Asghartabar Kashi, P. (2023). Multimaterial 3D printing of self-assembling smart thermo-responsive polymers into 4D printed objects: A review. *Additive Manufacturing*, *71*, 103598. DOI: 10.1016/j.addma.2023.103598

Shahrubudin, N., Chuan, L. T., & Ramlan, R. (2019). An Overview on 3D Printing Technology: Technological, Materials, and Applications. *Procedia Manufacturing*, *35*, 1286–1296. DOI: 10.1016/j.promfg.2019.06.089

Shang, Z., & Zhang, L. (2022). The Sustainable Digitalization in the Manufacturing Industry: A Bibliometric Analysis and Research Trend. *Mobile Information Systems*, *1*, 1–11. Advance online publication. DOI: 10.1155/2022/1451705

Sharma, A., & Patel, D. (2023). Barriers to energy efficiency in rural microgrids. *Energy and Environment Research*, *15*(1), 50–62.

Sharma, K., Kumar, V., Chaudhary, B., Kaith, B. S., Kalia, S., & Swart, H. C. (2016). Application of biodegradable superabsorbent hydrogel composite based on Gum ghatti-co-poly(acrylic acid-aniline) for controlled drug delivery. *Polymer Degradation & Stability*, *124*, 101–111. https://doi.org/https://doi.org/10.1016/j.polymdegradstab.2015.12.021. DOI: 10.1016/j.polymdegradstab.2015.12.021

Sharma, R., & Pillai, A. (2022). AI-driven energy dashboards for smart cities. *Urban Intelligence and Energy*, *4*(3), 105–118.

Sherwin, C. (2004). Design and sustainability: A discussion paper based on personal experience and observations. *The Journal of Sustainable Product Design*, *4*(1-2), 21–31. DOI: 10.1007/s10970-006-0003-x

Shirazi, S. F. S., Gharehkhani, S., Mehrali, M., Yarmand, H., Metselaar, H. S. C., Adib Kadri, N., & Osman, N. A. A. (2015). A review on powder-based additive manufacturing for tissue engineering: Selective laser sintering and inkjet 3D printing. In *Science and Technology of Advanced Materials* (Vol. 16, Issue 3). Institute of Physics Publishing. DOI: 10.1088/1468-6996/16/3/033502

Shree, M. V., Dhinakaran, V., Rajkumar, V., Ram, P. M. B., Vijayakumar, M., & Sathish, T. (2019). Effect of 3D printing on supply chain management. *Materials Today: Proceedings*, *21*, 958–963. DOI: 10.1016/j.matpr.2019.09.060

Siddiqi, K., Chatman, D., & Cook, G. (2008). Role of education and industry towards more sustainable construction. *International Journal of Environmental Technology and Management*, *8*(2/3), 310. DOI: 10.1504/IJETM.2008.017336

Silva, A., Gaspar, F., & Бакатович, A. A. (2023). Composite Materials of Rice Husk and Reed Fibers for Thermal Insulation Plates Using Sodium Silicate as a Binder. *Sustainability (Basel)*, *15*(14), 11273. DOI: 10.3390/su151411273

Silva, J. F., & Moradi, H. (2020). Smart grid technologies for resource optimization. *IEEE Transactions on Smart Grid*, *11*(5), 2300–2310.

Silva, M., & Rodrigues, A. (2020). ML-enhanced biomass energy prediction models. *Bioenergy Science and Analytics*, *5*(1), 60–73.

Simchi, A. (2006). Direct laser sintering of metal powders : Mechanism, kinetics, and microstructural features. *Materials Science and Engineering A*, *428*(1-2), 148–158. DOI: 10.1016/j.msea.2006.04.117

Sims, C. T., Stoloff, N. S., & Hagel, W. C. (1987). *Superalloys II: High-temperature materials for aerospace and industrial power*. Wiley.

Singh, A. K., & Patel, D. (2020). Optimization of Air Flow Over a Car by Wind Tunnel. *2020 International Conference on Computation, Automation and Knowledge Management (ICCAKM*, 1-4. Smith, J. R., & Chen, L. (2020). *Energy efficiency strategies for smart cities*. Springer. DOI: 10.1109/ICCAKM46823.2020.9051457

Singh, C. S. (2018). Green Construction: Analysis on Green and Sustainable Building Techniques. *Civil Engineering Research Journal*, *4*(3). Advance online publication. DOI: 10.19080/CERJ.2018.04.555638

Singh, M., Fuenmayor, E., Hinchy, E. P., Qiao, Y., Murray, N., & Devine, D. M. (2021). Digital Twin : Origin to Future. *Applied System Innovation*, *4*(2), 36–36. DOI: 10.3390/asi4020036

Singh, P. K. K. (2024). Measuring the Broader Value Proposition of Digital Transformation in Supply Chains. *International Journal of Supply Chain Management*, *13*(1), 16–24. DOI: 10.59160/ijscm.v13i1.6222

Sira, M. (2024). Potential of Advanced Technologies for Environmental Management Systems. *Management Systems in Production Engineering*, *32*(1), 33–44. DOI: 10.2478/mspe-2024-0004

Slotwinski, J. A., Garboczi, E. J., & Hebenstreit, K. M. (2014). Characterization of metal powders used for additive manufacturing. *Journal of Research of the National Institute of Standards and Technology*, *119*(), 460–493. DOI: 10.6028/jres.119.018

Smith, T., & Williams, R. (2022). Competitive advantage through sustainability in manufacturing. *Sustainable Production and Consumption*, *31*, 101–110. DOI: 10.1016/j.spc.2022.01.004

Sobin, N., Molenaar, K. R., & Gransberg, D. D. (2010). Sustainability by..." A Synthesis of Procurement Approaches for High Performance Buildings. Construction Research Congress 2022, 1366. DOI: 10.1061/41109(373)137

Song, M., Zhu, G., & Guo, J. (2023). 4D printing of biodegradable shape memory double-network hydrogel for highly bionic devices. *Journal of Materials Research and Technology*, *24*, 2935–2945. https://doi.org/https://doi.org/10.1016/j.jmrt.2023.03.180. DOI: 10.1016/j.jmrt.2023.03.180

Song, Y., & Zhang, H. (2018). Research on sustainability of building materials. *IOP Conference Series. Materials Science and Engineering*, *452*, 22169. DOI: 10.1088/1757-899X/452/2/022169

Soori, M., Jough, F. K. G., Dastres, R., & Arezoo, B. (2024). AI-Based Decision Support Systems in Industry 4.0, A Review [Review of *AI-Based Decision Support Systems in Industry 4.0, A Review*]. *Journal of Economy and Technology*. Elsevier BV. DOI: 10.1016/j.ject.2024.08.005

Soppimath, K. S., Aminabhavi, T. M., Dave, A. M., Kumbar, S. G., & Rudzinski, W. E. (2002). Stimulus-Responsive "Smart" Hydrogels as Novel Drug Delivery Systems. *Drug Development and Industrial Pharmacy*, *28*(8), 957–974. DOI: 10.1081/DDC-120006428

Sovacool, B. K. (2009). The cultural barriers to renewable energy and energy efficiency in the United States. *Technology in Society*, *31*(4), 365–373. DOI: 10.1016/j.techsoc.2009.10.009

Spierings, A. B., Voegtlin, M., Bauer, T., Wegener, K., & Kerres, M. (2013). Powder flowability characterization methodology for powder-bed-based metal additive manufacturing. *Additive Manufacturing*, *1–4*, 87–98. DOI: 10.1007/s40964-015-0001-4

Stahel, W. R. (2016). The circular economy. *Nature*, *531*(7595), 435–438. DOI: 10.1038/531435a PMID: 27008952

Stern, P. C. (2000). New environmental theories: Toward a coherent theory of environmentally significant behavior. *The Journal of Social Issues*, *56*(3), 407–424. DOI: 10.1111/0022-4537.00175

Sulaiman, N. N., Omar, A. J., Pakir, F., & Mohamad, M. (2024). Evaluating Environmental, Social and Governance (ESG) Practice Among Malaysian Public Listed Construction Companies Using FTSE Russell Rating Model. *International Journal of Sustainable Construction Engineering Technology*, *15*(1). Advance online publication. DOI: 10.30880/ijscet.2024.15.01.003

Sun, Z., Jiang, X., Dong, J., Wang, L., Xu, Z., Li, K., & Wang, M. (2025). Recovery of creep performance at different temperature and stress levels of a directionally solidified Ni-based superalloy by hot isostatic pressing and rejuvenation heat treatments. *Metallurgical and Materials Transactions A*. DOI: 10.1007/s11661-025-07855-9

Sung, H., Kim, K., & Kim, C. K. (2011). Case Study on Selective Demolition Method for Refurbishing Deteriorated Residential Apartments. *Journal of Construction Engineering and Management*, *138*(2), 294–303. DOI: 10.1061/(ASCE)CO.1943-7862.0000424

Sunter, D. A., Castellanos, S., & Kammen, D. M. (2019). Disparities in rooftop photovoltaics deployment in the United States by race and ethnicity. *Nature Sustainability*, *2*(1), 71–76. DOI: 10.1038/s41893-018-0204-z

Szmelter-Jarosz, A. (2026). Energy-efficient management policies in smart cities. In *Energy-efficient transformative technologies for data-driven smart cities* (pp. 161–178). Elsevier., DOI: 10.1016/B978-0-443-27618-7.00007-8

Taha, F. F. (2020). Adoption of BIM technology to develop sustainable buildings in the Iraqi construction sector. *Indian Journal of Science and Technology*, *13*(15), 1596–1606. DOI: 10.17485/IJST/v13i15.223

Tan, L. P., Choe, K. L., Ng, Y. K., Lai, K. F., & Khor, H. C. (2023). The Environmental, Social, and Governance (ESG) In Malaysia SME: A Case Study of Vitrox. In *Advances in economics, business and management research/Advances in Economics, Business and Management Research* (p. 344). Atlantis Press. DOI: 10.2991/978-94-6463-342-9_26

Tanaka, Y., & Singh, R. (2019). Economic barriers to solar adoption in urban areas. *Energy Economics Journal*, *41*(3), 310–322.

Tao, F., Zhang, M., Cheng, J., & Qi, Q. (2019). Digital Twin and Big Data Towards Smart Manufacturing and Industry 4.0: A 360-Degree Comparison. *IEEE Access : Practical Innovations, Open Solutions*, *6*, 3585–3593.

Teixeira, A. C. B., Ferreira, J. V., & Ramos, A. L. (2025). Intelligent Supply Chain Management: A Systematic Literature Review on Artificial Intelligence Contributions. *Information (Basel)*, *16*(5), 399. DOI: 10.3390/info16050399

Tekinerdoğan, B., & Verdouw, C. N. (2020). Systems Architecture Design Pattern Catalog for Developing Digital Twins. *Sensors (Basel)*, *20*(18), 5103. DOI: 10.3390/s20185103

Terry, D. J., & Hogg, M. A. (2000). *Attitudes, behavior, and social context: The role of norms and group membership*. Lawrence Erlbaum Associates.

Tesla, I. (2020). *Impact Report 2020*. Retrieved from https://www.tesla.com/impact

Tesla. (2021). *The Tesla Gigafactory: Revolutionising the Global Energy Ecosystem*. Retrieved from https://www.tesla.com/gigafactory

Thaler, R. (1985). Mental accounting and consumer choice. *Marketing Science*, *4*(3), 199–214. DOI: 10.1287/mksc.4.3.199

Thaler, R. H., & Sunstein, C. R. (2008). *Nudge: Improving decisions about health, wealth, and happiness*. Yale University Press.

Thomas, E., & Singh, D. (2021). Battery storage for sustainable energy systems. *Renewables Review*, *13*(3), 75–84.

Thompson, M. K., Moroni, G., Vaneker, T., Fadel, G., Campbell, R. I., Gibson, I., & Martina, F. (2016). Design for Additive Manufacturing : Trends, opportunities, considerations, and constraints. *CIRP Annals*, *65*(2), 737–760. DOI: 10.1016/j.cirp.2016.05.004

Tibbits, S., McKnelly, C., Olguin, C., Dikovsky, D., & Hirsch, S. (2014). *4D Printing and Universal Transformation*. 539–548. DOI: 10.52842/conf.acadia.2014.539

Tibbits, S. (2012). Design to Self-Assembly. *Architectural Design*, *82*(2), 68–73. DOI: 10.1002/ad.1381

Tiwari, A., & Turner, T. A. (2019). Additive manufacturing and sustainability : A review. *Materials Today: Proceedings*, *18*, 2121–2128. DOI: 10.1016/j.matpr.2019.07.166

Tofail, S. A. M., Koumoulos, E. P., Bandyopadhyay, A., Bose, S., O'Donoghue, L., & Charitidis, C. (2018). Additive manufacturing: Scientific and technological challenges, market uptake and opportunities. *Materials Today*, *21*(1), 22–37. https://doi.org/https://doi.org/10.1016/j.mattod.2017.07.001. DOI: 10.1016/j.mattod.2017.07.001

Tokbolat, S., Karaca, F., Durdyev, S., & Calay, R. K. (2019). Construction professionals' perspectives on drivers and barriers of sustainable construction. *Environment, Development and Sustainability*, *22*(5), 4361–4378. DOI: 10.1007/s10668-019-00388-3

Tran, T. S., Balu, R., Mettu, S., Roy Choudhury, N., & Dutta, N. K. (2022). 4D Printing of Hydrogels: Innovation in Material Design and Emerging Smart Systems for Drug Delivery. In *Pharmaceuticals* (Vol. 15, Issue 10). MDPI. DOI: 10.3390/ph15101282

Triandis, H. C. (1977). *Interpersonal behavior*. Brooks/Cole Publishing Company.

Trienens, M., Rasor, R., Kharatyan, A., Dumitrescu, R., & Anacker, H. (2024). Digital twins to increase sustainability throughout the system life cycle: A systematic literature review. *Proceedings of the Design Society*, *4*, 2277–2286. DOI: 10.1017/pds.2024.230

Tsolakis, N., Niedenzu, D., Simonetto, M., Dora, M., & Kumar, M. (2020). Supply network design to address United Nations Sustainable Development Goals: A case study of blockchain implementation in Thai fish industry. *Journal of Business Research, 131*, 495–519. DOI: 10.1016/j.jbusres.2020.08.003

Tsolakis, N., Schumacher, R., Dora, M., & Kumar, M. (2022). Artificial intelligence and blockchain implementation in supply chains: A pathway to sustainability and data monetisation? *Annals of Operations Research*, *327*(1), 157–210. DOI: 10.1007/s10479-022-04785-2

U.S. Department of Energy. (2020). *Better Buildings Challenge: Impact evaluation of residential energy efficiency programs*. U.S. Department of Energy.

United Nations Framework Convention on Climate Change. (2021). *Nationally determined contributions under the Paris Agreement: Synthesis report by the secretariat*. United Nations.

United Nations. (2018). *World urbanization prospects: The 2023 revision*. United Nations Department of Economic and Social Affairs.

Uriarte-Gallastegi, N., Landín, G. A., Landeta-Manzano, B., & Laskurain-Iturbe, I. (2024). The Role of AI in Improving Environmental Sustainability: A Focus on Energy Management. *Energies*, *17*(3), 649. DOI: 10.3390/en17030649

Uyan, M., & Celiktas, M. S. (2024). Novel constitutive models, challenges and opportunities of shape memory polymer composites. *Emerging Materials Research*, *13*(3), 1–20. DOI: 10.1680/jemmr.23.00116

Van Rooij, M., Lusardi, A., & Alessie, R. (2011). Financial literacy and stock market participation. *Journal of Financial Economics*, *101*(2), 449–472. DOI: 10.1016/j.jfineco.2011.03.006

Varghese, A. (2022). AI-Driven Solutions for Energy Optimization and Environmental Conservation in Digital Business Environments. *Asia Pacific Journal of Energy and Environment*, *9*(1), 49–60. DOI: 10.18034/apjee.v9i1.736

Vorkapić, M., Hasan, M. A., Tanović, D., Baltić, M., & Tomić, B. (2020). Implementation of 6R strategy in FDM printing process: Case: Small electronic enclosure box. *Journal of Engineering Management and Competitiveness*, *10*(2), 141–150. DOI: 10.5937/jemc2002141V

Vu, N. V., Nazari, M. A., Dang, T., Muralev, Y., Mohanraj, M., Tran, T., & Quoc, H. A. (2025). Type of the Paper*: Article*. DOI: 10.2139/ssrn.5384374

Výtisk, J., Kočí, V., Honus, S., & Vrtek, M. (2019). Current options in the life cycle assessment of additive manufacturing products. *Open Engineering*, *9*(1), 674–682. DOI: 10.1515/eng-2019-0073

Wang, J., & Xue, F. (2024). Emerging Trends of ESG in the Construction Sector: A Promising Pathway to Sustainable and Responsible Development. In Lecture notes in operations research (p. 2033). Springer International Publishing. DOI: 10.1007/978-981-97-1949-5_142

Wang, D., & Yang, T. Y. (2025). Research on the Promotion Effect of the Marketization of Data Elements on the Digital Transformation of Manufacturing Enterprises: An Empirical Evaluation of a Multiperiod DID Model. *Sustainability (Basel)*, *17*(7), 3199. DOI: 10.3390/su17073199

Wang, J., Wang, Z., Sui, Q., Xu, S., Yuan, Q., Zhang, D., & Liu, J. (2023). A comparison of the microstructure, mechanical properties, and corrosion resistance of the K213 superalloy after conventional casting and selective laser melting. *Materials (Basel)*, *16*(4), 1331. DOI: 10.3390/ma16041331

Wang, L., & Gupta, N. (2021). Artificial intelligence in wind farm efficiency optimization. *Renewable Energy Insights*, *18*(2), 105–117.

Wang, Y., Wang, S., & Wang, J. (2021). Digital twin model of equipment maintenance management in modern enterprises. *Journal of Physics: Conference Series*, *1986*(1), 12088. DOI: 10.1088/1742-6596/1986/1/012088

Wan, X., He, Y., Liu, Y., & Leng, J. (2022). 4D printing of multiple shape memory polymer and nanocomposites with biocompatible, programmable and selectively actuated properties. *Additive Manufacturing*, *53*, 102689. DOI: 10.1016/j.addma.2022.102689

Wasti, S., & Adhikari, S. (2020). Use of Biomaterials for 3D Printing by Fused Deposition Modeling Technique: A Review [Review of *Use of Biomaterials for 3D Printing by Fused Deposition Modeling Technique: A Review*]. *Frontiers in Chemistry, 8*. Frontiers Media. DOI: 10.3389/fchem.2020.00315

Waters, G. (2025). *Testing, Evaluation, Verification and Validation (TEVV) of Digital Twins: A Comprehensive Framework*. DOI: 10.48550/ARXIV.2507.04555

Weil, C., Bibri, S. E., Longchamp, R., Golay, F., & Alahi, A. (2023). Urban Digital Twin Challenges: A Systematic Review and Perspectives for Sustainable Smart Cities. *Sustainable Cities and Society*, *99*, 104862. DOI: 10.1016/j.scs.2023.104862

White, S. R., Sottos, N. R., Geubelle, P. H., Moore, J. S., Kessler, M. R., Sriram, S. R., Brown, E. N., & Viswanathan, S. (2001). Autonomic healing of polymer composites. *Nature*, *409*(6822), 794–797. DOI: 10.1038/35057232

Wong, K. V., & Hernandez, A. (2012). A review of additive manufacturing. *ISRN Mechanical Engineering*, *2012*, 1–10. DOI: 10.5402/2012/208760

Wu, Y., Wu, S., Tian, X., Wang, X., Wu, W., Zou, G., & Zhang, Q. (2011). Photoinduced reversible gel–sol transitions of dicholesterol-linked azobenzene derivatives through breaking and reforming of van der Waals interactions. *Soft Matter*, *7*(2), 716–721. DOI: 10.1039/C0SM00330A

Xiao, L., Cao, J., Liu, Z. G., & Luo, X. (2020). Sustainable Business Model Based on Digital Twin Platform Network: The Inspiration from Haier's Case Study in China. *Sustainability (Basel)*, *12*(3), 936. DOI: 10.3390/su12030936

Xiao, Y. (2022). *Effect and Influencing Factors of Digital Transformation of Manufacturing Industry*. Advances in Economics, Business and Management Research/ Advances in Economics, Business and Management Research., DOI: 10.2991/ aebmr.k.220405.072

Xiros, V., Castro, J. M. G., Fernandez-Pelaez, F., Magoutas, B., & Christidis, K. (2025). Supply Chain Data Analytics for Digital Twins: A Comprehensive Framework. *Applied Sciences (Basel, Switzerland)*, *15*(12), 6939. DOI: 10.3390/app15126939

Xu, X. (2023). *Vat photopolymerisation 3D printing of controlled drug delivery devices*.

Xue, X., Dhumras, H., Thakur, G., & Shukla, V. (2025). Integrating artificial intelligence and sustainable materials for smart eco innovation in production. *Scientific Reports*, *15*(1), 36942. Advance online publication. DOI: 10.1038/s41598-025-20803-2

Xu, W., Xu, J., & Zhang, W. (2025). Driving Green Transformation in Equipment Manufacturing Enterprises: The Role of Digital Technologies. *Systems*, *13*(5), 332. DOI: 10.3390/systems13050332

Yakovenko, Y., & Shaptala, R. (2024). Study of digital twins as the driving force of digital transformation and achieving the goals of sustainable development. *Technology Audit and Production Reserves*, *2*(4(76)), 11–20. DOI: 10.15587/2706-5448.2024.301423

Yang, G., Yang, G., & Yang, W. (2024). The impact of AI on enterprise energy management: From the perspective of carbon emissions. *Science and Technology for Energy Transition*, *80*, 8. DOI: 10.2516/stet/2024096

Yang, L., Martina, F., Colegrove, P., Williams, S., Wang, F., & Serres, H. (2019). Microstructural evolution and mechanical properties of Inconel 718 alloy processed by directed energy deposition additive manufacturing. *Materials Science and Engineering A*, *755*, 170–180. DOI: 10.1016/j.msea.2019.03.110

Yang, M., Fu, M., & Zhang, Z. (2021). The adoption of digital technologies in supply chains: Drivers, process and impact. *Technological Forecasting and Social Change*, *169*, 120795. DOI: 10.1016/j.techfore.2021.120795

Yang, M., Yuan, Z., Liu, J., Fang, Z., Fang, L., Yu, D., & Li, Q. (2019). Photoresponsive Actuators Built from Carbon-Based Soft Materials. *Advanced Optical Materials*, *7*(16), 1900069. https://doi.org/https://doi.org/10.1002/adom.201900069. DOI: 10.1002/adom.201900069

Yang, Y. (2014). Integrating BIM Technology to Achieve Green Buildings' Sustainable Performance. *Advanced Materials Research*, *1271*. Advance online publication. DOI: 10.4028/www.scientific.net/amr.1073-1076.1271

Yan, M.-R. (2015). Project-Based Market Competition and Policy Implications for Sustainable Developments in Building and Construction Sectors. *Sustainability (Basel)*, *7*(11), 15423–15448. DOI: 10.3390/su71115423

Yao, L., & Krause, S. (2003). Electromechanical Responses of Strong Acid Polymer Gels in DC Electric Fields. *Macromolecules*, *36*(6), 2055–2065. DOI: 10.1021/ma021326q

Yap, C. K., Leow, C. S., & Goh, B. (2024). Sustainable construction materials under ESG: a literature review and synthesis [Review of Sustainable construction materials under ESG: a literature review and synthesis]. MOJ Biology and Medicine, 9(1), 1. MedCrave Group. DOI: 10.15406/mojbm.2024.09.00208

Yoshimura, T., Uchikoshi, I., Yoshiura, Y., & Fujioka, R. (2005). Synthesis and characterization of novel biodegradable superabsorbent hydrogels based on chitin and succinic anhydride. *Carbohydrate Polymers*, *61*(3), 322–326. https://doi.org/https://doi.org/10.1016/j.carbpol.2005.06.014. DOI: 10.1016/j.carbpol.2005.06.014

Yunus, J. N., Fahrurazi, L. S., Hamid, M. S. A., Noor, R. N. H. R. M., & Majid, M. R. A. (2024). The Implementation of Sustainability Practices in Construction Site and its Effectiveness in Klang Valley. *International Journal of Academic Research in Business & Social Sciences*, *14*(9). Advance online publication. DOI: 10.6007/IJARBSS/v14-i9/23020

Yusuf, S. O., Ikhine, I. V., Nyamekeh, R., Oluwadare, O. E., Afoakwah, B., & Yusuf, N. (2025). The Impact of AI on Supply Chain Operations: A comparative analysis of traditional vs AI-enabled Processes. *World Journal of Advanced Research and Reviews*, *27*(2), 1688–1700. DOI: 10.30574/wjarr.2025.27.2.3027

Yu, Z., Umar, M., & Rehman, S. A. (2022). Adoption of technological innovation and recycling practices in automobile sector: Under the Covid-19 pandemic. *Operations Management Research : Advancing Practice Through Research*, *15*(1-2), 298–306. DOI: 10.1007/s12063-022-00263-x

Zafar, M. Q., Sajjad, R., Anwar, M. T., & Mustafa, M. S. (2025). A review on metal additive manufacturing: Types, applications and future trends. *Recent Progress in Materials*, *7*(1). Advance online publication. DOI: 10.21926/rpm.2501006

Zajemska, M., Wojtyto, D., Michalik, J., & Berski, S. (2025). Review of Current Trends in Sustainable Construction. *Energies*, *18*(10), 2559. DOI: 10.3390/en18102559

Zejjari, I., & Benhayoun, I. (2024). The use of artificial intelligence to advance sustainable supply chain: Retrospective and future avenues explored through bibliometric analysis. *Discover Sustainability*, *5*(1), 174. Advance online publication. DOI: 10.1007/s43621-024-00364-6

Zhang, H., Huang, S., Sheng, J., Wang, C., Zhang, J., & Zhu, M. Agyenim-Boateng, Emmanuel., Liang, C., Xue, B., Yang, H., Zhou, J., & Lu, J. (2023). 4D printing of shape memory polymer nanocomposites for enhanced performances and tunable response behavior. *European Polymer Journal, 201*, 112568. https://doi.org/https://doi.org/10.1016/j.eurpolymj.2023.112568

Zhang, S., & Xiao, T. (2013). Building Information Modeling and Sustainable Architecture Design Analysis. DOI: 10.2991/icaicte.2013.15

Zhang, D. (2024). AI integration in supply chain and operations management: Enhancing efficiency and resilience. *Applied and Computational Engineering*, *90*(1), 8–13. DOI: 10.54254/2755-2721/90/2024MELB0060

Zhang, H., Huang, S., Qian, Z., Zhang, Y., Sheng, J., Zhang, J., Agyenim-Boateng, E., Gao, Y., & Lu, J. (2024). Tunable deformation behavior of 4D printed shape memory polymers with embedded pre-stretched elastomers. *Chemical Engineering Journal*, *492*, 152187. https://doi.org/https://doi.org/10.1016/j.cej.2024.152187. DOI: 10.1016/j.cej.2024.152187

Zhang, M., & Li, X. (2018). Demand-side energy management with AI optimization. *Energy and Building*, *167*, 350–360.

Zhang, Y., Gao, Y., Michelin, L., Josien, L., Vidal, L., Schrodj, G., Simon-Masseron, A., & Lalevée, J. (2022a). Photopolymerization of ceramic/zeolite reinforced photopolymers: Towards 3D/4D printing and gas adsorption applications. *European Polymer Journal*, *179*, 111552. https://doi.org/https://doi.org/10.1016/j.eurpolymj.2022.111552. DOI: 10.1016/j.eurpolymj.2022.111552

Zhang, Y., & Luo, X. (2022). Intelligent fault detection in solar inverters using ML. *Journal of Electrical Energy Systems*, *9*(4), 201–214.

Zhang, Z., Tan, Z. J., Wang, Y. F., Ren, D. X., & Li, J. Y. (2023). The relationship between microstructures and mechanical properties in friction stir lap welding of titanium alloy. *Materials Chemistry and Physics*, *296*, 127251. DOI: 10.1016/j.matchemphys.2022.127251

Zhao, X., & Zhang, Y. (2024). Integrated management of urban resources toward Net-Zero smart cities considering renewable energies uncertainty and modeling in Digital Twin. *Sustainable Energy Technologies and Assessments*, *64*, 103656. DOI: 10.1016/j.seta.2024.103656

Zhao, Y., He, S., & Li, L. (2022). Application of hot isostatic pressing in nickel-based single crystal superalloys. *Crystals*, *12*(6), 805. DOI: 10.3390/cryst12060805

Zhou, C., Xu, J., Miller-Hooks, E., Zhou, W., Chen, C., Lee, L. H., Chew, E. P., & Li, H. (2021). Analytics with digital-twinning: A decision support system for maintaining a resilient port. *Decision Support Systems*, *143*, 113496. DOI: 10.1016/j.dss.2021.113496

Zhou, X., Ren, L., Song, Z., Li, G., Zhang, J., Li, B., Wu, Q., Li, W., Ren, L., & Liu, Q. (2023). Advances in 3D/4D printing of mechanical metamaterials: From manufacturing to applications. *Composites. Part B, Engineering*, *254*, 110585. DOI: 10.1016/j.compositesb.2023.110585

Zhu, H. H., Fuh, J. Y. H., & Lu, L. (2007). The influence of powder apparent density on the density in direct laser-sintered metallic parts. *International Journal of Machine Tools & Manufacture*, *47*(2), 294–298. DOI: 10.1016/j.ijmachtools.2006.03.019

Zhu, W., & Dhokia, V. (2021). A review of hybrid manufacturing processes—State-of-the-art and future perspectives. *Journal of Manufacturing Processes*, *68*, 279–297. DOI: 10.1016/j.jmapro.2021.05.042

Zhu, Z.-Q., Sun, H.-X., Qin, X.-J., Jiang, L., Pei, C.-J., Wang, L., Zeng, Y.-Q., Wen, S.-H., La, P.-Q., Li, A., & Deng, W.-Q. (2012). Preparation of poly(acrylic acid)–graphite oxide superabsorbent nanocomposites. *Journal of Materials Chemistry*, *22*(11), 4811–4817. DOI: 10.1039/c2jm14210d

Zuo, T., & Zhao, H. (2019). Life cycle assessment and external audit in aerospace manufacturing : Enhancing sustainability. *Journal of Cleaner Production*, *237*, 117729. DOI: 10.1016/j.jclepro.2019.117729

About the Contributors

Romdhane Ben Khalifa is assistant professor, academic and researcher in the field of mechanical engineering, primarily associated with the University of Gabes. He holds a position at the National engineering school of Gabes (ENIG) where his research and teaching focus on Mechanical Engineering (fatigue and fracture of materials, hydraulic systems, mechanical behavior, plastic and composite materials, CAD/CAM/CAE Modeling and simulation and Sustainable Prototyping) and His work is well-regarded for its contributions to the improvement of manufacturing processes and the application of artificial intelligence in engineering. He obtained his PhD Doctoral thesis in Mechanical Engineering at National Engineering School of Tunisia in 2012. In 2005, he obtained the Master degree of Mechanical Engineering from the National Engineering School of Tunisia. He has published numerous research papers in leading international journals in the field of mechanical engineering and has served as the editor of several academic books.

***

Csongor Czipf, Ph.D. Honorary Associate Professor, Assistant Professor Széchenyi István University, Győr, Hungary Gyula Kautz Faculty of Economics, Department of International Applied Economics Ongoing teaching experience at Széchenyi István University since 1999. Appointed as Regional Economic Manager of the Piarist Order Hungarian Province in Budapest as of January 2022. Master's degree in Religious Education and Dormitory Teaching obtained from the College of Theology in Győr in 2017. Doctoral degree awarded in 2011 by Janus Pannonius University of Pécs. Research specialization: corporate evaluation from a human resources perspective. Bachelor's degree in Economics with a specialization in Comparative Economic Policy completed in 1999 at Széchenyi István College.

Madhukumar K is an accomplished academician and researcher in the field of Mechanical Engineering, specializing in metal matrix composites (MMCs), advanced materials, surface coatings, additive manufacturing, and thermo-mechanical processing. He earned his Ph.D. from Visvesvaraya Technological University (VTU) in 2024, with doctoral research focused on the development and property evaluation of Aluminum Alloy 6082 reinforced with micro-Yttria Stabilized Zirconia particulates, subjected to T6 heat treatment. His work contributes to the development of high-performance composites for structural and marine applications. With over 13 years of academic experience and industry exposure in injection mold design, Dr. Madhukumar has cultivated expertise in stir casting, MMCs, T6 heat treatment, nanocomposite coatings, rapid prototyping, and mold system design. He is proficient in advanced design and simulation tools such as Pro-E with EMX 4.1, SolidWorks, AutoCAD, and MoldFlow. His research portfolio includes 16 patents, 3 design grants, publications in Scopus and Web of Science-indexed journals, 11 UGC-CARE-listed articles, 8 book chapters, 2 authored textbooks, and presentations at multiple national and international conferences. His ongoing research interests span lightweight MMCs, AI/ML integration in material process optimization, eco-friendly corrosion-resistant nanocoatings, bio-nanocomposites for biomedical applications, and mold design. Dr. Madhukumar's work aims to bridge academic innovation with industrial application, fostering engineering solutions that are sustainable, future-focused, and technologically advanced.

. **Leevesh Kumar**-Associate Professor Civil Engineering Dr. Leevesh Kumar is a distinguished academician and researcher with over 13 years of progressive experience in the field of civil and construction engineering across India, Ethiopia, the United Kingdom, and Bangladesh. Currently serving as an associate professor at the World University of Bangladesh and holding a concurrent part-time role with Liverpool John Moores University and the University of East London (UK), Dr. Kumar has built a diverse and global portfolio in higher education and construction technology research. He holds a Ph.D. in Civil Engineering from Career Point University (India), preceded by both M.Tech and B.Tech degrees in Civil Engineering from Bhagwant University. His educational foundation is complemented by a postgraduate diploma and several certifications from prestigious institutions like IIT Madras and NITTTR Chandigarh. Throughout his academic journey, Dr. Kumar has remained committed to advancing sustainable practices in the construction industry. Dr. Kumar's research focuses on sustainable construction materials, with specific interest in the mechanical and durability

K. K. Mishra is working as a Professor of Marketing at the University School of Business, Chandigarh University, Mohali, Punjab. He has published

54 research papers in Scopus-indexed journals and book chapters, and presented 31 papers at national and international conferences. He has four patent rights and three copyrights. He has 25 years of experience in teaching and research. He has supervised two PhD scholars. He has twice qualified for the UGC NET in Management. He has received the Best Faculty Award three times from the Indian Institute of Materials Management, National Headquarters, Mumbai, in 2009, 2016 and 2018, respectively. He has been conferred the Best Paper Award for the paper titled "Harnessing AI: Unveiling Educators' Intentions and Synergy in Education", presented at the 8th International Conference on Information System Design and Intelligent Applications (ISDIS 2024) in Dubai, UAE, in 2024. He has chaired technical sessions of international conferences organised by the Indian Institute of Materials Management, Mumbai, in 2016 and 2018 and OP Jindal University, Raigarh, in 2022 and 2023. He did his PhD in "Strategy formulation for marketing services in rural Uttar Pradesh" from Uttar Pradesh Rajarshi Tandon University, Allahabad.

Pawan Pant is working as an Associate Professor of Management at the University School of Business, Chandigarh University, Mohali, Punjab. He has published eight research papers, one in a national and one in an international journal. He has 11years of teaching experience and worked as an Assistant Manager (Purchase) at Pacific Industries Ltd., Udaipur (Rajasthan) for one year. His area of specialisation is Economics.

Pushpak Sharma is an Associate Professor at the University School of Business, Chandigarh University, Punjab. He did his Ph.D from Panjab University, Chandigarh. His research interests include Development Economics and Labour Economics. Pushpak Sharma has authored over 20 research papers in journals listed in SCOPUS and has published book chapters with prominent publishers such as Routledge and Palgrave Macmillan. Additionally, he serves as the Ph.D Program Coordinator (Economics) in the department.

Sitesh Kumar Singh is an accomplished academic and Chartered Engineer with a Ph.D. in Civil Engineering and over nine years of diverse experience in education, research, consultancy, and educational administration. Currently serving as an Assistant Professor in the Department of Civil and Environmental Engineering at the National University of Science and Technology, Muscat, Sultanate of Oman, he specializes in Civil and Transportation Engineering, Project Management, and Data Analytics. Dr. Singh holds multiple advanced degrees including M.Tech, B.Tech, PGDIPR, PGDUPDL. His career encompasses curriculum development, research supervision, quality assurance, and strategic academic planning. He has supervised over 40 postgraduate and 2 doctoral students, contributing to academic leadership through workshops, conferences, and editorial roles, and has published over 40 research articles, several patents, and authored books and book chapters.

Azmee Zaheer is an Assistant Professor at the University School of Business, MBA department, Chandigarh University, Gharuan, Punjab. She has completed her PhD and MPhil in Management from Babasaheb Bhimrao Ambedkar University (A Central University), Lucknow, Uttar Pradesh. She received a Gold Medal for her M.Phil's exemplary work. She has four years of research experience, 3.5 years of corporate marketing experience, and 2.5 years of academic marketing experience. She has also contributed to various research journals, including those indexed in Scopus, Web of Science, and UGC Care, as well as edited books. She has participated in multiple national and international seminars.

Index

Symbols

A

C

D

E

F

U

www.ingramcontent.com/pod-product-compliance
Ingram Content Group UK Ltd.
Pitfield, Milton Keynes, MK11 3LW, UK
UKHW051038091026
15624UKWH00011B/516

9 798337 354736